MATHEMATICAL METHODS FOR
SCIENCE STUDENTS

Mathematical Methods for Science Students

G. STEPHENSON

B.SC., PH.D., D.I.C.

Reader in Mathematics, Imperial College, London

SECOND EDITION

Longman Scientific & Technical
Longman Group UK Limited
Longman House, Burnt Mill, Harlow
Essex CM20 2JE, England
and Associated Companies throughout the world

FIRST PUBLISHED 1961
SECOND EDITION 1973
REPRINTED 1975, 1977, 1978, 1981, 1982, 1984, 1986, 1988, 1989

British Library Cataloguing in Publication Data

Stephenson, G.
Mathematical methods for science students — 2nd ed.
1. Calculus
I. Title
515   QH303
ISBN 0-582-44416-0

Produced by Longman Singapore Publishers Pte Ltd.
Printed in Singapore.

CONTENTS

Contents

PREFACE TO THE FIRST EDITION

THIS book presents a course of mathematics suitable for under-graduate students of the physical sciences, and has grown out of a course of lectures given by the author at Imperial College during the past five years. It largely covers the mathematical requirements (excluding mechanics) of the Cambridge Natural Science Tripos (Part 1) and the various London University ancillary mathematics examinations for physicists and chemists, and should also be useful as an introduction to mathematical methods for the B.Sc. General and B.Sc. Special Mathematics degrees. However, no attempt has been made to follow any particular examination syllabus. Throughout it is assumed that the reader has a good knowledge of mathematics up to the Advanced Level of G.C.E. and that he is therefore fairly familiar with such topics as analytical geometry, simple differentiation and integration, and the use of the sine, cosine, exponential and logarithmic functions.

The method of illustrating mathematics by various physical problems (as is the tendency in many books of this type) is not adopted here, as in the author's experience the understanding of the physics of such problems very often seems to present more difficulties to the student than the mathematical technique itself. Furthermore, problems which are of interest to physicists, for example, are often of little interest to chemists, and vice-versa. Consequently the large number of worked examples given in the text, and the problems to be solved by the reader, are all of the mathematical (as distinct from physical) type.

Various topics (such as group theory and integral equations), which are usually considered too difficult to be mentioned in an undergraduate course, have nevertheless been introduced here, not so much in an attempt to give the reader a deep understanding of them, but rather to stimulate his interest and to make him aware of the existence of such things.

Apart from a chapter on numerical integration, little attention has been given to the subject of numerical methods in general. This is to some extent unfortunate since much more can be done numerically in solving differential and integral equations than can be done by formal methods. Nevertheless the author feels that the subject is too vast and too important these days to be dealt with in a few pages, and the reader is well advised to undertake a study of numerical methods in their own right.

Many of the problems set at the end of each chapter are taken from recent London University and Cambridge University examination papers. Such problems are denoted by (L.U.) and (C.U.) respectively, and the author is grateful to the authorities of both Universities for permission to use them here.

The author wishes to express his grateful thanks to Dr. T. Kovari, Mr. D. Dunn and Dr. C. W. Kilmister for their careful readings of the manuscript, and for many criticisms and suggestions which have substantially improved the text. His thanks are also due to Dr. A. N. Gordon for reading the proofs and for checking the answers to the problems.

Department of Mathematics,   1960
Imperial College,
London.

PREFACE TO THE SECOND EDITION

WITH the opportunity to provide a new edition, it was tempting to introduce a large amount of new material. However, the usefulness and widespread adoption of the book seems to be due mainly to the compact presentation of essential mathematics, and to its very modest price. For both these reasons, therefore, it was decided to keep the amount of new material down to an absolute minimum, extending only where necessary to give a more complete and balanced account. Additional topics and problems have been included in existing chapters as follows:

    Linear inequalities (Ch. 2)
    Maxima and minima (Ch. 3)
    Functions of a complex variable (Ch. 7)
    Lagrange multipliers (Ch. 9)
    Orthogonal and unitary matrices; eigenvalues and eigenvectors
    (Ch. 17)

Three-dimensional geometry of lines and planes (Ch. 19)
Vector operators in cylindrical polar coordinates (Ch. 19)
Matrix formalism for simultaneous differential equations (Ch. 21)
Further properties of Bessel functions (Ch. 22)

The author wishes to thank all those colleagues and students who have, over the past thirteen years, recommended the book, and assisted by noting errors and misprints.

Department of Mathematics,   G. S.
Imperial College,
London.
1973

CHAPTER 1

Real Numbers and Functions of a Real Variable

1.1 Real Numbers

The concept of a number is introduced to most people at a very early age, and the existence of this concept is implicit in the development of all mathematical analysis. Any measurement carried out in the physical world leads directly to a number.

Of the various types of numbers that exist we first meet the rational number system which consists of the positive and negative integers (and zero) and numbers of the form p/q, where p and q are integers. Division by zero is not permitted and p/q is not defined therefore for $q = 0$. The addition, subtraction, multiplication and division of two rational numbers always leads to another rational number.

However, not all numbers can be written in the form p/q. For example, there is no rational number x which satisfies the equation $x^2 = 2$. To see that this is so we suppose $x = p/q$, where p and q are in their lowest form (i.e. have no common factor). Then if $x^2 = 2$, $p^2 = 2q^2$. Hence p is even and is therefore of the form $p = 2k$, where k is an integer. But then $4k^2 = 2q^2$ or $q^2 = 2k^2$, whence q is even also. We have now reached a contradiction since if p and q are both even then (contrary to assumption) p/q is not in its lowest form. Hence we conclude that there is no number x of the form p/q satisfying $x^2 = 2$. In order that equations of this type should have a solution a new class of numbers is needed. These numbers are called irrational numbers in the sense that they are not expressible in rational form p/q; they are in fact infinite non-recurring decimals. The number $x = \sqrt{2}$ ($= 1 \cdot 414 \ldots$), which is the solution of the equation just given, is an example of an irrational number. Another well-known irrational number is π ($= 3 \cdot 1415 \ldots$) which, in geometrical language, is the ratio of the circumference of a circle to its diameter. However, whereas the proof that $\sqrt{2}$ is not a rational number is easy, the proof of the irrationality of π requires much deeper mathematical analysis. Moreover, for some numbers, notably Euler's

1

constant γ (see Chapter 5, 5.3) the rationality or irrationality is still undecided. We may, of course, approximate as closely as we please to irrational numbers by choosing a rational number with suitable p and q values (for example, π can be approximately represented by $\frac{22}{7}$, or, to a better accuracy, by $\frac{355}{113}$). The rational and irrational numbers together make up the real number system. Real numbers, however, are a special class of numbers and in later chapters (particularly Chapter 7) we shall meet another number system involving complex numbers. Such numbers are extensions of real numbers (and contain real numbers as a special case). They are important in many topics in physics, chemistry and engineering and, in particular, enable many pure mathematical theorems to be proved in a concise way. In this chapter, however, we shall only be concerned with real numbers like 2, π, $\frac{3}{4}$, $-6\cdot183$, or, in general, just x.

It is worth remarking here that ∞ is not a number but only a symbol to denote that x may take on values as large as we please without limit.

1.2 Operations with Real Numbers

In order to make use of a real number x we must know how to combine it with other real numbers y, z ... etc. The operations that we may perform are restricted by a set of rules which we call Laws of (Elementary) Algebra. These rules are as follows:

(1) $x+y = y+x$,   (commutative law of addition)
(2) $x+(y+z) = (x+y)+z$,   (associative law of addition)
(3) $xy = yx$,   (commutative law of multiplication)
(4) $x(yz) = (xy)z$,   (associative law of multiplication)
(5) $x(y+z) = xy+xz$,   (distributive law)
(6) $x^n x^m = x^{n+m}$,   (index law)
(7) $x+0 = x$, $x0 = 0$, $x1 = x$.

It is important to realise that these rules only apply to real numbers. Later on we shall meet other mathematical entities (for example, matrices) which, in general, do not obey all of these rules.

The modulus or absolute value of a real number x is defined to be $+\sqrt{x^2}$ and is denoted by $|x|$ (read as ' mod x '). This implies that mod x is just the numerical value of x. In other words, mod x is

x itself if x is positive and is $-x$ if x is negative. For example, $|-6| = 6$, $|3/4| = 3/4$. Similarly, statements such as $|x| \leqq 2$ imply that the numerical value of x is always less than or equal to 2, and may be written as $-2 \leqq x \leqq 2$. This introduces the idea of an interval or range; we say that x lies within a closed interval of length $b-a$ if $a \leqq x \leqq b$, where a and b are given real numbers. If, however, x is not allowed to take the values a and b, but lies between them such that $a < x < b$, the interval is said to be open.

In both cases a and b are called the end points of the interval. When one or both of the end points are infinite we have either a semi-infinite interval $a < x < \infty$, or $-\infty < x < b$, or an unbounded interval $-\infty < x < \infty$. We note here that it would be wrong to write, for example, $-\infty \leqq x \leqq \infty$ since, as remarked in 1.1, ∞ is not a number and the question of equality therefore cannot arise.

The relation $|x-a| < b$ means, using the definition of the modulus, that $x-a < b$ and $a-x < b$. Hence $a-b < x < a+b$, which implies that x is not greater than b units away from a.

1.3 Functions of a Real Variable

If a relation between two real variables y and x is such that when x is given y is determined, then y is said to be a function of x. This is usually denoted by $y = f(x)$. We call x the independent variable (or argument of the function) and y the dependent variable. For example, $y = 6 + x^3$ and $y = \sin x$ are simple functions which give unique values of y for every value of x; for this reason they are called single-valued functions of x.

(a) Many-valued Functions

It is not always the case that one value of x leads to just one value of y. For example, $y^2 = x+2$ (with $x > -2$) gives two real values of y for every x. Such functions are called two-valued functions of x. In general, if one value of x leads to n values of y then the function is said to be an n-valued (or many-valued) function of x.

(b) Interval of Definition

The range of values of x for which y is defined as a function of x is called the interval of definition. For example, provided x and y are real numbers, the function

3

$$y = \frac{1}{\sqrt{(4 - x^2)}} \qquad (1)$$

is only defined in the open interval $-2 < x < 2$. Outside this interval (or range) y is either infinite (at $x = \pm 2$) or meaningless.

(c) *Functions of More than One Variable*

It is not necessary that y should be a function of only one independent variable. If

$$y = f(x_1, x_2) \quad (= \sin x_1 + x_1 \cos x_2, \text{ say}) \qquad (2)$$

then y is said to be a function of the two independent variables x_1 and x_2; its value is only determined when definite values are given to both x_1 and x_2. In general, if y is a function of n-independent variables $x_1, x_2, x_3 \ldots x_n$ it is written as

$$y = f(x_1, x_2, \ldots x_n). \qquad (3)$$

Functions of two or more independent variables will be discussed in detail in Chapter 9.

(d) *Polynomials*

A function of the type

$$f(x) = a_0 x^n + a_1 x^{n-1} + \ldots + a_{n-1} x + a_n, \qquad (4)$$

where $a_0, a_1, a_2, \ldots a_n$ are real constants and n is a positive integer, is said to be a polynomial in x of degree n. For example, $x^3 - 2x + 6$ is a polynomial of degree 3.

(e) *Rational Functions*

A rational function is defined to be of the type

$$f(x) = \frac{P(x)}{Q(x)}, \qquad (5)$$

where $P(x)$ and $Q(x)$ are polynomials in x of any degree. Such functions, however, are not defined for those values of x which make $Q(x) = 0$ since division by zero is not permitted. The function

4

$$y = \frac{x^3 + 2x + 3}{x + 6} \tag{6}$$

is a rational function, but it is not defined at $x = -6$. A particular type of rational function which we shall meet again later on when dealing with complex numbers is the bilinear function

$$f(x) = \frac{ax + b}{cx + d} \quad (c \neq 0). \tag{7}$$

The name arises from the fact that both the denominator and numerator are linear (i.e. of degree one) in x.

(f) Algebraic Functions

Any relation of the type

$$P(x)y^m + Q(x)y^{m-1} + \ldots + U(x)y + V(x) = 0, \tag{8}$$

where $P(x)$, $Q(x)$ ... $U(x)$, $V(x)$ are polynomials of any degree in x, defines y as an algebraic function of x.

(g) Transcendental Functions

Certain functions like $\sin x$, e^x, $\log_e x$ are not definable in terms of a finite number of polynomials and are not, therefore, of algebraic type. They bear a close analogy to the irrational numbers discussed earlier which are not expressible in terms of integers. In Chapter 6 we shall show how these functions may be expressed as power series in x.

(h) Even and Odd Functions

A function $f(x)$ is said to be an even function of x if $f(x) = f(-x)$, and to be an odd function of x if $f(x) = -f(-x)$. For example, $y = x^2$ is an even function of x since the value of y is the same when x is changed to $-x$. When considered graphically, even functions clearly are symmetrical about the y-axis. The function $y = x^3$ is an example of an odd function since y changes to $-y$ when x changes to $-x$. Such functions are not symmetrical about the y-axis, but have the origin of the Cartesian coordinate system as their centre of symmetry (see Fig. 1.1).

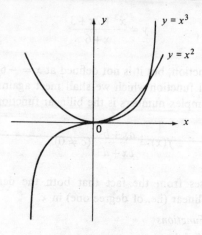

Fig. 1.1

Most functions are neither even nor odd, as for example

$$y = x^3 + x^2.$$

However, every function may be expressed as the sum of an even and an odd function. For suppose we take a function which is neither even nor odd; then $f(x)$ may be written as

$$f(x) = \tfrac{1}{2}\{f(x) + f(-x)\} + \tfrac{1}{2}\{f(x) - f(-x)\}. \tag{9}$$

The first bracket is unaltered by changing the sign of x and is therefore an even function, whilst the second bracket changes sign when x changes sign and is therefore an odd function. Hence $f(x)$ is the sum of an even function and an odd function.

For example, if $f(x) = e^{2x} \sin x$ we may write

$$e^{2x} \sin x = \tfrac{1}{2}(e^{2x} - e^{-2x}) \sin x + \tfrac{1}{2}(e^{2x} + e^{-2x}) \sin x, \tag{10}$$

the first term being even and the second odd.

The functions $\dfrac{e^{ax} + e^{-ax}}{2}$, and $\dfrac{e^{ax} - e^{-ax}}{2}$, where a is a positive constant, are called hyperbolic functions and are written as cosh ax and sinh ax, respectively. We shall deal with these functions in detail in Chapter 8.

6

(i) Periodic Functions

If the value of a function repeats itself at regular intervals of x, we say that the function is periodic. For example, $y = \sin x$ is periodic in x with period 2π since $y = \sin(x+2n\pi) = \sin x, n = 0, 1, 2 \ldots$. In general, a function $f(x)$ is periodic in x with period T if

$$f(x+nT) = f(x) \quad \text{(for all } x\text{), where } n = 1, 2 \ldots. \tag{11}$$

As another example we introduce the quantity $[x]$, which means the greatest possible integer $\leqq x$, and consider the function $x-[x]$. The graph of this function (Fig. 1.2) shows that the function is periodic in x with a period equal to unity.

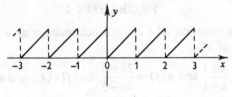

Fig. 1.2

(j) Monotonic Functions

A function $f(x)$ is said to be a strictly monotonic increasing function in an interval (a, b) if, for all pairs of numbers x_1 and x_2 in this interval,

$$f(x_1) < f(x_2) \quad \text{when} \quad x_1 < x_2. \tag{12}$$

Conversely $f(x)$ is said to be a strictly monotonic decreasing function in (a, b) if, for all pairs of numbers x_1 and x_2 in this interval,

$$f(x_1) > f(x_2) \quad \text{when} \quad x_1 < x_2. \tag{13}$$

If, however, $f(x_1) \leqq f(x_2)$, or $f(x_1) \geqq f(x_2)$ for $x_1 < x_2$ the function is said to be monotonic increasing or monotonic decreasing, respectively, the word ' strictly ' being omitted to allow for the possibility that the function may be constant in value over part of the range (a, b). For example, e^{ax} is a strictly monotonically increasing function for all x when a is positive, and a strictly monotonically decreasing function for all x when a is negative. However, the function $y = e^{ax} \cos bx$ (a, b constants, $b \neq 0$) is not monotonic for all x as may be seen by considering its graph with, say, $a > 0$ (see Fig. 1.3).

7

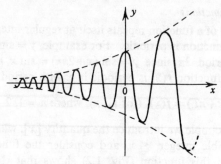

Fig. 1.3

PROBLEMS 1

1. Does the equation $x^2 + 2y^3 = 3$ determine y as a single-valued function of x?

2. If $f(x) = \dfrac{x+1}{x-1}$ and $g(y) = \dfrac{2y+5}{4y-3}$, find $f\{g(x)\}$ and $g\{f(x)\}$.

3. If $f(x) = 2x^2 - 1$, show that $f(\cos \theta) = \cos 2\theta$.

4. Express the function $f(x) = \dfrac{1}{x+1} + \dfrac{x}{4-x} - \dfrac{1}{x-2}$ in the form $\dfrac{P(x)}{Q(x)}$, where $P(x)$ and $Q(x)$ are polynomials. State the degree of these polynomials and the values of x for which the function is not defined.

5. Determine whether the following functions are either even or odd, or neither

    (a) $x^3 + 6x$,   (b) $x^2 + 2 \sin x$,   (c) $e^{x^2} \cos 3x$,

    (d) $\displaystyle\int_0^x \sin^2 y \, dy$,   (e) $\tan^{-1} x$.

6. Write the functions $e^{-ax} \cos bx$ (a positive), and $\sqrt{\left(\dfrac{1-x}{1+x}\right)}$ as the sums of an even and an odd function.

7. Find which of the following functions are periodic and state their periods

    (a) $\cos 2x$,   (b) $\dfrac{\sin x}{x}$,   (c) $|\sin x|$,

8

(d) | sin x cos x |,   (e) cos $(wx+\alpha)$; w, α constant,

(f) $\log_e \sin^2 x$.

Draw rough graphs of (b), (d) and (f).

8. Show by drawing a graph, or otherwise, that if $f(x)$ is a periodic function of x, with period T, then

(a) $\int_0^{nT} f(x)\, dx = n \int_0^T f(x)\, dx$, where n is an integer,

(b) $\int_a^{T+a} f(x)\, dx = \int_0^T f(x)\, dx$, where a is a constant.

9. Draw rough graphs of the following functions and state which of them are monotonic increasing functions for all x within their intervals of definition

(a) $\log_e x$, $(x>0)$,   (b) $\dfrac{e^x}{x}$,   (c) $x^2 \sin x$,

(d) $\tan^{-1} x$,   (e) e^{-2x^2}.

CHAPTER 2

Inequalities

2.1 Definitions

We now discuss the meaning and use of the inequality signs $>$, $\geqq$, $<$, $\leqq$, $\not>$, $\not<$.

If x and y are real numbers, then $x > y$ implies that x is algebraically greater than y; or in other words that $(x-y)$ is positive. For example, $-3 > -4$ since $(-3)-(-4) = 1$ (which is > 0). Similarly $(x+1)^2 > 2x$ for all x, since $(x+1)^2 - 2x = x^2 + 1 > 0$. The relation $x \geqq y$ is to be read as ' x is either algebraically greater than or equal to y '. Conversely, $x < y$ means x is algebraically less than y; that is $(x-y)$ is negative, whilst $x \leqq y$ again allows for the possibility of x and y being equal.

The signs $\not>$ and $\not<$ mean ' not greater than ' and ' not less than ', respectively.

2.2 Operations with Inequalities

(a) Transposition

If x, y, u, v are real numbers and if

$$x+y > u+v, \tag{1}$$

then
$$x > u+v-y. \tag{2}$$

This result is easily proved since (1) and (2) both imply (according to the definitions) that $(x+y-u-v) > 0$ and are therefore equivalent statements.

(b) Multiplication by a Constant

If $x > y$, then $ax > ay$ if a is a positive number. This is true since $ax > ay$ implies $a(x-y) > 0$ which is satisfied by positive a. When a is negative, however, $ax < ay$, since $a(x-y) < 0$. Hence multiplication of an inequality by a negative number reverses the inequality sign. For example, since $7 > 5$, multiplication by 3 gives $21 > 15$, and multiplication by -3, $-21 < -15$.

10

(c) Addition of Inequalities

If $x > y$, and $u > v$, then $x+u > y+v$. This is proved since $x+u > y+v$ implies $(x-y)+(u-v) > 0$, which is true since both brackets are positive.

(d) Subtraction of Inequalities

If $x > y$ and $u > v$, we cannot deduce that $(x-u) > (y-v)$ since $(x-u)-(y-v) = (x-y)-(u-v)$ is not necessarily positive. Clearly $3 > 2$ and $1 > -1$, but $(3-1) \ngtr (2-(-1))$.

(e) Multiplication of Inequalities

If x, y, u, v are positive, then $x > y$, $u > v$ imply $xu > yv$, since $xu-yv = x(u-v)+v(x-y)$ is positive. This result is not necessarily valid when some of the numbers are negative.

(f) Division of Inequalities

If $x > y$, and $u > v$, these do not imply $x/u > y/v$. For example, $3 > 1$ and $2 > \frac{1}{4}$, but $\frac{3}{2} \ngtr 1/\frac{1}{4}(= 4)$.

We now give two simple examples on the use of inequalities.

Example 1. Find the values of x for which the inequality

$$\frac{5}{5x-1} > \frac{2}{2x+1} \tag{3}$$

is satisfied.

If the inequality is satisfied then $\dfrac{5}{5x-1} - \dfrac{2}{2x+1}$ must be positive.

Hence we must have

$$y = \frac{7}{(5x-1)(2x+1)} > 0. \tag{4}$$

This is positive for $x > \frac{1}{5}$ and $x < -\frac{1}{2}$ only. Hence the original inequality is only satisfied with these conditions on x.

This result can also be deduced by considering the graph of y against x as shown in Fig. 2.1. The reader is warned here against the common mistake of cross multiplying in (3), an operation which is **not** permitted in inequality relations of this type.

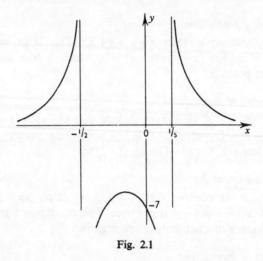

Fig. 2.1

Example 2. Find the values of x for which the polynomial

$$f(x) = x^3 - 7x^2 + 21x - 27 \qquad (5)$$

is greater than zero. Factorising the polynomial, the inequality may be written as $(x-3)\{(x-2)^2+5\} > 0$. If $x < 3$, the first bracket is negative and hence, for the left-hand side to be positive, $(x-2)^2+5$ must be negative. But this is impossible for any x; hence the inequality cannot be satisfied if $x < 3$. For $x > 3$, however, both brackets are positive, and the inequality is only satisfied therefore for such values of x.

Example 3. Find the values of x which satisfy the inequality

$$|7 - 3x| < 2. \qquad (6)$$

Using the definition of the modulus given in Chapter 1, (6) requires that $7 - 3x < 2$ and $3x - 7 < 2$. Hence $3x > 5$ and $3x < 9$. Accordingly the inequality is satisfied by $x > \frac{5}{3}$ and $x < 3$; that is $\frac{5}{3} < x < 3$.

2.3 Properties of Arithmetic, Geometric and Harmonic Means

If $x_1, x_2, x_3 \ldots x_n$ are n positive numbers, their arithmetic mean is defined by

$$A = \frac{x_1 + x_2 + \ldots + x_n}{n}, \qquad (7)$$

12

and their geometric mean by

$$G = \sqrt[n]{(x_1 x_2 x_3 \ldots x_n)}. \tag{8}$$

A further mean, the harmonic mean H, is defined by

$$\frac{1}{H} = \frac{1}{n}\left(\frac{1}{x_1} + \frac{1}{x_2} + \ldots + \frac{1}{x_n}\right) \tag{9}$$

These definitions may be put into simpler forms by using the summation and product signs $\sum$ and $\prod$, where

$$\sum_{r=1}^{n} x_r = x_1 + x_2 + \ldots + x_n, \tag{10}$$

and

$$\prod_{r=1}^{n} x_r = x_1 x_2 \ldots x_n. \tag{11}$$

Hence

$$A = \frac{1}{n}\sum_{r=1}^{n} x_r, \quad G = \sqrt[n]{\left(\prod_{r=1}^{n} x_r\right)}, \quad \text{and} \quad \frac{1}{H} = \frac{1}{n}\sum_{r=1}^{n}\frac{1}{x_r}. \tag{12}$$

For example, if $x_1 = 2$, $x_2 = 4$, $x_3 = 1$, then

$$A = \tfrac{7}{3}, \quad G = 8^{1/3} = 2, \quad \text{and} \quad H = \tfrac{12}{7}.$$

It can be proved, if $x_1, x_2 \ldots x_n$ are positive numbers, that $A \geqq G$ for all n, and that $A = G$ only when

$$x_1 = x_2 = x_3 = \ldots = x_n.$$

For example, for any two positive numbers x_1 and x_2 we have

$$A = \frac{x_1 + x_2}{2}, \quad G = \sqrt{(x_1 x_2)}. \tag{13}$$

Hence for $A \geqq G$ to be satisfied we must have

$$(x_1 + x_2) \geqq 2\sqrt{(x_1 x_2)}. \tag{14}$$

Squaring both sides (since x_1 and x_2 are positive) we have

$$(x_1 + x_2)^2 \geqq 4x_1 x_2,$$

which gives

$$(x_1 - x_2)^2 \geqq 0. \tag{15}$$

13

Since (15) is always satisfied for any values of x_1 and x_2, we have $A \geq G$, the equality sign occurring when $x_1 = x_2$.

Now $\dfrac{1}{H}$ is the arithmetic mean of $\dfrac{1}{x_1}, \dfrac{1}{x_2} \ldots \dfrac{1}{x_n}$. Hence, using the result that $A \geq G$, we have

$$\frac{1}{H} \geq \sqrt[n]{\left(\frac{1}{x_1}\frac{1}{x_2}\frac{1}{x_3}\ldots\frac{1}{x_n}\right)} = \frac{1}{G}. \tag{16}$$

Accordingly $G \geq H$, equality occurring only when $x_1 = x_2 = \ldots = x_n$. Hence finally we have the basic inequality

$$A \geq G \geq H. \tag{17}$$

Example 4. Find the value of x for which the function

$$y = \frac{a+bx^4}{x^2} \tag{18}$$

has its least value, where a and b are positive numbers.

Now a/x^2 and bx^2 are positive numbers. Hence, using the result that $A \geq G$, we have

$$\frac{1}{2}\left(\frac{a}{x^2}+bx^2\right) \geq \sqrt{\left(\frac{a}{x^2}\cdot bx^2\right)} \tag{19}$$

or

$$\frac{a}{x^2}+bx^2 \geq 2\sqrt{(ab)}. \tag{20}$$

The least value of the left-hand side of equation (20) occurs when the equality sign holds; that is, when

$$\frac{a}{x^2} = bx^2. \tag{21}$$

Hence $x = \sqrt[4]{\left(\dfrac{a}{b}\right)}$ gives the least value of y.

2.4 Inequalities involving Simple Functions

The question of whether the value of one function (for example, $\sin x$) is greater or less than the value of another (say, $\log_e x$) in a given interval of x will be dealt with more fully in the next chapter as this requires some of the results obtained there. Nevertheless we may obtain an important set of inequalities between the circular functions (sine, cosine, etc.,) using the following geometrical

approach. In Fig. 2.2, the area of the
triangle OBD is clearly greater than the
area of the sector of the circle OAB,
which in turn is greater than the area of
the triangle OAB.

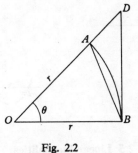

Expressed mathematically we therefore have

$$\tfrac{1}{2}r^2 \tan \theta > \tfrac{1}{2}r^2\theta > \tfrac{1}{2}r^2 \sin \theta \quad (22)$$

for $0 < \theta < \pi/2$.

Fig. 2.2

Hence $\qquad\qquad \tan \theta > \theta > \sin \theta.$ $\qquad\qquad$ (23)

These inequalities may be also written (by dividing through by
$\sin \theta$) as

$$\frac{1}{\cos \theta} > \frac{\theta}{\sin \theta} > 1. \qquad (24)$$

As θ approaches zero, $\cos \theta$ approaches unity and consequently, for
(24) to be satisfied, $\dfrac{\sin \theta}{\theta}$ must also approach unity. We express this

result by saying that the limit of $\dfrac{\sin \theta}{\theta}$ as θ tends to zero is unity, or

$$\lim_{\theta \to 0} \left(\frac{\sin \theta}{\theta} \right) = 1. \qquad (25)$$

This result could not have been obtained by simply putting $\theta = 0$
in the function $\dfrac{\sin \theta}{\theta}$ for this gives $\dfrac{0}{0}$ which is not defined in the real

number system. The function $\dfrac{\sin \theta}{\theta}$ is also an example of an even

function (see Chapter 1, 1.3 (h)) since under the transformation
$\theta \to -\theta$ the function remains unaltered; its graph is shown in Fig.
2.3.

In the next chapter we shall discuss the concept of a limit in more
detail.

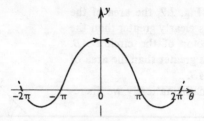

Fig. 2.3

2.5 Linear Inequalities

Consider first the linear equation

$$ax + by = c, \qquad (26)$$

where a, b and c are constants. This equation defines a straight line in the plane (see Fig. 2.4), which divides the plane into three distinct regions

(i) the points in the plane for which

$$ax + by > c,$$

(ii) the points in the plane for which

$$ax + by < c,$$

and

(iii) the points on the line itself.

The inequalities $ax + by \geq c$, and $ax + by \leq c$, define half-planes in the sense that they divide the whole plane into two parts. Half planes are convex in the sense that, if A and B are any two points in the half-plane, the straight line joining them always lies within the half-plane. Hence the region formed by the intersection of any number of half-planes is also a convex region.

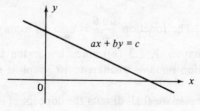

Fig. 2-4

The following examples show how regions can be defined by linear inequalities.

Example 5 (see Fig. 2.5).
The inequalities

$$2x - 3y \leq 6,$$
$$x + y \leq 4,$$

define the infinite (shaded) region.

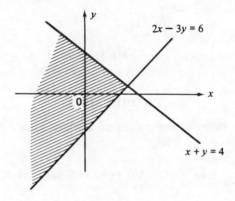

Fig. 2.5

Example 6 (see Fig. 2.6).
The inequalities

$$x + y \leq 4,$$
$$2x - 3y \leq 6,$$
$$3x - y \geq -3,$$
$$x \leq 2,$$

define the finite (shaded) region.

The coordinates of the corners of the convex regions so formed are found by letting the equality signs hold in the various relations and then solving for x and y.

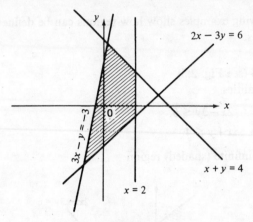

Fig. 2.6

PROBLEMS 2

1. Find the range of real values of x which satisfy the following inequalities

   (a) $\dfrac{1}{2-x} < 1$,   (b) $x(4-x) \leqq 4$,   (c) $x^2 + 8 < 2x$,

   (d) $x^2 - 3x > 10$,   (e) $-1 < \dfrac{3x+4}{x-6} < 1$.

2. Verify Cauchy's inequality

$$\sum_{r=1}^{n} (x_r^2) \sum_{r=1}^{n} (y_r^2) > \left(\sum_{r=1}^{n} x_r y_r \right)^2$$

   for $n = 2$. Show that the inequality sign must be replaced by an equality sign when $\dfrac{x_1}{y_1} = \dfrac{x_2}{y_2}$.

3. Verify the Weierstrass inequality

$$\prod_{r=1}^{n} (1+x_r) > 1 + \sum_{r=1}^{n} x_r,$$

   (where x_r all have the same sign, and the factors $(1+x_r)$ are positive) for $n = 2$ and $n = 3$.

18

4. If x, y, z are real numbers, show that

$$(y+z-x)^2+(z+x-y)^2+(x+y-z)^2 \geq yz+zx+xy.$$

5. If x_1, x_2, x_3 are positive numbers show that

$$(x_1+x_2+x_3)\left(\frac{1}{x_1}+\frac{1}{x_2}+\frac{1}{x_3}\right) \geq 9.$$

6. Show that, for $0 < x < 1$,

$$1+x < e^x < \frac{1}{1-x}.$$

7. By considering the graph of $\log_e x$, show that

$$\int_m^{m+1} \log_e x\, dx > \log_e m > \int_{m-1}^m \log_e x\, dx,$$

for all integers $m > 1$.

Hence show that $\quad \int_2^{n+1} \log_e x\, dx > \log_e n! > \int_1^n \log_e x\, dx,$

and deduce, by evaluating the integrals, that

$$e < n!\left(\frac{e}{n}\right)^n < \frac{en}{4}\left(1+\frac{1}{n}\right)^{n+1}.$$

Using $e^3 \simeq 20$, check this inequality for the case $n = 4$. (C.U.)

8. The quadratic form $P(x, y) = ax^2+2hxy+by^2$ (a, b, h not all zero) is said to be positive definite if $P(x, y) > 0$ for all (x, y) other than $(0, 0)$. Show that the condition for $P(x, y)$ to be positive definite is that $h^2 < ab$, and that $P(x, y)$ has the sign of a.

9. Show that of all rectangles having a specified perimeter, the square encloses the largest area.

10. Sketch the region defined by the linear inequalities

$$3x+2y \leq 6, \quad x-y \leq 2, \quad x \leq 1,$$

and determine the coordinates of the corners.

CHAPTER 3

Limits, Continuity and Differentiability

3.1 Limits

In the last chapter the idea of a limit was introduced by showing that as θ becomes small $\dfrac{\sin \theta}{\theta}$ approaches unity. We now consider more carefully what is meant by a limit. Suppose $f(x)$ is a given function of x. Then if we can make $f(x)$ as near as we please to a given number l by choosing x sufficiently near to a number a, l is said to be the limit of $f(x)$ as $x \to a$, and is written as

$$\lim_{x \to a} f(x) = l. \tag{1}$$

It is important to emphasise the following points:

(a) the independent variable x may approach the point a either from left to right (that is, from $-\infty$ to a) or from right to left (from ∞ to a). In many cases the limits of the function obtained in these two ways are different, and when this is the case we write them as

$$\lim_{x \to a-} f(x) = l_1, \quad \lim_{x \to a+} f(x) = l_2,$$

respectively.

For example, the function $y = \tan^{-1}\left(\dfrac{1}{x}\right)$ tends to $\dfrac{\pi}{2}$ when x approaches zero from the positive side, and to $-\dfrac{\pi}{2}$ when x approaches zero from the negative side.

Consequently we write

$$\lim_{x \to 0+} \tan^{-1}\left(\frac{1}{x}\right) = \frac{\pi}{2}, \quad \lim_{x \to 0-} \tan^{-1}\left(\frac{1}{x}\right) = -\frac{\pi}{2}.$$

Sometimes we are faced with a function which becomes arbitrarily large when x is chosen sufficiently close to a number a. When this happens we write

$$\lim_{x \to a} f(x) = \infty. \tag{2}$$

For example, the function $y = \dfrac{1}{x}$ tends to ∞ when $x \to 0$ from the positive side, and to $-\infty$ when $x \to 0$ from the negative side (see Fig. 3.1). Accordingly

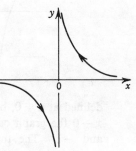

$$\lim_{x \to 0+}\left(\frac{1}{x}\right) = \infty, \quad \lim_{x \to 0-}\left(\frac{1}{x}\right) = -\infty.$$

In all cases when the limits as $x \to a$ from both directions are equal (say l) we simply write

$$\lim_{x \to a} f(x) = l. \tag{3}$$

Fig. 3.1

(b) in proceeding to the limit of $f(x)$ as $x \to a$ we have to exclude x from becoming equal to a for two reasons. Firstly, the value of the function may not be defined at $x = a$, as, for example. $\dfrac{\sin x}{x}$ at $x = 0$. Secondly, if $f(x)$ is defined at $x = a$ its value may not be equal to $\lim_{x \to a} f(x)$. For example, if $f(x)$ is defined by

$$f(x) = \begin{cases} 1 & \text{for } x \leqq 1, \\ \tfrac{1}{2} & \text{for } x > 1, \end{cases} \tag{4}$$

(see Fig. 3.2) then

$$\lim_{x \to 1+} f(x) = \tfrac{1}{2}.$$

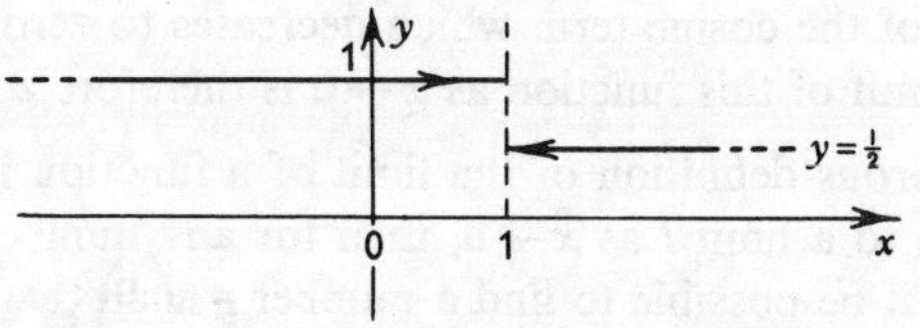

Fig. 3.2

21

This is not equal to the value of the function at $x = 1$, which by (4) is equal to unity.

The function $y = \cos(1/x)$ (see Fig. 3.3) is not only un-

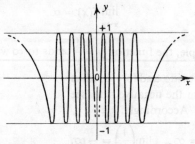

Fig. 3.3

defined at $x = 0$, but possesses no limit there either, since as $x \to 0$ the graph oscillates infinitely many times between $+1$ and -1. The function therefore does not approach any particular value as $x \to 0$. However, $y = x \cos \dfrac{1}{x}$ (Fig. 3.4)

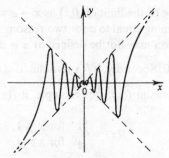

Fig. 3.4

although again oscillating infinitely many times as $x \to 0$ nevertheless does possess a limit in virtue of the factor x in front of the cosine term which decreases to zero in the limit. The limit of this function as $x \to 0$ is therefore zero.

A more rigorous definition of the limit of a function is as follows: if $f(x)$ tends to a limit l as $x \to a$, then for any number ε (however small) it must be possible to find a number η such that

$$|f(x) - l| < \varepsilon \quad \text{when} \quad |x - a| < \eta \qquad (5)$$

In general the value of η depends on the value of ε. Consider $f(x) = 1 - \dfrac{1}{x+2}$. According to (5) we are permitted to say that $\lim_{x \to 1} f(x) = \frac{2}{3}$ provided a value of η exists such that, for any ε,

$$\left| \left(1 - \frac{1}{x+2} \right) - \frac{2}{3} \right| < \varepsilon,$$

when
$$|x-1| < \eta. \tag{6}$$

Suppose we take $\varepsilon = 10^{-3}$. Then

$$0.334 > \frac{1}{x+2} > 0.332 \quad \text{(to 3 decimals)},$$

which gives
$$0.994 < x < 1.010.$$

For (6) to be satisfied we need therefore only take

$$(1.010 - 1) < \eta, \tag{7}$$

or $\eta > 0.010$. Hence, since the conditions (5) can be satisfied, the limit of $f(x)$ as $x \to 1$ exists and is equal to $\frac{2}{3}$.

We now state without proof three important theorems on limits. If $f(x)$ and $g(x)$ are two functions of x such that $\lim_{x \to a} f(x)$ and $\lim_{x \to a} g(x)$ exist, then

Theorem 1.
$$\lim_{x \to a} \left\{ f(x) + g(x) \right\} = \lim_{x \to a} f(x) + \lim_{x \to a} g(x),$$

Theorem 2.
$$\lim_{x \to a} \left\{ f(x)g(x) \right\} = \lim_{x \to a} f(x) \cdot \lim_{x \to a} g(x),$$

Theorem 3.
$$\lim_{x \to a} \left\{ \frac{f(x)}{g(x)} \right\} = \frac{\lim_{x \to a} f(x)}{\lim_{x \to a} g(x)}$$

provided $\lim_{x \to a} g(x) \neq 0.$

23

These theorems may be readily extended to any finite number of functions.

The following example illustrates the use of these theorems.

Example 1. Suppose we wish to evaluate

$$\lim_{x \to 1} \left(\frac{x^m - 1}{x - 1} \right),$$

where m is a positive integer. Dividing the denominator into the numerator, the limit may be written as

$$\lim_{x \to 1} (1 + x + x^2 + \dots + x^{m-1}), \tag{8}$$

which by Theorem 1 is the same as

$$\lim_{x \to 1} 1 + \lim_{x \to 1} x + \lim_{x \to 1} x^2 + \dots + \lim_{x \to 1} x^{m-1}. \tag{9}$$

The value of each of these limits is unity and since there are m of them, the sum is m. Similarly if m is a negative integer, say $-k$, where k is a positive integer, then

$$\lim_{x \to 1} \left(\frac{x^m - 1}{x - 1} \right) = \lim_{x \to 1} \left(\frac{x^{-k} - 1}{x - 1} \right) = \lim_{x \to 1} \left(\frac{1/x^k - 1}{x - 1} \right) = \lim_{x \to 1} \left(\frac{1 - x^k}{(x - 1)x^k} \right)$$

$$= -\lim_{x \to 1} \left\{ \left(\frac{x^k - 1}{x - 1} \right) \frac{1}{x^k} \right\}. \tag{10}$$

By Theorem 2, (10) may be written as

$$\lim_{x \to 1} \left(\frac{x^m - 1}{x - 1} \right) = -\lim_{x \to 1} \left(\frac{x^k - 1}{x - 1} \right) \lim_{x \to 1} \frac{1}{x^k} = -k = m, \tag{11}$$

(making use of the result for a positive integer).

Likewise if m is fractional, say p/q, where p and q are integers, then

$$\lim_{x \to 1} \left(\frac{x^m - 1}{x - 1} \right) = \lim_{x \to 1} \left(\frac{x^{p/q} - 1}{x - 1} \right). \tag{12}$$

Now putting $x^{1/q} = y$ so that $x = y^q$ we have

$$\lim_{x \to 1} \left(\frac{x^m - 1}{x - 1} \right) = \lim_{y \to 1} \left(\frac{y^p - 1}{y^q - 1} \right) = \lim_{y \to 1} \left\{ \frac{\left(\dfrac{y^p - 1}{y - 1} \right)}{\left(\dfrac{y^q - 1}{y - 1} \right)} \right\}. \tag{13}$$

By Theorem 3, therefore

$$\lim_{x \to 1} \left(\frac{x^m - 1}{x - 1} \right) = \frac{\lim_{y \to 1} \left(\dfrac{y^p - 1}{y - 1} \right)}{\lim_{y \to 1} \left(\dfrac{y^q - 1}{y - 1} \right)} = \frac{p}{q} = m \tag{14}$$

as before. Hence for all rational values of m

$$\lim_{x \to 1} \left(\frac{x^m - 1}{x - 1} \right) = m. \tag{15}$$

3.2 Continuous and Discontinuous Functions
A single-valued function of x is said to be continuous at $x = a$ if

(a) $\lim_{x \to a} f(x)$ exists,

(b) the function is defined for the value $x = a$,

and (c) if $\lim_{x \to a} f(x) = f(a)$.

When a function does not satisfy these conditions it is said to be
discontinuous and $x = a$ is called a point of discontinuity. In general
if the graph of a function has a break in it at a particular value of x
it is discontinuous at that point. For example, the function $y = 1/x$
represented in Fig. 3.1 is discontinuous at $x = 0$, whilst the function
defined by 3.1 (3) and represented in Fig. 3.2 is discontinuous at
$x = 1$. There is, however, a slight difference between these two
examples. The first function $(y = 1/x)$ becomes infinite at the point
of discontinuity and is said to have an infinite discontinuity at
$x = 0$; the second function remains finite at the discontinuity and
is therefore said to have a finite discontinuity at $x = 1$. Another
example of a function with discontinuities is $y = x - [x]$, whose

25

graph is shown in Chapter 1, Fig. 1.2. This function has an infinite number of finite discontinuities which occur at all integral values of x (positive, negative and zero).

Functions like $\dfrac{\sin x}{x}$ and $\dfrac{\tan x}{x}$ are discontinuous at $x = 0$ since they are not defined there (see condition (b) above).

It is an important result (and one that we shall need later on) that every polynomial of any degree is continuous for all x.

To prove this consider a polynomial of degree n

$$P_n(x) = a_0 x^n + a_1 x^{n-1} + \ldots + a_{n-1} x + a_n \tag{16}$$

and take as a function $f(x)$ any typical term $x^m (m \leqq n)$ in the polynomial · Then for any arbitrary value of x, say $x = a$, $f(a) = a^m$. Now by Theorem 2, (3.1)

$$\lim_{x \to a} f(x) = \lim_{x \to a} x^m = \left(\lim_{x \to a} x \right)^m = a^m = f(a). \tag{17}$$

Hence the function x^m is continuous at $x = a$, and since a is arbitrary, it must be continuous for all x. This result applies to every term of the polynomial, and hence every polynomial is continuous for all x. An immediate consequence of this result is that every rational function (see Chapter 1, 1.3 (e)) is continuous everywhere except at the points where the denominator vanishes. For example,

$$y = \frac{5x^2 + 3}{(x-1)(x-2)} \tag{18}$$

is continuous everywhere except at $x = 1$ and $x = 2$. The discontinuities are shown graphically by the existence of asymptotes at these values of x.

In general, the sums, differences, products and quotients of continuous functions are continuous functions (except, of course, at the zeros of the denominator in the case of a quotient).

3.3 Differentiability

Consider a function $y = f(x)$ whose graph is represented in Fig. 3.5, and let P be a typical point on the curve with coordinates (x, y). The coordinates of a neighbouring point Q can be written

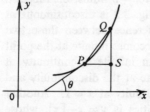

Fig. 3.5

as $(x+\delta x, y+\delta y)$, where the small change δx in x produces the small change δy in y. The expression

$$\frac{f(x+\delta x)-f(x)}{\delta x} = \tan QPS \qquad (19)$$

is then the slope of the straight line joining the points P and Q, and may be thought of as the mean value of the gradient of the curve $y = f(x)$ in the range $(x, x+\delta x)$. As the point Q approaches P, (19) may approach a limiting value given by

$$\lim_{\delta x \to 0} \left(\frac{f(x+\delta x)-f(x)}{\delta x} \right) = l \quad \text{(say)}. \qquad (20)$$

If this limit exists then geometrically this implies the existence of a tangent such that $l = \tan \theta$, where θ is the angle between the tangent at P and the x-axis. We refer to (20) as the differential coefficient of y with respect to x and denote it by the symbol $\frac{dy}{dx}$. Sometimes, however, it is convenient to denote $\frac{dy}{dx}$ by $f'(x)$, $\frac{df}{dx}$, or by Dy or Df, where D is the operator $\frac{d}{dx}$. We shall meet the idea of an operator in later chapters of this book, and, in particular, the D-operator in Chapter 21.

A function $y = f(x)$ is said to be differentiable if it possesses a differential coefficient, and to be differentiable at a point $x = a$ if $\frac{dy}{dx}$ (or $f'(x)$) exists at that point.

From the definition of the differential coefficient as a limit we may obtain the differential coefficient of any function of one variable. In the same way we may also derive the well-known rules for differentiating the product and quotient of two functions. It is assumed here that the reader is familiar with these ideas, and that the following examples will be sufficient to illustrate the technique of differentiating from first principles.

Example 2. The differential coefficient of $y = \sin x$ is obtained by evaluating

$$\frac{d(\sin x)}{dx} = \lim_{\delta x \to 0} \left(\frac{\sin (x+\delta x)-\sin x}{\delta x} \right) \qquad (21)$$

$$= \lim_{\delta x \to 0} \left\{ \frac{2 \sin \dfrac{\delta x}{2} \cos \left(x + \dfrac{\delta x}{2} \right)}{\delta x} \right\} \tag{22}$$

$$= \lim_{\delta x \to 0} \left(\frac{\sin \dfrac{\delta x}{2}}{\delta x/2} \right) \cdot \lim_{\delta x \to 0} \left\{ \cos \left(x + \frac{\delta x}{2} \right) \right\} \tag{23}$$

(by Theorem 2, 3.1).

As $\delta x \to 0$, the first limit becomes equal to unity, and the second to $\cos x$. Hence

$$\frac{d}{dx} (\sin x) = \cos x. \tag{24}$$

Example 3. If f and g are two functions of x, then

$$\frac{d}{dx} (fg) = f \frac{dg}{dx} + g \frac{df}{dx}, \tag{25}$$

and

$$\frac{d}{dx} \left(\frac{f}{g} \right) = \frac{g \dfrac{df}{dx} - f \dfrac{dg}{dx}}{g^2}. \tag{26}$$

Both of these well-known formulae can be proved from first principles, and we illustrate this statement by deriving (26).

Now

$$\frac{d}{dx} \left(\frac{f}{g} \right) = \lim_{\delta x \to 0} \left\{ \frac{\dfrac{f(x+\delta x)}{g(x+\delta x)} - \dfrac{f(x)}{g(x)}}{\delta x} \right\} \tag{27}$$

$$= \lim_{\delta x \to 0} \left\{ \frac{f(x+\delta x)g(x) - f(x)g(x+\delta x)}{g(x)g(x+\delta x)\,\delta x} \right\} \tag{28}$$

$$= \lim_{\delta x \to 0} \left\{ \frac{1}{g(x)g(x+\delta x)} \right.$$
$$\left. \times \left[g(x) \cdot \frac{f(x+\delta x)-f(x)}{\delta x} - f(x) \frac{g(x+\delta x)-g(x)}{\delta x} \right] \right\}, \tag{29}$$

28

which, by using the theorems on limits stated in 3.1, and the definition of the differential coefficient, reduces to

$$\frac{d}{dx}\left(\frac{f}{g}\right) = \frac{g\frac{df}{dx} - f\frac{dg}{dx}}{g^2}, \qquad (30)$$

as required.

Example 4. The differential coefficients of the inverse circular functions $\sin^{-1} x$, $\cos^{-1} x$, (sometimes written as arc sin x, arc cos x) may be obtained as follows:

If $y = \sin^{-1} x$, then $x = \sin y$.

Hence
$$\frac{dx}{dy} = \cos y \qquad (31)$$

and
$$\frac{dy}{dx} = \frac{1}{\cos y} = \frac{1}{\sqrt{(1-\sin^2 y)}} = \frac{1}{\sqrt{(1-x^2)}}. \qquad (32)$$

It is usual to take the positive sign of the square root in (32) to define the differential coefficient of the principal value of $\sin^{-1} x$, the principal value being such that $-\pi/2 \leq \sin^{-1} x \leq \pi/2$. When principal values of many-valued functions are implied it is usual to write the functions with capital letters. For example,

$$\frac{d}{dx}(\mathrm{Sin}^{-1} x) = \frac{1}{\sqrt{(1-x^2)}} \quad \text{and} \quad \frac{d}{dx}(\mathrm{Cos}^{-1} x) = -\frac{1}{\sqrt{(1-x^2)}},$$

where the principal value of $\cos^{-1} x$ is such that $0 \leq \cos^{-1} x \leq \pi$.

In the next chapter we shall consider the operation of indefinite integration. This is the inverse operation to differentiation in that the differential coefficient of the indefinite integral of a function is the function itself.

3.4 Continuity and Differentiability

Continuity and differentiability are closely related in the sense that, if $f(x)$ is a function of x and $\frac{df}{dx}$ exists at $x = a$, then $f(x)$ is continuous at $x = a$. This follows since, if $f(x)$ were not continuous at $x = a$, $f(a+\delta x)-f(a)$ would not tend to zero as $\delta x \to 0$, and consequently

$$\lim_{\delta x \to 0}\left\{\frac{f(a+\delta x)-f(a)}{\delta x}\right\} \qquad (33)$$

(which is the differential coefficient at $x = a$) could not exist. Hence differentiability at a point implies continuity, whilst discontinuity implies non-differentiability. The converse, however, is not true; continuity does not imply differentiability. This may be easily seen by considering the function represented graphically in Fig. 3.6. At

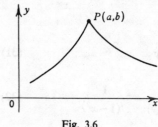

Fig. 3.6

the point P (a, b) the curve is continuous despite the ' kink ' since the function is defined and the limit of the function as $x \to a$ from either direction is equal to $f(a)$. The differential coefficient, however, is not uniquely defined at P (a, b) since a definite tangent to the curve at this point does not exist. The function is not differentiable therefore at this point, although (as shown) it is differentiable everywhere else. As an example, we mention the function $f(x) = x \sin\frac{1}{x}, f(0) = 0$, which is continuous at $x = 0$ but not differentiable there. Certain functions, moreover, are known to be continuous for all x and yet differentiable at none. Such functions are usually termed ' pathological ' (i.e. ill) and are not often of any great interest in physical applications.

3.5 Rolle's Theorem and the Mean-Value Theorem
(i) Rolle's Theorem

If $f(x)$ is continuous in the interval $a \leqq x \leqq b$ and differentiable in $a < x < b$, and if $f(a) = f(b) = 0$, then, provided $f(x)$ is not identically zero for $a < x < b$, there exists at least one value of x (say $x = c$) such that $f'(c) = 0$, where $a < c < b$. In other words, there must exist at least one maximum or minimum in the interval (a, b).

The validity of this theorem may be easily illustrated geometrically (see Fig. 3.7).

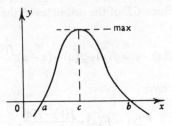

Fig. 3.7

(ii) *First Mean-Value Theorem*

If $f(x)$ is a continuous function of x in the interval $a \leqq x \leqq b$ and is differentiable in $a < x < b$ then there exists at least one value of x (say $x = c$) lying in the interval (a, b) such that

$$f'(c) = \frac{f(b) - f(a)}{b - a}. \tag{34}$$

In other words, considered graphically (see Fig. 3.8), there exists a value $x = c$ such that the tangent to the curve at this point is parallel to the chord AB.

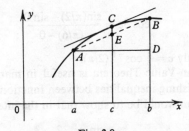

Fig. 3.8

We may prove this theorem geometrically in the following way: the equation of the line AB is

$$y = f(a) + (x - a)\frac{f(b) - f(a)}{b - a}, \tag{35}$$

since $BD = f(b) - f(a)$ and $AD = b - a$.

31

Hence the difference CE of the ordinates of the curve AB and the straight line AB is

$$F(x) = f(x) - y = f(x) - f(a) - (x-a)\frac{f(b)-f(a)}{b-a}. \tag{36}$$

Differentiating we have

$$F'(x) = f'(x) - \frac{f(b)-f(a)}{b-a}, \tag{37}$$

which is a defined quantity in $a < x < b$.

Also $F(a) = F(b) = 0$ since the curve AB and the straight line AB intersect at these points. Hence the function $F(x)$ satisfies Rolle's Theorem and consequently there exists a value of x (say $x = c$) such that $F'(c) = 0$. This implies (from (37)) that there exists a value $(x = c)$ such that

$$f'(c) = \frac{f(b)-f(a)}{b-a}, \tag{38}$$

which proves (34).

Example 5. If $f(x) = \sin 3x$, and $a = 0$, $b = \pi/6$, c can be found from the equation (see (34) or (38))

$$3\cos 3c = \frac{\sin(\pi/2) - \sin 0}{(\pi/6) - 0}. \tag{39}$$

This gives directly $c = \frac{1}{3}\cos^{-1}(2/\pi)$.

The first Mean-Value Theorem is useful in many ways; in particular in establishing inequalities between functions. For example, a typical problem would be to show that in the interval $0 < x < \pi/2$

$$1 > \frac{\sin x}{x} > \frac{2}{\pi}. \tag{40}$$

This is an extension of the inequality relation already obtained graphically in Chapter 2. Problems like this may be conveniently dealt with by using the following result:

If $f(x)$ is continuous in the range $a \leqq x \leqq b$, and differentiable in $a < x < b$, and if $f'(x) > 0$ in $a < x < b$, then for $a < x_1 < x_2 < b$

$$f(a) < f(x_1) < f(x_2) < f(b).$$

32

Similarly if $f'(x) < 0$ in $a < x < b$, then

$$f(a) > f(x_1) > f(x_2) > f(b)$$

for $a < x_1 < x_2 < b$.

These statements are obvious when represented graphically (see Figs. 3.9 and 3.10), but nevertheless we indicate an analytical proof here.

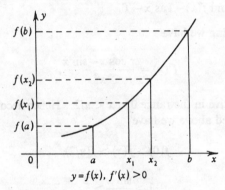

Fig. 3.9

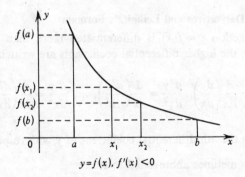

Fig. 3.10

Consider the case when $f'(x) > 0$. The first Mean-Value Theorem gives

$$f(x_1) - f(a) = (x_1 - a)f'(c), \qquad (41)$$

where $a < c < x_1$. But if $f'(x) > 0$, then $f'(c) > 0$. Also, by

33

assumption, $x_1 > a$. Hence

$$f(x_1) > f(a). \tag{42}$$

Similarly $f(x_2) > f(x_1)$ and $f(b) > f(x_2)$, and hence the statement is proved. A similar proof exists when $f'(x) < 0$.

Example 6. Consider now the inequality relation (40). Here $f(x) = \dfrac{\sin x}{x}$, and $f(x) \to 1$ as $x \to 0$.

Differentiating we have

$$f'(x) = \frac{x \cos x - \sin x}{x^2}, \tag{43}$$

which is negative in the range $0 < x < \pi/2$. Hence according to the results obtained above we have

$$f(0) > f(x) > f(\pi/2), \tag{44}$$

which gives
$$1 > \frac{\sin x}{x} > \frac{2}{\pi}. \tag{45}$$

3.6 Higher Derivatives and Leibnitz's Formula

When a function $y = f(x)$ is differentiated more than once with respect to x, the higher differential coefficients are written as

$$\frac{d^2 y}{dx^2} = \frac{d}{dx}\left(\frac{dy}{dx}\right), \; \frac{d^3 y}{dx^3} = \frac{d}{dx}\left(\frac{d^2 y}{dx^2}\right), \; ..., \; \frac{d^n y}{dx^n} = \frac{d}{dx}\left(\frac{d^{n-1} y}{dx^{n-1}}\right),$$

where $\dfrac{d^n y}{dx^n}$ is the nth differential coefficient of y with respect to x.

(These are sometimes abbreviated to either

$$f''(x), \; f'''(x) \; ... \; f^{(n)}(x)$$

or
$$D^2 y, \quad D^3 y \quad ... \quad D^n y,$$

where $D \equiv d/dx$.)

We now give a few examples showing how the nth differential coefficients of some simple functions may be obtained.

34

Example 7. If $y = \sin x$, then

$$Dy \equiv \frac{dy}{dx} = \cos x = \sin\left(\frac{\pi}{2} + x\right),$$

$$D^2 y \equiv \frac{d^2 y}{dx^2} = -\sin x = \sin(\pi + x),$$

$$D^3 y \equiv \frac{d^3 y}{dx^3} = -\cos x = \sin\left(\frac{3\pi}{2} + x\right),$$

and in general

$$D^n y \equiv \frac{d^n y}{dx^n} = \sin\left(\frac{n\pi}{2} + x\right). \tag{46}$$

Example 8. If $y = \log_e x$, then

$$Dy = 1/x,$$

$$D^2 y = -1/x^2,$$

$$D^3 y = 2/x^3,$$

and

$$D^n y = (-1)^{n-1} \frac{(n-1)!}{x^n}. \tag{47}$$

(Equation (47) is valid for all n, including $n = 1$, if by 0! we mean unity.)

In these two examples the functions have been simple enough to enable the nth differential coefficient to be written down in a few lines. When, however, the nth differential coefficient of a product of two functions $u(x)$ and $v(x)$ is required it is better to proceed as follows:

We have shown earlier from first principles that

$$D(uv) = u\,Dv + v\,Du. \tag{48}$$

Differentiating (48) now gives

$$D^2(uv) = u\,D^2 v + 2Du \cdot Dv + v\,D^2 u. \tag{49}$$

Similarly we obtain

$$D^3(uv) = u\,D^3v + 3Du\,.\,D^2v + 3D^2u\,.\,Dv + v\,D^3u, \tag{50}$$

$$D^4(uv) = u\,D^4v + 4Du\,.\,D^3v + 6D^2u\,.\,D^2v + 4D^3u\,.\,Dv + v\,D^4u, \tag{51}$$

and so on.

By inspection of these results the following formula (due to Leibnitz) may be written down for the nth differential coefficient of uv:

$$D^n(uv) = u\,D^nv + {}^nC_1Du\,.\,D^{n-1}v + {}^nC_2D^2u\,.\,D^{n-2}v + \ldots$$
$$+ {}^nC_{n-1}D^{n-1}u\,.\,Dv + v\,D^nu, \tag{52}$$

where $\quad {}^nC_r = \dfrac{n!}{(n-r)!\,r!}$.

This may be written more concisely as

$$D^n(uv) = \sum_{r=0}^{n} {}^nC_r D^{n-r}v\,.\,D^ru. \tag{53}$$

Leibnitz's formula (52) may be proved by induction as follows. Suppose (52) is true for one value of n, say m; then by differentiating we find

$$D^{m+1}(uv) = (uD^{m+1}v + Du\,.\,D^mv) + {}^mC_1(Du\,.\,D^mv + D^2u\,.\,D^{m-1}v)$$
$$+ {}^mC_2(D^2u\,.\,D^{m-1}v + D^3u\,.\,D^{m-2}v) + \ldots$$
$$+ (Dv\,.\,D^mu + v\dot{D}^{m+1}u), \tag{54}$$

$$= uD^{m+1}v + (1 + {}^mC_1)Du\,.\,D^mv + ({}^mC_1 + {}^mC_2)\,D^2u\,.\,D^{m-1}v$$
$$+ \ldots + vD^{m+1}u. \tag{55}$$

Now $\qquad\qquad\qquad {}^mC_{r-1} + {}^mC_r = {}^{m+1}C_r \tag{56}$

and hence (55) becomes

$$D^{m+1}(uv) = uD^{m+1}v + {}^{m+1}C_1Du\,.\,D^mv + {}^{m+1}C_2D^2u\,.\,D^{m-1}v$$
$$+ \ldots + vD^{m+1}u. \tag{57}$$

This again is the Leibnitz formula (52) with $m+1$ in place of m. Hence if the formula is true for $n = m$, it is certainly true for $n = m+1$.

However, we know (from first principles) that it is true for $n = 1$, and therefore it is true for $n = 2, 3, \ldots$, and consequently for all positive integral values of n.

Example 9. To obtain the nth differential coefficient of $y = (x^2 + 1)e^{2x}$ we put $x^2 + 1 = u$ and $e^{2x} = v$. Then by (52)

$$D^n y = D^n(uv) = (x^2 + 1)2^n e^{2x} + 2nx \cdot 2^{n-1} e^{2x} + n(n-1)2^{n-2} e^{2x} \quad (58)$$

$$= 2^{n-2} e^{2x}(4x^2 + 4nx + n^2 - n + 4). \quad (59)$$

Example 10. The nth differential coefficient of $y = x \log_e x$ may be obtained by putting $x = u$ and $\log_e x = v$, and using (47) for the nth differential coefficient of $\log_e x$. Equation (52) then gives

$$D^n y = D^n(uv) = x(-1)^{n-1} \frac{(n-1)!}{x^n} + n(-1)^{n-2} \frac{(n-2)!}{x^{n-1}} \quad (60)$$

$$= (-1)^{n-2} \frac{(n-2)!}{x^{n-1}}, \quad (n \geq 2). \quad (61)$$

Example 11. Leibnitz's formula may also be applied to a differential equation to obtain a relation between successive differential coefficients. As this forms a step towards finding power series solutions of certain types of differential equations (see Chapters 6 and 22) we now consider the following problem. Suppose y satisfies the equation

$$\frac{d^2 y}{dx^2} + x^2 y = \sin x. \quad (62)$$

Then differentiating each term n times (using Leibnitz's formula for the product term $x^2 y$), we obtain (using (46))

$$D^{n+2} y + (x^2 D^n y + 2nx\, D^{n-1} y + n(n-1)D^{n-2} y) = \sin\left(\frac{n\pi}{2} + x\right), \quad (63)$$

which is a relation between the $(n-2)$th, $(n-1)$th, nth and $(n+2)$th differential coefficients of y for all x. If we now put $x = 0$ in (63) we find

$$D^{n+2} y + n(n-1)D^{n-2} y = \begin{cases} 0 & \text{if } n \text{ is even,} \\ \pm 1 & \text{if } n \text{ is odd.} \end{cases} \quad (64)$$

Relations of this type between differential coefficients at $x = 0$ are useful in developing power series solutions of differential equations as we shall see in Chapter 22.

3.7 Maxima and Minima

A particularly important application of differentiation is to the problem of finding the maxima and minima values of a given function $f(x)$ in some interval $a \leqq x \leqq b$. Purely on geometrical grounds we can see (Fig. 3.11) that provided $f(x)$ is differentiable in the range (a, b) then at a maximum or minimum the tangent to the curve must be parallel to the x-axis. Accordingly a necessary condition for a point x_0 (say) to be a maximum or a minimum is that

$$f'(x_0) = 0 \quad \left(\text{i.e. } \frac{df(x)}{dx} = 0 \quad \text{at } x = x_0 \right). \tag{65}$$

Such points are called critical points.

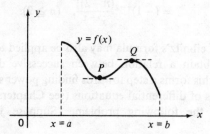

Fig. 3.11

Although at the end points $x = a$ and $x = b$ it would seem that the function possesses larger and smaller values respectively than at the maximum point Q and the minimum point P, we do not count these as true maxima and minima but note that they are just the greatest and least values of the function in the range $a \leqq x \leqq b$.

Now suppose $f'(x) > 0$. Then the function $y = f(x)$ increases with increasing x. If $f''(x) > 0$ then $f'(x)$ is also increasing and hence the curve is concave upwards (as near the minimum P). If, however, $f''(x) < 0$, the curve is concave downwards (as near the maximum point Q). Hence if $f'(x_0) = 0$ and $f''(x_0) > 0$ the point x_0 is a minimum point, whilst if $f'(x_0) = 0$ and $f''(x_0) < 0$ the point x_0 is a maximum point. It may happen that both $f'(x_0)$ and $f''(x_0)$

vanish (for example, $f(x) = x^3$ has a critical point at $x = x_0 = 0$, and $f''(0) = 0$). Such points are called points of inflection. A more detailed theory based on Taylor series (see Chapter 6) enables the nature of a critical point to be determined when the first n (say) derivatives vanish at the critical point. However, we shall not deal with this situation here.

Finally, we note that we have so far assumed that the function is continuous and has a continuous first derivative. If the function is not differentiable then it may still possess maxima and minima but they cannot be found by differentiation. For example, the function $y = |x|$ is shown in Fig. 3.12. This is not a differentiable function for the range $-a \leqq x \leqq a$ (say). However, a true minimum does exist at $x = 0$.

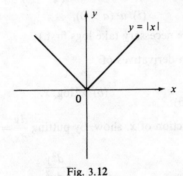

Fig. 3.12

PROBLEMS 3

1. Evaluate the following limits

(a) $\displaystyle\lim_{x \to \infty} \frac{x+1}{x+2}$, (b) $\displaystyle\lim_{x \to \infty} \frac{x^3+3}{2x^3+4x+1}$, (c) $\displaystyle\lim_{x \to 0} \frac{\sin^{-1} x}{x}$,

(d) $\displaystyle\lim_{x \to 0} \frac{\sqrt{(x+2)}-\sqrt{2}}{x}$, (e) $\displaystyle\lim_{x \to \infty} [\sqrt{(x+1)}-\sqrt{x}]$,

(f) $\displaystyle\lim_{x \to 0} \frac{1-\cos x}{x^2}$, (g) $\displaystyle\lim_{x \to 0} \frac{\sin^{-1} x}{\sin x}$,

(h) $\displaystyle\lim_{x \to 0} \frac{\sec x - \cos x}{\sin x}$, (i) $\displaystyle\lim_{x \to 1} \frac{1+\cos \pi x}{\tan^2 \pi x}$.

2. Find the points of discontinuity of the following functions

   (a) $\dfrac{x^3+4x+6}{x^2-6x+8}$,   (b) $\sec x$,   (c) $\dfrac{\sin x}{\sqrt{x}}$.

3. Differentiate from first principles $\sqrt{\left(\dfrac{1-x}{1+x}\right)}$ and $\sqrt{(a^2-x^2)}$.

4. Differentiate

   (a) $\log_e \cos\left(\dfrac{1}{x}\right)$,   (b) e^{3x^2},   (c) $\sin^{-1}\dfrac{x}{x+1}$,   (d) $e^{\sin^2 x}$,

   (e) $\sin(\cos x)$,   (f) $\log_e \cos\left(\dfrac{\pi}{4}-x^2\right)$,   (g) $x^{\cos x}$.

   (h) x^x,   (i) $a^x (a>0)$,   (j) $\log_e(\sin^{-1} x^2)$.

   (Hint: where necessary take logs first.)

5. Find the nth derivatives of

   (a) $\dfrac{1}{\sqrt{(ax+b)}}$,   (b) $x^4 \log_e x$.

6. If y is a function of x, show, by putting $\dfrac{dy}{dx}=p$, that

$$\frac{d^2x}{dy^2}=\frac{-\dfrac{d^2y}{dx^2}}{\left(\dfrac{dy}{dx}\right)^3}.$$

7. A function $f(x)$ is defined by

$$f(x)=\begin{cases}(x-1)^3, & x\leqq 1 \\ (x-1)^4, & x\geqq 1.\end{cases}$$

   How many times is $f(x)$ differentiable at $x=1$?

8. Sketch the graph of the function $f(x)$ defined by

$$f(x)=\begin{cases}e^{-x}+2x & \text{for } x\geqq 0, \\ e^x & \text{for } x<0.\end{cases}$$

   Sketch the curves of the first, second and third derivatives and show that the third derivative is discontinuous at $x=0$.

9. Show that the nth differential coefficients of

$$e^{-ax} \cos (bx+c), (a > 0) \text{ and } e^{-x} \sin x$$

are

$$r^n e^{-ax} \cos (bx+c+n\varepsilon),$$

where $r \cos \varepsilon = -a, \ r \sin \varepsilon = b$, and $r = +\sqrt{(a^2+b^2)}$, and

$$(\sqrt{2})^n e^{-x} \sin \left(\frac{3n\pi}{4}+x\right),$$

respectively.

10. If $y = \sin (\pi\sqrt{(x+1)})$, prove that

$$4(x+1)\frac{d^2y}{dx^2}+2\frac{dy}{dx}+\pi^2 y = 0,$$

and hence by differentiating n times using Leibnitz's theorem obtain a relation between any three successive derivatives of y.

(L.U.)

11. Prove that the nth derivative of the function y/x is

$$(-1)^n\frac{n!}{x^{n+1}}\left(y-xDy+\frac{x^2}{2!}D^2y-\frac{x^3}{3!}D^3y+...+(-1)^n\frac{x^n}{n!}D^ny\right),$$

where y is any function of x.
    Prove that
$$D^n\left(\frac{e^{-x}}{x}\right) = \frac{(-1)^n}{x^{n+1}}\left(n!-\int_0^x t^n e^{-t}\,dt\right).$$

(L.U.)

12. If $y = \sin^{-1}x+(\sin^{-1}x)^2$, prove that

$$(1-x^2)\frac{d^2y}{dx^2}-x\frac{dy}{dx}$$

is independent of x and deduce that, for $n > 1$,

$$(1-x^2)\frac{d^{n+2}y}{dx^{n+2}}-x(2n+1)\frac{d^{n+1}y}{dx^{n+1}}-n^2\frac{d^ny}{dx^n} = 0.$$

Show that the value of $\dfrac{d^{2r+1}y}{dx^{2r+1}}$ when $x = 0$ is $\dfrac{1}{2^{2r}}\left\{\dfrac{(2r)!}{r!}\right\}^2$.

(L.U.)

13. Determine the maxima and minima values (if any) of

    (a) $\sin^{-1}(x^2+2)$,   (b) $1+x^{2/3}$.

14. Find the critical points of $y = x^2 e^{-x}$, and determine whether they are maxima or minima.

CHAPTER 4

Indefinite, Definite and Improper Integrals

4.1 Definitions

We have already mentioned in Chapter 3 that integration may be regarded as the inverse operation to differentiation. However, this is not the only way of introducing the concept of an integral and the reader is no doubt familiar to some extent with the definition of an integral as the area under a curve. These two approaches lead, respectively, to definitions of (a) the indefinite integral and (b) the definite Riemann integral.

We now discuss these two types of integrals in more detail.

(a) Indefinite Integrals

If $F(x)$ and $f(x)$ are two functions of x such that

$$\frac{dF(x)}{dx} = f(x), \tag{1}$$

then $F(x)$ is said to be an indefinite integral of $f(x)$ and is written as

$$F(x) = \int f(x)\, dx. \tag{2}$$

The function $f(x)$ is called the integrand and is said to be integrable if $F(x)$ exists.

There is always a certain degree of freedom in defining the indefinite integral of a function since if (2) is an indefinite integral of $f(x)$, then (by (1)) so also is $F(x) + c$, where c is an arbitrary constant. It is usual to refer to c as a constant of integration and to always include it in evaluating indefinite integrals. For example,

$$\int \sin x\, dx = c - \cos x, \text{ since } \frac{d}{dx}(c - \cos x) = \sin x, \tag{3}$$

43

$$\int \frac{1}{x}\,dx = \log cx, \text{ since } \frac{d}{dx}(\log cx) = \frac{1}{x}, \tag{4}$$

$$\text{and } \int \frac{dx}{\sqrt{(4-x^2)}} = c + \sin^{-1}\frac{x}{2}, \text{ since } \frac{d}{dx}\left(c + \sin^{-1}\frac{x}{2}\right) = \frac{1}{\sqrt{(4-x^2)}}. \tag{5}$$

A short list of indefinite integrals of elementary functions is given in Table 1 (integration constants being omitted). (For a comprehensive list see H. B. Dwight, Tables of Integrals and Other Mathematical Data, Macmillan, 1957.)

TABLE 1

$f(x)$	$F(x) = \int f(x)\,dx$	$f(x)$	$F(x) = \int f(x)\,dx$
$x^n\ (n \neq -1)$	$\dfrac{x^{n+1}}{n+1}$	$\cot x$	$\log|\sin x|$
$\dfrac{1}{x}$	$\log_e |x|$	$\sec^2 x$	$\tan x$
$e^{ax},\ (a \neq 0)$	$\dfrac{1}{a} e^{ax}$	$\operatorname{cosec}^2 x$	$-\cot x$
$a^x,\ (a > 0)$	$\dfrac{1}{\log_e a} a^x$	$\dfrac{1}{\sqrt{(a^2-x^2)}}$	$\sin^{-1}\dfrac{x}{a},\ (|x| < a)$
$\sin x$	$-\cos x$	$\dfrac{1}{a^2+x^2}$	$\dfrac{1}{a}\tan^{-1}\dfrac{x}{a}$
$\cos x$	$\sin x$	$\log_e x$	$x\log_e x - x$
$\tan x$	$\log_e|\sec x|$	$\sec x$	$\log_e|\sec x + \tan x|$

(b) The Definite Riemann Integral

A rigorous analytical definition of this integral would lead us too far into real analysis to be of any great value here and we therefore proceed geometrically. Consider a function $y = f(x)$ which is assumed to be positive and continuous in the range $a \leq x \leq b$ (see Fig. 4.1). Let the interval $(b-a)$ be divided into n parts by arbitrarily

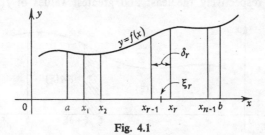

Fig. 4.1

choosing points $x_1, x_2, \ldots x_{n-1}$ on the x-axis such that

$$a(= x_0) < x_1 < x_2 < \ldots < x_{n-2} < x_{n-1} < b(= x_n). \tag{6}$$

Suppose further that ξ_r represents a point on the x-axis lying between the $(r-1)$th and rth points such that

$$x_{r-1} < \xi_r < x_r, \tag{7}$$

and that

$$\delta_r = x_r - x_{r-1}. \tag{8}$$

Then an approximation to the area of the rth strip is given by

$$f(\xi_r)\delta_r. \tag{9}$$

The whole area bounded by the curve, the x-axis and the lines $x = a$, $x = b$ is approximately represented by the sum of n expressions like (9) and is denoted by S_n, where

$$S_n = \sum_{r=1}^{n} f(\xi_r)\delta_r. \tag{10}$$

If the limit of this sum as $n \to \infty$ exists independently of the way of choosing x_r and ξ_r, and all $\delta_r \to 0$ then this limit-sum is called the definite integral of $f(x)$ from $x = a$ to $x = b$ and is written as

$$I = \lim_{n \to \infty} \sum_{r=1}^{n} f(\xi_r)\,\delta_r = \int_a^b f(x)\,dx. \tag{11}$$

The relation between this integral and the indefinite integral discussed in (a) may be found by considering the variation of the function $f(x)$ over a small interval $x = \xi$ to $x = \xi + \delta\xi$ (see Fig. 4.2). If f_{min} and f_{max} are respectively the least and greatest values of $f(x)$ in this

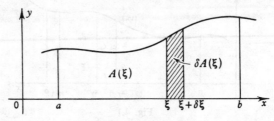

Fig. 4.2

interval, the elemental area δA (shown shaded) must satisfy the inequality

$$f_{min}\,\delta\xi < \delta A < f_{max}\,\delta\xi. \tag{12}$$

As $\delta\xi \to 0$, f_{min} and f_{max} both approach $f(\xi)$ and hence

$$\frac{dA}{d\xi} = f(\xi). \tag{13}$$

The area $A(\xi)$ under the curve from $x = a$ to $x = \xi$ is therefore (by (13)) an indefinite integral of $f(\xi)$. Suppose now that $F(\xi)$ is any indefinite integral of $f(\xi)$; then

$$F(\xi) = A(\xi) + c, \tag{14}$$

where c is an arbitrary constant. Since $A(a) = 0$, we have (with $\xi = b$)

$$\int_a^b f(x)\,dx = F(b) - F(a). \tag{15}$$

This is an important formula for computing the definite Riemann integral and we have demonstrated here that it is equivalent to the area under a continuous curve as given by (11). However, certain discontinuous functions which do not bound a finite area nevertheless possess a Riemann integral as defined by (15). The Riemannian theory of integration gives a detailed account of the relation between the integral regarded as a limit-sum and as defined by (15). We shall not go into these questions here, but shall always adopt (15) as the definition of the definite integral and where necessary invoke the equivalence with an area with the convention that areas above the x-axis are positive and those below, negative.

Finally, we give a simple example of definite integration as defined by (15):

$$\int_0^2 \frac{dx}{4+x^2} = \left(\frac{1}{2}\tan^{-1}\frac{x}{2}\right)_0^2 = \frac{1}{2}\tan^{-1}1 - \frac{1}{2}\tan^{-1}0 = \frac{\pi}{8}, \tag{16}$$

since $\dfrac{d}{dx}\left(\dfrac{1}{2}\tan^{-1}\dfrac{x}{2}\right) = \dfrac{1}{4+x^2}$.

4.2 Properties of Definite Integrals

In what follows, $f(x)$ and $g(x)$ are supposed to be integrable functions of x in the specified range of integration:

(a)
$$\int_a^b f(x)\,dx = -\int_b^a f(x)\,dx. \tag{17}$$

This is proved (using (15)) since

$$-\int_b^a f(x)\,dx = -\{F(a)-F(b)\} = F(b)-F(a) = \int_a^b f(x)\,dx.$$

(b) If $a \leq c \leq b$, then

$$\int_a^c f(x)\,dx + \int_c^b f(x)\,dx = \int_a^b f(x)\,dx. \tag{18}$$

Again this follows from (15) since

$$\{F(c)-F(a)\} + \{F(b)-F(c)\} = F(b)-F(a).$$

(c)
$$\int_a^b \{f(x)+g(x)\}\,dx = \int_a^b f(x)\,dx + \int_a^b g(x)\,dx. \tag{19}$$

This is proved using (15) again since the first integral is $F(b)+G(b)-F(a)-G(a)$, which when rearranged becomes

$$\{F(b)-F(a)\} + \{G(b)-G(a)\}.$$

47

(d) If k is a constant, then

$$\int_a^b kf(x) \, dx = k \int_a^b f(x) \, dx. \tag{20}$$

(e)

$$\int_0^a f(x) \, dx = \int_0^a f(a-x) \, dx. \tag{21}$$

This can be proved by putting $x = a - y$ in the first integral; then, using (17),

$$\int_0^a f(x) \, dx = -\int_a^0 f(a-y) \, dy = \int_0^a f(a-y) \, dy.$$

Whether we write y or x in the last integral is quite irrelevant, since the value of the integral depends only on the form of the integrand and the values of the limits of integration. Hence (21) is proved.

Example 1.

$$\int_0^{\pi/2} \sin^3 x \, dx = \int_0^{\pi/2} \sin^3 \{(\pi/2) - x\} dx = \int_0^{\pi/2} \cos^3 x \, dx,$$

and in general

$$\int_0^{\pi/2} f(\sin x) \, dx = \int_0^{\pi/2} f(\cos x) \, dx.$$

(f) If x is a variable upper limit of integration, a is a constant, and $f(x)$ is a continuous function then

$$\frac{d}{dx} \int_a^x f(t) \, dt = f(x). \tag{22}$$

This follows from the definition of the definite integral, for

$$\frac{d}{dx} \int_a^x f(t) \, dt = \frac{d}{dx} \{F(x) - F(a)\} = f(x).$$

Example 2.

$$\frac{d}{dx} \int_{\pi/4}^x \sin^2 t \, dt = \sin^2 x.$$

4.3 Particular Types of Integrands

(a) If $f(x)$ is an even function of x (see Chapter 1, 1.3 (h)), then $f(x) = f(-x)$ and

$$\int_{-a}^{a} f(x)\, dx = 2\int_{0}^{a} f(x)\, dx. \qquad (23)$$

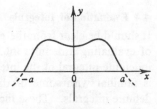

Fig. 4.3

This is obvious graphically since $f(x)$ is represented by a curve which is symmetrical about the y-axis (see Fig. 4.3). The areas under the curve from $x = -a$ to $x = 0$, and from $x = 0$ to $x = a$ are then equal in magnitude and sign.

(b) If $f(x)$ is an odd function of x, then $f(x) = -f(-x)$ and

$$\int_{-a}^{a} f(x)\, dx = 0. \qquad (24)$$

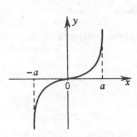

Fig. 4.4

This again is obvious graphically (see Fig. 4.4) since, although the areas from $-a$ to 0 and from 0 to a are numerically equal in virtue of the symmetry of $f(x)$ through the origin, they are of opposite sign and hence cancel out.

(c) If $f(x)$ is a periodic function of x of period T (see Chapter 1, 1.3 (i)), then $f(x+nT) = f(x)$ for all x ($n = 1, 2 \dots$) and we have

$$\int_{0}^{a} f(x)\, dx = \int_{T}^{T+a} f(x)\, dx, \qquad (25)$$

$$\int_{0}^{nT} f(x)\, dx = n\int_{0}^{T} f(x)\, dx. \qquad (26)$$

Again these results are obvious when considered graphically.

49

Example 3.

$$\int_0^{\pi/2} |\sin x| \, dx = \int_\pi^{3\pi/2} |\sin x| \, dx, \qquad (27)$$

and

$$\int_0^{2\pi} |\sin x| \, dx = 2\int_0^\pi |\sin x| \, dx. \qquad (28)$$

4.4 Evaluation of Integrals

It should be clear from the earlier parts of this chapter that one way of evaluating a definite integral is to find, in the first instance, an indefinite integral of the integrand. We assume here that the reader is familiar with some of the simple techniques for evaluating indefinite integrals. These include

(a) reducing the integral to standard form by either change of variable or by partial fractions,

(b) by directly integrating by parts,

and (c) by the use of reduction formulae.

The following examples illustrate the use of these methods.

Example 4. To evaluate

$$\int \frac{\sin \sqrt{x}}{\sqrt{x}} \, dx \qquad (29)$$

we put $\sqrt{x} = y$. Since $dx = 2y \, dy$, (29) now becomes

$$2\int \sin y \, dy = c - 2\cos y = c - 2\cos \sqrt{x}, \qquad (30)$$

where c is the constant of integration.

Example 5. To evaluate

$$\int \frac{dx}{5 + 3\cos x} \qquad (31)$$

we use the ' t ' substitution

$$\tan (x/2) = t, \qquad (32)$$

which gives

$$dx = \frac{2\,dt}{1+t^2}, \quad \sin x = \frac{2t}{1+t^2}, \quad \cos x = \frac{1-t^2}{1+t^2}. \tag{33}$$

Using the first and last of these results in (31), the integral becomes

$$\int \frac{dt}{4+t^2} = \frac{1}{2}\tan^{-1}\frac{t}{2}+c = \frac{1}{2}\tan^{-1}\left(\frac{1}{2}\tan\frac{x}{2}\right)+c, \tag{34}$$

where c is the constant of integration.

The integrals of cosec x and sec x given in Table 2 are easily found in a similar way.

Example 6. The integral

$$\int \sqrt{(a^2-x^2)}\,dx, \tag{35}$$

where a is a constant and $|x| < a$, may be evaluated by putting $x = a\sin\theta$, for (35) then becomes

$$a^2\int \cos^2\theta\,d\theta = \frac{a^2}{2}\int(1+\cos 2\theta)\,d\theta = \frac{a^2}{2}\left(\theta+\frac{\sin 2\theta}{2}\right)+c. \tag{36}$$

Hence, in terms of x,

$$\int \sqrt{(a^2-x^2)}\,dx = \frac{1}{2}\left\{a^2\sin^{-1}\frac{x}{a}+x\sqrt{(a^2-x^2)}\right\}+c. \tag{37}$$

Similarly the integrals $\int \sqrt{(x^2-a^2)}\,dx$ and $\int \sqrt{(x^2+a^2)}\,dx$ may be evaluated by putting $x = a\sec\theta$, and $x = a\tan\theta$, respectively.

Example 7. The integral

$$\int \frac{dx}{x(x^2+1)} \tag{38}$$

may best be evaluated by writing the integrand as partial fractions such that

$$\frac{1}{x(x^2+1)} = \frac{A}{x}+\frac{Bx+C}{x^2+1}. \tag{39}$$

Comparing the coefficients of identical powers of x on each side of (39), we find

$$A = 1, B = -1, C = 0,$$

and hence

$$\int \frac{dx}{x(x^2+1)} = \int \frac{dx}{x} - \int \frac{x\,dx}{x^2+1} = \log_e x - \tfrac{1}{2}\log_e(x^2+1)+c. \quad (40)$$

Example 8. In the case of integrals of the type

$$\int \frac{dx}{\sqrt{Q}}, \quad (41)$$

where Q is a polynomial in x of degree two, it is usually convenient to write Q as a perfect square so that the integral takes on a standard form. For example,

$$\int \frac{dx}{\sqrt{(3+2x-x^2)}} = \int \frac{dx}{\sqrt{\{4-(x-1)^2\}}} = \int \frac{du}{\sqrt{(4-u^2)}}, \quad (42)$$

where $u = (x-1)$. Putting $u = 2\sin\theta$, the last integral can be evaluated to give $\sin^{-1}\dfrac{u}{2} + c$. Hence

$$\int \frac{dx}{\sqrt{(3+2x-x^2)}} = \sin^{-1}\left(\frac{x-1}{2}\right)+c. \quad (43)$$

Example 9. To evaluate the integral

$$\int \frac{dx}{P\sqrt{Q}}, \quad (44)$$

where P is a linear function of x, and Q is quadratic as in Example 8, we put $P = 1/u$. For example,

$$\int \frac{dx}{(x+1)\sqrt{(x^2+4)}} \quad (45)$$

becomes, on putting $x+1 = 1/u$,

$$-\frac{1}{\sqrt{5}} \int \frac{du}{\sqrt{(u^2 - \frac{2}{5}u + \frac{1}{5})}} = -\frac{1}{\sqrt{5}} \int \frac{du}{\sqrt{\{(u - \frac{1}{5})^2 + \frac{4}{25}\}}}, \qquad (46)$$

which is of standard type. Finally we obtain

$$\int \frac{dx}{(x+1)\sqrt{(x^2+4)}} = c - \frac{1}{\sqrt{5}} \log_e \tfrac{1}{2}\{y + \sqrt{(y^2+4)}\}, \qquad (47)$$

here $y = \dfrac{5}{x+1} - 1$, and c is the constant of integration.

Example 10. Since (from Chapter 3) we have

$$\frac{d}{dx}(uv) = u\frac{dv}{dx} + v\frac{du}{dx}, \qquad (48)$$

where u and v are functions of x, it follows that

$$\int u\left(\frac{dv}{dx}\right) dx = uv - \int v\left(\frac{du}{dx}\right) dx. \qquad (49)$$

This is the formula for integration by parts. For example, to evaluate

$$\int x^2 e^{ax}\, dx, \qquad (50)$$

where a is a constant, we put $x^2 = u$ and $e^{ax} = \dfrac{dv}{dx}$. Then (49) gives

$$\int x^2 e^{ax}\, dx = \frac{x^2 e^{ax}}{a} - \frac{2}{a}\int x e^{ax}\, dx. \qquad (51)$$

In the last integral, we now put $x = u$, $e^{ax} = \dfrac{dv}{dx}$ and again integrate by parts; then

$$\int x e^{ax}\, dx = \frac{x e^{ax}}{a} - \frac{1}{a}\int e^{ax}\, dx = \frac{x e^{ax}}{a} - \frac{e^{ax}}{a^2} + c. \qquad (52)$$

Hence from (51) and (52)

$$\int x^2 e^{ax}\,dx = e^{ax}\left(\frac{x^2}{a} - \frac{2x}{a^2} + \frac{2}{a^3}\right) + c. \tag{53}$$

Similarly we may evaluate

$$\int x^n \log_e x\,dx \tag{54}$$

by putting $\log_e x = u$ and $x^n = \dfrac{dv}{dx}$. Then

$$\int x^n \log_e x\,dx = \frac{x^{n+1}}{n+1}\log_e x - \frac{1}{n+1}\int x^n\,dx \tag{55}$$

$$= \frac{x^{n+1}}{n+1}\left(\log_e x - \frac{1}{n+1}\right) + c, \tag{56}$$

provided $n \neq -1$.

Example 11. As an example of the use of reduction formulae in evaluating integrals we take

$$I_n = \int x^n e^x\,dx, \tag{57}$$

where n is a positive integer.

Integrating (57) by parts gives

$$I_n = x^n e^x - n\int x^{n-1} e^x\,dx,$$

or

$$I_n = x^n e^x - nI_{n-1}. \tag{58}$$

This result is termed a reduction formula for I_n (or a recurrence relation between I_n and I_{n-1}) and enables I_n to be expressed in terms of I and so on. For example, if $n = 4$, then by (58)

$$\begin{aligned}
I_4 &= x^4 e^x - 4I_3,\\
I_3 &= x^3 e^x - 3I_2,\\
I_2 &= x^2 e^x - 2I_1,\\
I_1 &= xe^x - I_0,
\end{aligned} \tag{59}$$

where
$$I_0 = \int e^x \, dx = e^x + c. \tag{60}$$

Hence

$$\int x^4 e^x \, dx = (x^4 - 4x^3 + 12x^2 - 24x + 24)e^x + c. \tag{61}$$

An important integral which may be evaluated by finding a reduction formula is

$$I_n = \int_0^{\pi/2} \cos^n x \, dx, \tag{62}$$

where n is a positive integer.

Writing I_n as $\int_0^{\pi/2} \cos x \cdot \cos^{n-1} x \, dx$ and integrating by parts we have

$$I_n = \left[\sin x \cos^{n-1} x \right]_0^{\pi/2} + (n-1) \int_0^{\pi/2} \sin^2 x \cos^{n-2} x \, dx, \tag{63}$$

or, since the first term on the right-hand side of (63) vanishes at both limits of integration,

$$I_n = (n-1)I_{n-2} - (n-1)I_n. \tag{64}$$

From (64) we find the relation

$$I_n = \left(\frac{n-1}{n} \right) I_{n-2}. \tag{65}$$

Hence, if n is even

$$\int_0^{\pi/2} \cos^n x \, dx = \frac{(n-1)(n-3)(n-5) \dots 3 \cdot 1}{n(n-2)(n-4) \dots 4 \cdot 2 \cdot} \, I_0, \tag{66}$$

where $I_0 = \int_0^{\pi/2} d\theta = \pi/2$, whilst, if n is odd

$$\int_0^{\pi/2} \cos^n x \, dx = \frac{(n-1)(n-3)(n-5) \dots 4 \cdot 2}{n(n-2)(n-4) \dots 5 \cdot 3} \, I_1, \tag{67}$$

where $I_1 = \int_0^{\pi/2} \cos x \, dx = 1.$

Since by 4.2 (e)

$$\int_0^{\pi/2} f(\cos x)\, dx = \int_0^{\pi/2} f(\sin x)\, dx, \tag{68}$$

the results (66) and (67) (which are usually known as Wallis's formulae) also apply to $\int_0^{\pi/2} \sin^n x\, dx$.

4.5 Other Methods of Integration

The integrals evaluated so far are all simple in the sense that they can be expressed explicitly in terms of the well-known elementary functions ($\sin x$, $\log_e x$, and so on). However, many functions can be integrated analytically only with the help of more complicated functions such as the elliptic functions and the Γ-function which are discussed in Chapters 11 and 13 respectively. In other cases it is possible to evaluate a difficult integral by differentiating a known integral a finite number of times. The conditions under which this operation may be carried out and its use in evaluating integrals are discussed in Chapter 10.

For those functions which are either too complicated to be integrated in terms of known functions or are defined only by a discrete set of numbers we must proceed numerically as shown in Chapter 14.

A more sophisticated and powerful method of evaluating definite integrals called 'contour integration' which requires an extensive knowledge of the properties of functions of a complex variable will not be dealt with in this book.

4.6 Inequalities for Integrals

If $f(x)$ and $g(x)$ are two integrable functions in the range $x = a$ to $x = b$ then from the definitions of an integral as an area we have

$$(a) \qquad \left| \int_a^b f(x)\, dx \right| \leqq \int_a^b |f(x)|\, dx, \quad \text{if } b > a, \tag{69}$$

and

$$(b) \qquad \int_a^b f(x)\, dx \leqq \int_a^b g(x)\, dx, \quad \text{if } f(x) \leqq g(x) \text{ and } b > a. \tag{70}$$

For example, if $f(x) = \cos x$ and $a = 0$, $b = 3\pi/2$, then (69) gives

$$\left| \int_0^{3\pi/2} \cos x \, dx \right| = 1 < \int_0^{3\pi/2} |\cos x| \, dx = 3, \tag{71}$$

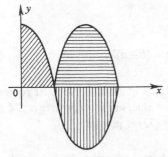

Fig. 4.5   Fig. 4.6

the first integral representing (see Fig. 4.5) the sum of the areas shaded with vertical and crosswise lines (areas below the x-axis counted as negative) and the second integral representing the area shaded with horizontal and crosswise lines. As an example of (b) we take $f(x) = \cos x$ and $g(x) = e^x$ in the interval $0 \leqq x \leqq \pi/2$. Then (70) gives

$$\int_0^{\pi/2} \cos x \, dx = 1 < \int_0^{\pi/2} e^x \, dx = e^{\pi/2} - 1, \tag{72}$$

since $\cos x \leqq e^x$ in the range of integration. This result is obvious when the integrals are interpreted graphically as the areas under the curves (see Fig. 4.6).

4.7 Mean-Value Theorems and Estimation of Integrals

Theorem 1. If $f(x)$ is continuous in $a \leqq x \leqq b$, then

$$\int_a^b f(x) \, dx = (b-a)f(\xi), \tag{73}$$

where $a \leqq \xi \leqq b$.

The proof of this theorem is as follows:

If $f(x)$ is continuous in $a \leqq x \leqq b$, it must have a least and a greatest value. If these values are f_1 and f_2 respectively, then

$$\int_a^b f_1 \, dx \leqq \int_a^b f(x) \, dx \leqq \int_a^b f_2 \, dx, \tag{74}$$

57

which gives

$$f_1(b-a) \leqq \int_a^b f(x)\,dx \leqq f_2(b-a), \tag{75}$$

or

$$\int_a^b f(x)\,dx = (b-a)k, \tag{76}$$

where $f_1 \leqq k \leqq f_2$. Now since $f(x)$ is continuous, k must represent the value of the function at some x-value, say $x = \xi$, where $a \leqq \xi \leqq b$ Hence $k = f(\xi)$, and consequently (73) is proved.

Theorem 2. If $f(x)$ and $g(x)$ are both continuous in $a \leqq x \leqq b$, and if $g(x)$ is positive in this range, then

$$\int_a^b f(x)g(x)\,dx = f(\xi)\int_a^b g(x)\,dx, \tag{77}$$

where $a \leqq \xi \leqq b$.

We may prove this theorem by considering (as in Theorem 1) the least and greatest values of $f(x)$ in $a \leqq x \leqq b$. Now since

$$f_1 g(x) \leqq f(x)g(x) \leqq f_2 g(x), \tag{78}$$

we have

$$f_1 \int_a^b g(x)\,dx \leqq \int_a^b f(x)g(x)\,dx \leqq f_2 \int_a^b g(x)\,dx, \tag{79}$$

or

$$\int_a^b f(x)g(x)\,dx = k\int_a^b g(x)\,dx, \tag{80}$$

where $f_1 \leqq k \leqq f_2$. Since $k = f(\xi)$ for $a \leqq \xi \leqq b$, (77) is proved.

Clearly the first mean-value theorem is a special case of the second with $g(x) = 1$.

The second mean-value theorem may be used to obtain a numerical estimate of the value of an integral as shown by the following examples.

58

Example 12. To obtain a numerical estimate of

$$I = \int_0^{\pi/2} e^{-x/15} \sin^2 x \, dx \tag{81}$$

we put $e^{-x/15} = f(x)$ and $\sin^2 x = g(x)$, and use (77). Then

$$I = e^{-\xi/15} \int_0^{\pi/2} \sin^2 x \, dx = (\pi/4) e^{-\xi/15}, \tag{82}$$

where $0 \leq \xi \leq \pi/2$. Hence I must satisfy the inequality relation

$$(\pi/4) e^0 \geqq I \geqq (\pi/4) e^{-\pi/30}, \tag{83}$$

or $\qquad\qquad \pi/4 \geqq I \geqq 0{\cdot}90(\pi/4). \tag{84}$

Example 13. If

$$I = \int_0^{\pi/4} \frac{d\theta}{\sqrt{(1 - \tfrac{7}{16} \sin^2 \theta)}}, \tag{85}$$

then (77) gives (with $g(\theta) = 1$)

$$I = \frac{1}{\sqrt{(1 - \tfrac{7}{16} \sin^2 \xi)}} \int_0^{\pi/4} d\theta = \frac{\pi}{4\sqrt{(1 - \tfrac{7}{16} \sin^2 \xi)}}, \tag{86}$$

where $0 \leqslant \xi \leqslant \pi/4$.

Hence

$$\frac{\pi}{4} \leqq I \leqq \frac{\pi}{4\sqrt{(1 - \tfrac{7}{32})}} = \frac{\pi\sqrt{2}}{5}. \tag{87}$$

The exact evaluation of (85), which is an elliptic integral, will be carried out in Chapter 11.

4.8 Improper Integrals

In defining the Riemann definite integral and showing its equivalence with an area under a curve we have assumed that the limits of integration a and b are finite and that the integrand remains finite in the range of integration. If either or both of these conditions are not satisfied the Riemann integral may not exist. We now consider separately the conditions for the Riemann integral to exist when

(a) the range of integration is infinite,

(b) the integrand becomes infinite in the range of integration.

(a) Infinite Range

Suppose $I = \int_a^b f(x)\, dx$ exists when a and b are finite.

Then we define the improper integral

$$\int_a^\infty f(x)\, dx = \lim_{b \to \infty} I, \tag{88}$$

so that if $\lim_{b \to \infty} I = l$, where l is finite, (88) is said to converge and to be equal to l. If $l \to \pm\infty$ (or does not exist) as $b \to \infty$, the integral is said to diverge.

Example 14. Consider

$$I = \int_1^\infty \frac{dx}{x^n}, \quad \text{where } n > 1. \tag{89}$$

Then $\int_1^\infty \frac{dx}{x^n} = \lim_{b \to \infty} \int_1^b \frac{dx}{x^n} = \lim_{b \to \infty} \frac{1}{(n-1)}\left(1 - \frac{1}{b^{n-1}}\right) = \frac{1}{n-1},$ (90)

which is finite. Hence $\int_1^\infty \frac{dx}{x^n}$ is convergent for $n > 1$.

Example 15. The integral

$$\int_a^\infty e^{kx}\, dx, \quad (k > 0) \tag{91}$$

is divergent since

$$\lim_{b \to \infty} \int_a^b e^{kx}\, dx = \frac{1}{k} \lim_{b \to \infty} (e^{kb} - e^{ka}) \tag{92}$$

is infinite.

Example 16. The integral

$$I = \int_0^\infty \cos x \, dx \qquad (93)$$

is divergent since

$$\lim_{b \to \infty} \int_0^b \cos x \, dx = \lim_{b \to \infty} (\sin b) \qquad (94)$$

does not exist, $\sin b$ oscillating between ± 1. In special circumstances a definite value may be assigned to (93) by considering the integral

$$J = \int_0^\infty e^{-kx} \cos x \, dx = \frac{k}{1+k^2}. \qquad (95)$$

As $k \to 0$, $J \to I$ and $J \to 0$. Hence, in the limit, $I = 0$. The term e^{-kx} introduced into the integral is called a convergence factor. However, the reader is warned against this procedure, since the integral defined by the limit of (95) is not strictly identical with (93), which is truly divergent.

A simple test of convergence for any integral with an infinite range of integration is to consider the form of the integrand for large x. Suppose

$$I = \int_a^\infty f(x) \, dx, \qquad (96)$$

where $f(x)$ is assumed to be continuous in the range of integration. Then if, for large x,

$$f(x) = \frac{g(x)}{x^n}, \qquad (97)$$

where $g(x)$ is non-zero, $|g(x)| < \beta$ (where β is a constant) and $n > 1$, the integral (96) is convergent. This test is only a test of sufficiency in that many integrands not satisfying these conditions have convergent integrals.

Example 17. If

$$I = \int_0^\infty \frac{\sin^2 x}{1+x^2}\, dx, \tag{98}$$

the integrand may be written for large x as $\dfrac{\sin^2 x}{x^2}$. Comparing with (97) we have $g(x) = \sin^2 x$, $|g(x)| \leq 1$ and $n = 2$. Hence (98) is convergent.

Example 18. The integral

$$\int_0^\infty \frac{\sin x}{x}\, dx \tag{99}$$

does not satisfy the simple test for convergence since $n = 1$. Nevertheless the integral is convergent and equal to $\pi/2$ (see Chapter 10).

(b) Integrand Infinite in Range of Integration

Suppose

$$I = \int_a^b f(x)\, dx, \quad (a \text{ and } b \text{ finite}) \tag{100}$$

and $f(x) \to \pm \infty$ as $x \to a$. Then by (100) we mean

$$I = \lim_{\varepsilon \to 0} \int_{a+\varepsilon}^b f(x)\, dx. \tag{101}$$

If this limit is finite (and equal to l) the integral I is said to converge and to be equal to l. If the limit is infinite or does not exist, the integral is said to be divergent.

Example 19. The integral

$$\int_0^2 x^{-1/2}\, dx \tag{102}$$

has an integrand which becomes infinite at $x = 0$. We therefore consider

$$\int_0^2 x^{-1/2}\, dx = \lim_{\varepsilon \to 0} \int_\varepsilon^2 x^{-1/2}\, dx = \lim_{\varepsilon \to 0} 2(\sqrt{2} - \sqrt{\varepsilon}) = 2\sqrt{2}, \tag{103}$$

which is finite.

Hence (102) is convergent.

Example 20. To examine the convergence of

$$\int_0^1 \frac{dx}{x^2} \tag{104}$$

we consider

$$\lim_{\varepsilon \to 0} \int_\varepsilon^1 \frac{dx}{x^2} = \lim_{\varepsilon \to 0} \left(\frac{1}{\varepsilon} - 1 \right), \tag{105}$$

which becomes infinite as $\varepsilon \to 0$. Hence (104) diverges.

If the integrand $f(x)$ becomes infinite at the upper limit of integration similar arguments hold, and we define

$$\int_a^b f(x)\, dx = \lim_{\varepsilon \to 0} \int_a^{b-\varepsilon} f(x)\, dx. \tag{106}$$

Similarly if $f(x)$ becomes infinite at points $c_1, c_2, c_3 \ldots c_n$, where

$$a < c_1 < c_2 < c_3 < \ldots c_n < b,$$

then we write

$$\int_a^b f(x)\, dx = \int_a^{c_1} f(x)\, dx + \int_{c_1}^{c_2} f(x)\, dx + \ldots + \int_{c_n}^b f(x)\, dx, \tag{107}$$

and test each integral on the right-hand side of (107) for convergence.

A simple test of convergence will be stated here without proof. If $f(x)$ can be written as

$$f(x) = \frac{g(x)}{(x-a)^n}, \tag{108}$$

where $g(x)$ is non-zero when $x = a$ and bounded in (a, b), the integral

$$\int_a^b f(x)\, dx \tag{109}$$

converges when $n < 1$. We speak of n as the order of the infinity (or singularity) at $x = a$, and consequently if the order of every infinity of the integrand in the range of integration is less than unity the integral converges.

Example 21. The integral

$$I = \int_0^1 \frac{\sin x}{\sqrt{(1-x^2)}}\, dx = \int_0^1 \frac{\sin x}{(1+x)^{1/2}(1-x)^{1/2}}\, dx \qquad (110)$$

has an infinity at $x = 1$ of order $\frac{1}{2}$ only, and is therefore convergent.

Example 22. The integral

$$I = \int_0^4 \frac{dx}{(x-3)\sqrt{(1-x^2)}} = \int_0^4 \frac{dx}{(x-3)(1-x)^{1/2}(1+x)^{1/2}} \qquad (111)$$

has infinities at $x = 1$ and $x = 3$ of orders $\frac{1}{2}$ and 1, respectively, and is therefore divergent.

It should be remarked finally that if an integral has an infinite range of integration and an integrand with one or more infinities in this range, the convergence tests given in (a) and (b) must both be applied.

Suppose now that $f(x)$ becomes infinite at $x = c$, where $a < c < b$. Then

$$\int_a^b f(x)\, dx = \lim_{\varepsilon \to 0} \int_a^{c-\varepsilon} f(x)\, dx + \lim_{\eta \to 0} \int_{c+\eta}^b f(x)\, dx. \qquad (112)$$

It may happen, however, that although neither of the limits on the right-hand side of (112) exists, yet

$$\lim_{\varepsilon \to 0} \left\{ \int_a^{c-\varepsilon} f(x)\, dx + \int_{c+\varepsilon}^b f(x)\, dx \right\} \qquad (113)$$

exists. When this is so, (113) is called the Cauchy principal value of $\int_a^b f(x)\, dx$.

Example 23. The integral

$$\int_0^\infty \frac{dx}{x^2 - 4} \qquad (114)$$

does not exist in the sense of (112) but has a principal value (by (113)) equal to

$$\lim_{\varepsilon \to 0} \left\{ \int_0^{2-\varepsilon} \frac{dx}{x^2-4} + \int_{2+\varepsilon}^\infty \frac{dx}{x^2-4} \right\} = \lim_{\varepsilon \to 0} \left\{ \frac{1}{4} \log_e \left(\frac{4+\varepsilon}{4-\varepsilon} \right) \right\} = 0. \qquad (115)$$

PROBLEMS 4

1. Evaluate the following indefinite integrals

   (a) $\displaystyle\int x \sin x\, dx,$ (b) $\displaystyle\int \frac{x^5}{x^3-1}\, dx,$ (c) $\displaystyle\int \frac{dx}{4\cos x+3\sin x},$

   (d) $\displaystyle\int x^3 e^{x^2}\, dx,$ (e) $\displaystyle\int \frac{(x+a)\, dx}{(1+2ax+x^2)^{3/2}},$ (f) $\displaystyle\int \cos^{-1} x\, dx,$

   (g) $\displaystyle\int \frac{dx}{x^2-3x+2},$ (h) $\displaystyle\int \frac{x^{1/2}}{1+x}\, dx.$

2. Evaluate the following definite integrals

   (a) $\displaystyle\int_0^1 \frac{x^2}{\sqrt{(1-x^2)}}\, dx,$ (b) $\displaystyle\int_0^{\pi/2} \frac{dx}{1+\cos\alpha\cos x},$

   (c) $\displaystyle\int_0^\infty e^{-x}\sin x\, dx,$ (d) $\displaystyle\int_0^{\pi/6} \frac{\tan y}{1+\sin^2 y}\, dy,$

   (e) $\displaystyle\int_0^{\pi/2} \frac{dx}{(1+\cos x)^2},$ (f) $\displaystyle\int_0^{\pi/2} x\sin^2 x\, dx,$

   (g) $\displaystyle\int_0^{\pi/4} \sec^4 x\, dx,$ (h) $\displaystyle\int_2^\infty \frac{dx}{(x-1)\sqrt{(x^2-1)}},$

   (i) $\displaystyle\int_1^e x^{-1/2}\log_e x\, dx.$

3. Prove that $\displaystyle\int_0^\infty x^n e^{-x^2}\, dx = \tfrac{1}{2}(n-1)\int_0^\infty x^{n-2}e^{-x^2}\, dx,$ where $n>1$,

   and hence evaluate $\displaystyle\int_0^\infty x^5 e^{-x^2}\, dx.$ (C.U.)

4. If $\displaystyle I_{m,\,n} = \int_0^{\pi/2} \sin^m\theta\cos^n\theta\, d\theta,$ where m, n are positive integers,

65

show by setting up reduction formulae that

$$I_{m,n} = \left(\frac{m-1}{m+n}\right)I_{m-2,n} = \left(\frac{n-1}{m+n}\right)I_{m,n-2} = \left(\frac{n-1}{m+1}\right)I_{m+2,n-2}$$

$$= \left(\frac{m-1}{n+1}\right)I_{m-2,n+2}.$$

5. If $\log_e x$ is defined by the integral $\int_1^x \frac{1}{t}\, dt$, show that

$$\log_e x + \log_e y = \log_e xy.$$

6. Prove that, if $n > 2$,

$$\int \tan^n \theta\, d\theta = \frac{\tan^{n-1}\theta}{n-1} - \int \tan^{n-2}\theta\, d\theta.$$

Hence evaluate $\int_0^{\pi/3} \tan^4 \theta\, d\theta$ and $\int_0^{\pi/3} \tan^5 \theta\, d\theta.$ \hfill (C.U.)

7. Find the area enclosed by the curve $y = e^{-ax} \sin bx$ $(a > 0, b > 0)$ and the x-axis, between two successive roots of the equation $y = 0$. Hence evaluate

$$\int_0^\infty e^{-ax} |\sin bx|\, dx. \hfill \text{(C.U.)}$$

8. Prove that the nth differential coefficient with respect to x of $x^n e^{-x}$ is of the form $(-1)^n L_n(x)e^{-x}$, where $L_n(x)$ is a polynomial in x of degree n with the coefficient of x^n unity. If m, n are integers greater than or equal to unity prove that

$$\int_0^\infty L_n(x)L_m(x)e^{-x}\, dx = (-1)^{n-1}\int_0^\infty \frac{d^{n-1}}{dx^{n-1}}(x^n e^{-x})\frac{dL_m(x)}{dx}\, dx,$$

and hence, by repeated integration by parts, prove that the integral on the left is equal to $(n!)^2$ if $m = n$, and to zero if $m \neq n$.
\hfill (C.U.)

9. Verify the Cauchy–Schwarz inequality

$$\left|\int_a^b f(x)g(x)\, dx\right|^2 \leq \left\{\int_a^b |f(x)|^2\, dx . \int_a^b |g(x)|^2\, dx\right\}$$

when $f(x) = x^2$, $g(x) = x - 1$ and $a = 0$, $b = 2$.

10. Estimate the values of the following integrals using the second mean-value theorem

(a) $\int_4^9 \dfrac{dx}{(2+\sqrt{x})^2}$,

(b) $\int_0^{\pi/4} \sqrt{(1+\sin^4 \theta)}\, d\theta$,

(c) $\int_{\pi/3}^{\pi} \dfrac{x\, dx}{1+\cos^2 x}$,

(d) $\int_0^{\pi/2} x\sqrt{(\sin x)}\, dx$,

(e) $\int_0^1 x^{1/4} e^{-x}\, dx$.

11. Examine the convergence of the following integrals and evaluate those which converge

$$\int_0^{\pi/2} \dfrac{dx}{\sqrt{(1-\cos x)}}, \quad \int_0^1 (1-x)\log_e x\, dx, \quad \int_0^{\infty} x^3 e^{-x^2}\, dx. \qquad \text{(L.U.)}$$

12. Find the values of α and β (if any) for which

$$\int_0^{\infty} |\sin x^{\alpha}|\, dx, \quad \int_0^1 x^{\alpha-1}(1-x)^{\beta-1}\, dx \quad \text{and} \quad \int_0^{\infty} \dfrac{x^{\alpha} \sin x\, dx}{1+x^2}$$

converge.

13. Show that the principal value of

$$\int_0^2 \dfrac{dx}{1-x^2} = \dfrac{1}{2}\log_e 3.$$

CHAPTER 5

Convergence of Infinite Series

5.1 Definitions

If $a_1, a_2, a_3 \ldots$ is a given sequence of numbers, the sum of the first n numbers is called the nth partial sum and is represented by

$$S_n = a_1 + a_2 + a_3 + \ldots + a_n = \sum_{r=1}^{n} a_r. \tag{1}$$

If the partial sums $S_1, S_2, S_3 \ldots$ converge to a finite limit S, where

$$S = \lim_{n \to \infty} S_n, \tag{2}$$

then S is defined as the sum of the infinite series

$$a_1 + a_2 + \ldots = \sum_{r=1}^{\infty} a_r, \tag{3}$$

and the series is said to be convergent. When the sequence of partial sums tends to an infinite limit, or oscillates either finitely or infinitely, the series is said to be divergent.

Example 1. The series

$$\sum_{r=1}^{\infty} \frac{1}{r(r+1)} = \frac{1}{1 \cdot 2} + \frac{1}{2 \cdot 3} + \ldots, \tag{4}$$

has partial sums $S_1 = \frac{1}{2}$, $S_2 = \frac{2}{3}$, $S_3 = \frac{3}{4}$, $S_4 = \frac{4}{5}$, $S_5 = \frac{5}{6}$, ... which with increasing n tend to unity. Hence the series is convergent with a sum $S = 1$. This result can also be obtained by using the method of differences to sum the finite series

$$S_n = \sum_{r=1}^{n} \frac{1}{r(r+1)} \tag{5}$$

and then letting $n \to \infty$ in the result. For writing the rth term as

$$a_r = \frac{1}{r} - \frac{1}{r+1}, \tag{6}$$

we have

$$a_n = \frac{1}{n} - \frac{1}{n+1},$$

$$a_{n-1} = \frac{1}{n-1} - \frac{1}{n},$$

$$\vdots \qquad \vdots \qquad \vdots$$

$$a_2 = \tfrac{1}{2} - \tfrac{1}{3},$$

$$a_1 = 1 - \tfrac{1}{2},$$

$$\qquad (7)$$

which, on adding, give

$$S_n = \sum_{r=1}^{n} a_r = \sum_{r=1}^{n} \frac{1}{r(r+1)} = 1 - \frac{1}{n+1}. \qquad (8)$$

Hence $S = \lim_{n \to \infty} S_n = 1$, as before.

Example 2. The geometric series

$$\sum_{r=0}^{\infty} ak^r = a(1 + k + k^2 + \ldots) \qquad (9)$$

(where a is a constant) has an nth partial sum S_n given by

$$S_n = a \frac{1 - k^n}{1 - k}. \qquad (10)$$

Hence if $|k| < 1$,

$$S = \lim_{n \to \infty} S_n = \frac{a}{1-k} \qquad (11)$$

and the series is convergent.

The series is divergent, however, when $|k| \geqq 1$, since the partial sum S_n either increases without limit as $n \to \infty$, or oscillates either finitely ($k = -1$) or infinitely ($k < -1$).

5.2 Theorems on Series

Theorem 1. The series $\sum_{r=1}^{\infty} a_r$ cannot converge unless $\lim_{n \to \infty} a_n = 0$

69

This may be proved by considering the $(n-1)$th and nth partial sums given by

$$S_{n-1} = a_1 + a_2 + \ldots + a_{n-1}, \tag{12}$$

and

$$S_n = a_1 + a_2 + \ldots + a_n. \tag{13}$$

Subtracting (12) from (13) we have

$$S_n - S_{n-1} = a_n. \tag{14}$$

Now if the series converges to a sum S then

$$S = \lim_{n \to \infty} S_n = \lim_{n \to \infty} S_{n-1}, \tag{15}$$

and hence, from (14) and (15),

$$\lim_{n \to \infty} a_n = 0. \tag{16}$$

This condition is necessary but not sufficient for convergence in that there are many series satisfying (16) which nevertheless do not converge. The harmonic series

$$1 + \tfrac{1}{2} + \tfrac{1}{3} + \ldots = \sum_{r=1}^{\infty} \frac{1}{r} \tag{17}$$

is a good example of this since, although

$$\lim_{n \to \infty} a_n = \lim_{n \to \infty} \frac{1}{n} = 0, \tag{18}$$

the sum of the series is infinite (see next section). However, the series $\sum_{r=1}^{\infty} \cos \dfrac{\pi r}{4}$, for example, cannot converge since $\lim_{n \to \infty} \cos \dfrac{\pi n}{4} \neq 0$.

Theorem 2. If $\sum_{r=1}^{\infty} a_r = S$, then $\sum_{r=1}^{\infty} ka_r = kS$, where k is a constant.
This follows from the obvious identity

$$\sum_{r=1}^{n} ka_r = k \sum_{r=1}^{n} a_r \tag{19}$$

and proceeding to the limit $n \to \infty$.

Theorem 3. If $\sum\limits_{r=1}^{\infty} a_r = S$ and $\sum\limits_{r=1}^{\infty} b_r = T$, then

$$\sum_{r=1}^{\infty}(a_r+b_r)=S+T.$$

Again this theorem is proved by considering the identity

$$\sum_{r=1}^{n} (a_r+b_r) = \sum_{r=1}^{n} a_r + \sum_{r=1}^{n} b_r \qquad (20)$$

and then letting $n \to \infty$.

Theorem 4. If $\sum\limits_{r=1}^{\infty} a_r = S$, then $\sum\limits_{r=0}^{\infty} a_r = S + a_0$, where a_0 is any number.

Writing $\bar{S}_n = \sum\limits_{r=0}^{n} a_r$ and $S_n = \sum\limits_{r=1}^{n} a_r$,

we have $\qquad\qquad \bar{S}_n = S_n + a_0. \qquad (21)$

Hence, letting $n \to \infty$ in (21), the theorem is proved. This theorem shows that any new term may be introduced at the beginning of a series without affecting the convergence of the series. A simple extension of this result shows that the removal or insertion of a finite number of terms anywhere in the series does not affect its convergence.

5.3 Series of Positive Terms

When a series $\sum\limits_{r=1}^{\infty} a_r$ consists only of positive terms ($a_r \geqq 0$ for all r) it must either converge, or diverge to $+\infty$; it clearly cannot oscillate. Numerous tests of convergence are known for series of this type, and four such tests are given below:

(a) Comparison Test

If $\sum\limits_{r=1}^{\infty} a_r$ is a series of positive terms, and if $\sum\limits_{r=1}^{\infty} b_r$ is a series of positive terms that is known to converge, then $\sum\limits_{r=1}^{\infty} a_r$ is convergent

71

if $a_r \leqq b_r$ for all sufficiently large r. Similarly, if $\sum_{r=1}^{\infty} b_r$ is known to diverge then $\sum_{r=1}^{\infty} a_r$ is divergent if $a_r \geqq b_r$ for all sufficiently large r.

Since, by Theorem 4, the removal or insertion of a finite number of terms does not affect the convergence of a series, it can be assumed in the proof that follows that the condition $a_r \leqq b_r$ (and $a_r \geqq b_r$) holds for *all* r.

The proof of this test may easily be seen by a graphical argument. Suppose each term of a series represents the area of a rectangle of base equal to unity and height equal to the magnitude of the term (see Fig. 5.1). Then the sum of the series is represented by the sum of the

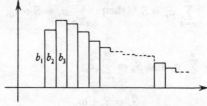

Fig. 5.1

areas of the rectangles. If now $\sum_{r=1}^{\infty} b_r$ converges to a sum then the total area of the rectangles must be finite, and if $a_r \leqq b_r$ for all r, the area of the rectangles representing the series a_r must also be finite. Hence $\sum_{r=1}^{\infty} a_r$ converges if $\sum_{r=1}^{\infty} b_r$ converges.

A similar argument applies to the second part of the test.

An analytic proof of the test may be obtained by considering the partial sums

$$S_n = \sum_{r=1}^{n} a_r, \quad T_r = \sum_{r=1}^{n} b_r. \tag{22}$$

Then $a_r \leqq b_r$ implies $S_n \leqq T_n$, and hence

$$\lim_{n \to \infty} S_n \leqq \lim_{n \to \infty} T_n = T, \tag{23}$$

where T is the sum of the convergent series $\sum_{r=1}^{\infty} b_r$. Now since $a_r \geqq 0$, S_n never decreases, and therefore

$$\lim_{n \to \infty} S_n = S \leqq T. \tag{24}$$

Hence, by (23), the first part of the test is proved. A similar argument exists for the second part of the test.

Example 3. The harmonic series

$$1+\tfrac{1}{2}+\tfrac{1}{3}+\tfrac{1}{4}+ \ldots = \sum_{r=1}^{\infty} \frac{1}{r} \qquad (25)$$

may be shown to be divergent by writing it as

$$1+\tfrac{1}{2}+(\tfrac{1}{3}+\tfrac{1}{4})+(\tfrac{1}{5}+\tfrac{1}{6}+\tfrac{1}{7}+\tfrac{1}{8})+ \ldots . \qquad (26)$$

The terms in brackets are now greater than $\tfrac{1}{2}$; by grouping terms together in this way throughout the series so that the value of each group exceeds $\tfrac{1}{2}$ we see, by comparison with the divergent series

$$1+\tfrac{1}{2}+\tfrac{1}{2}+\tfrac{1}{2}+ \ldots , \qquad (27)$$

that (26) is divergent. It should be noted, however, that bracketing of terms in series (as in (26)) is in general not possible without altering the character of the series (see 5.8).

Example 4. The p-series

$$1+\frac{1}{2^p}+\frac{1}{3^p}+\frac{1}{4^p}+ \ldots = \sum_{r=1}^{\infty} \frac{1}{r^p} \qquad (28)$$

converges if $p > 1$, and diverges if $p \leq 1$. We can prove these statements by taking the three cases $p \gtreqless 1$ separately:

(*a*) if $p = 1$, (28) becomes the harmonic series (25) and is consequently divergent;

(*b*) if $p < 1$, each term of (28) (apart from the first) is greater than the corresponding term of the harmonic series. The series is therefore divergent;

(*c*) if $p > 1$, we write the series as

$$1+\left(\frac{1}{2^p}+\frac{1}{3^p}\right)+\left(\frac{1}{4^p}+\frac{1}{5^p}+\frac{1}{6^p}+\frac{1}{7^p}\right)+ \ldots \qquad (29)$$

and continue grouping the terms throughout the series into brackets such that every bracket is less than the corresponding term of the series

$$1+\frac{2}{2^p}+\frac{4}{4^p}+ \dots. \tag{30}$$

Now (30) is a geometric series with a common ratio $k = 2^{1-p}$ which is known to be convergent for $|k| < 1$. Consequently (29) converges for $p > 1$.

(b) Ratio Comparison Test

If $\sum\limits_{r=1}^{\infty} a_r$ and $\sum\limits_{r=1}^{\infty} b_r$ are two series of positive terms and

$$\frac{a_{r+1}}{a_r} \leqq \frac{b_{r+1}}{b_r}$$

for all sufficiently large r, then $\sum\limits_{r=1}^{\infty} a_r$ converges when $\sum\limits_{r=1}^{\infty} b_r$ converges. Similarly, if

$$\frac{a_{r+1}}{a_r} \geqq \frac{b_{r+1}}{b_r},$$

then $\sum\limits_{r=1}^{\infty} a_r$ diverges when $\sum\limits_{r=1}^{\infty} b_r$ diverges.

To prove these results we assume first that $\frac{a_{r+1}}{a_r} \leqq \frac{b_{r+1}}{b_r}$ for *all* r (see the remarks in 5.3 *(a)*). Then writing

$$a_r = \frac{a_r}{a_{r-1}} \cdot \frac{a_{r-1}}{a_{r-2}} \cdot \frac{a_{r-2}}{a_{r-3}} \dots \frac{a_2}{a_1} \cdot a_1 \leqq \frac{b_r}{b_{r-1}} \cdot \frac{b_{r-1}}{b_{r-2}} \dots \frac{b_2}{b_1} \cdot a_1 \tag{31}$$

we have

$$a_r \leqq \frac{b_r a_1}{b_1}. \tag{32}$$

Since (32) is true for all r, the comparison test shows that $\sum\limits_{r=1}^{\infty} a_r$ converges when $\sum\limits_{r=1}^{\infty} b_r$ converges. The second part of the test may be proved in the same way.

Example 5. The series $\sum\limits_{r=1}^{\infty} a_r \equiv \sum\limits_{r=1}^{\infty} \dfrac{1}{r^3}$ is convergent since, using the convergent series $\sum\limits_{r=1}^{\infty} b_r \equiv \sum\limits_{r=1}^{\infty} \dfrac{1}{r^2}$, we have

$$\frac{a_{r+1}}{a_r} = \left(\frac{r}{r+1}\right)^3 < \frac{b_{r+1}}{b_r} = \left(\frac{r}{r+1}\right)^2 \tag{33}$$

for all r.

(c) d'Alembert's Ratio Test

The series of positive terms $\sum\limits_{r=1}^{\infty} a_r$ converges if $\lim\limits_{r \to \infty} \dfrac{a_{r+1}}{a_r} = k < 1$,

and diverges if $\lim\limits_{r \to \infty} \dfrac{a_{r+1}}{a_r} = k > 1$.   If $\lim\limits_{r \to \infty} \dfrac{a_{r+1}}{a_r} = 1$, the series may either converge or diverge.

To prove the first part of this test we assume that

$$\lim_{r \to \infty} \frac{a_{r+1}}{a_r} = k < 1 \tag{34}$$

and choose a number h such that $k < h < 1$.   Then for some sufficiently large value of r, say s, we have

$$\frac{a_{s+1}}{a_s} < h, \quad \frac{a_{s+2}}{a_{s+1}} < h, \quad \frac{a_{s+3}}{a_{s+2}} < h, \dots \tag{35}$$

and so on.

Therefore

$$\left.\begin{aligned}
a_{s+1} &< a_s h, \\
a_{s+2} &< a_{s+1}h < a_s h^2, \\
a_{s+3} &< a_{s+2}h < a_s h^3, \\
\cdot \qquad & \cdot \qquad \cdot \\
\cdot \qquad & \cdot \qquad \cdot \\
\cdot \qquad & \cdot \qquad \cdot
\end{aligned}\right\} \tag{36}$$

which give, on adding,

$$a_{s+1} + a_{s+2} + a_{s+3} + \dots < a_s(h + h^2 + h^3 + \dots). \tag{37}$$

The series on the right-hand side of (37) is a convergent geometric series since, by assumption, $h < 1$. Hence the series on the left-hand side of (37) also converges. Finally, therefore, if $k < 1$, $\sum\limits_{r=1}^{\infty} a_r$ is convergent.

The case of $k > 1$ may be proved in the same way. The ratio test clearly gives no information when $k = 1$ as can be seen by considering the p-series for which

$$\lim_{r \to \infty} \frac{a_{r+1}}{a_r} = \lim_{r \to \infty} \frac{r^p}{(r+1)^p} = \lim_{r \to \infty} \left(1 + \frac{1}{r}\right)^{-p} = 1 \qquad (38)$$

for *all p*.

Example 6. The series

$$\frac{1}{2} + \frac{2}{2^2} + \frac{3}{2^3} + \dots = \sum_{r=1}^{\infty} \frac{r}{2^r} \qquad (39)$$

converges since

$$\lim_{r \to \infty} \frac{a_{r+1}}{a_r} = \lim_{r \to \infty} \left\{ \left(\frac{r+1}{r}\right)\left(\frac{2^r}{2^{r+1}}\right) \right\} = \frac{1}{2} \lim_{r \to \infty} \left(1 + \frac{1}{r}\right) = \frac{1}{2}. \qquad (40)$$

(d) Cauchy's Integral Test

If $\sum\limits_{r=1}^{\infty} a_r$ is a series of positive decreasing terms and if there exists, for $x \geqq 1$, a positive, monotonic decreasing integrable function $f(x)$ such that $f(r) = a_r$ for $r = 1, 2, 3 \dots n$, then

$$0 < \sum_{r=1}^{n} a_r - \int_{1}^{n+1} f(x)\, dx < f(1). \qquad (41)$$

It may be further proved that

$$\lim_{n \to \infty} \left(S_n - \int_{1}^{n+1} f(x)\, dx \right) \qquad (42)$$

is finite. A direct consequence of (42) is that the series $\sum\limits_{r=1}^{\infty} a_r$ con-

76

verges when $\int_1^\infty f(x)\,dx$ converges (in the sense of Chapter 4, 4.8), and diverges when $\int_1^\infty f(x)\,dx$ diverges.

A simple proof of (41) may be easily obtained using the type of graphical argument given in proving the comparison test. Consider first the area $ABCD$ shown in Fig. 5.2. Then, since $AB = a_1$ and $AD = 1$, we have that

$$\text{area } ABCD = a_1.$$

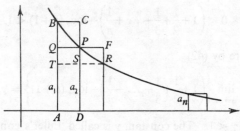

Fig. 5.2

The area under the curve $f(x)$ between A and D is $\int_1^2 f(x)\,dx$. Consequently

$$a_1 - \int_1^2 f(x)\,dx = \text{area } BCP < \text{area } BCPQ. \tag{43}$$

Similarly, considering the next rectangle of height a_2

$$a_2 - \int_2^3 f(x)\,dx = \text{area } PFR < \text{area } PFRS = \text{area } QPST. \tag{44}$$

After n such expressions we have, on adding,

$$0 < (a_1 + a_2 + a_3 + \ldots + a_n) - \int_1^{n+1} f(x)\,dx < \text{area } ABCD = f(1), \tag{45}$$

which proves the basic inequality (41) of the integral test.

Example 7. The series $\sum\limits_{r=2}^{\infty} \dfrac{1}{r(\log r)^p}$ converges only if $p > 1$. This follows from the integral test since

$$I = \lim_{b \to \infty} \int_2^b \frac{dx}{x(\log x)^p} = \lim_{b \to \infty} \int_{\log_e 2}^{\log_e b} \frac{du}{u^p} \qquad (46)$$

converges only if $p > 1$.

Example 8. Using the divergent harmonic series in the integral test we have (from (41))

$$0 < \left(1 + \frac{1}{2} + \frac{1}{3} + \dots + \frac{1}{n}\right) - \log_e (n+1) < 1. \qquad (47)$$

Furthermore by (42)

$$\lim_{n \to \infty} \left\{ \left(1 + \frac{1}{2} + \frac{1}{3} + \dots + \frac{1}{n}\right) - \log_e (n+1) \right\} = \gamma, \qquad (48)$$

where $0 < \gamma < 1$. The constant γ is called Euler's constant and is approximately equal to $0 \cdot 5772$.

5.4 Alternating Series

If $\sum\limits_{r=1}^{\infty} a_r$ is a series of terms which are alternately positive and negative, and if the terms *continually* decrease in magnitude and $\lim\limits_{n \to \infty} a_n = 0$, then the series converges.

Suppose

$$\sum_{r=1}^{\infty} a_r = a_1 - a_2 + a_3 - a_4 + a_5 \dots, \qquad (49)$$

where $a_1, a_2, a_3 \dots$ are positive decreasing terms. Plotting the values of the first few partial sums $S_1, S_2, S_3 \dots$ along the line Ox (see Fig. 5.3) it is clear that these partial sums approach more and more closely to a definite value S. Hence the series converges.

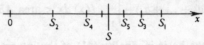

Fig. 5.3

Example 9. The series

$$1-\frac{1}{2}+\frac{1}{3}-\frac{1}{4}\dots = \sum_{r=1}^{\infty}\frac{(-1)^{r+1}}{r} \tag{50}$$

satisfies all the conditions stated above and therefore converges. As we shall see in 5.8 the sum of this series is $\log_e 2$.

5.5 Absolute Convergence and Conditional Convergence
Suppose

$$\sum_{r=1}^{\infty}a_r = a_1+a_2+a_3+\dots \tag{51}$$

is a series of positive and negative terms. Then

$$\sum_{r=1}^{\infty}|a_r| = |a_1| + |a_2| + |a_3| + \dots \tag{52}$$

is a series of positive terms which are just the absolute values of a_r. If (52) is convergent, (51) is said to be absolutely convergent, and it can be proved that any absolutely convergent series is also convergent.
If, however, $\sum_{r=1}^{\infty}|a_r|$ diverges, but $\sum_{r=1}^{\infty}a_r$ converges, then $\sum_{r=1}^{\infty}a_r$ is said to be conditionally convergent. For example, the series

$$1-\frac{1}{2}+\frac{1}{3}-\frac{1}{4}+\dots = \sum_{r=1}^{\infty}\frac{(-1)^{r+1}}{r} \tag{53}$$

discussed in Example 9 is conditionally convergent since the series formed from the absolute values of its terms

$$1+\frac{1}{2}+\frac{1}{3}+\frac{1}{4}+\dots = \sum_{r=1}^{\infty}\left|\frac{(-1)^{r+1}}{r}\right| \tag{54}$$

is the divergent harmonic series.
On the other hand, the series

$$1-\frac{1}{2^2}+\frac{1}{3^2}-\frac{1}{4^2}\dots = \sum_{r=1}^{\infty}\frac{(-1)^{r+1}}{r^2} \tag{55}$$

is absolutely convergent since

$$1+\frac{1}{2^2}+\frac{1}{3^2}+ \dots = \sum_{r=1}^{\infty} \left| \frac{(-1)^{r+1}}{r^2} \right| \tag{56}$$

is convergent (*p*-series with $p = 2$).

5.6 Absolute Convergence Tests

Since $\sum_{r=1}^{\infty} |a_r|$ is a series of positive terms its convergence may be discussed using any of the tests given in 5.3. For example, d'Alembert's ratio test for absolute convergence now takes the form:

the series of positive and negative terms $\sum_{r=1}^{\infty} a_r$ is absolutely convergent (and hence convergent) if

$$\lim_{r \to \infty} \left| \frac{a_{r+1}}{a_r} \right| = k < 1, \tag{57}$$

and is divergent if

$$\lim_{r \to \infty} \left| \frac{a_{r+1}}{a_r} \right| = k > 1. \tag{58}$$

As before, the test does not decide between absolute convergence and divergence when $k = 1$.

Example 10. If

$$\sum_{r=1}^{\infty} a_r = 1+2x+3x^2+ \dots , \tag{59}$$

then

$$\lim_{r \to \infty} \left| \frac{a_{r+1}}{a_r} \right| = \lim_{r \to \infty} \left| \frac{(r+1)x^r}{rx^{r-1}} \right| = |x| \lim_{r \to \infty} \left(\frac{r+1}{r} \right) = |x|. \tag{60}$$

Hence, when $|x| < 1$, (59) is absolutely convergent, and when $|x| > 1$, it is divergent. The question of what happens when $|x| = 1$ may be answered by taking the two possible cases $x = +1$ and $x = -1$ separately. When $x = +1$, (59) becomes

$$1+2+3+4+ \dots , \tag{61}$$

which is divergent; similarly when $x = -1$ the series becomes

$$1 - 2 + 3 - 4 \dots, \tag{62}$$

which is divergent since $\lim_{r \to \infty} a_r \neq 0$.

Hence (59) is absolutely convergent for $|x| < 1$, and divergent for $|x| \geqq 1$.

Example 11. The series

$$\frac{\sin x}{1^2} + \frac{\sin 2x}{2^2} + \frac{\sin 3x}{3^2} + \dots = \sum_{r=1}^{\infty} \frac{\sin rx}{r^2} \tag{63}$$

is absolutely convergent for all x, since, using the comparison test,

$$\left| \frac{\sin rx}{r^2} \right| \leq \frac{1}{r^2} \tag{64}$$

for all r, and $\sum_{r=1}^{\infty} \frac{1}{r^2}$ is known to converge.

5.7 The Product of Two Series

If $\sum_{r=1}^{\infty} a_r$ and $\sum_{r=1}^{\infty} b_r$ are two absolutely convergent series, the series $\sum_{r=1}^{\infty} c_r$, where

$$c_r = a_1 b_r + a_2 b_{r-1} + \dots + a_r b_1 \tag{65}$$

is called the Cauchy product of $\sum_{r=1}^{\infty} a_r$ and $\sum_{r=1}^{\infty} b_r$, and is itself absolutely convergent. Furthermore, if $\sum_{r=1}^{\infty} a_r$ converges to a sum S, and $\sum_{r=1}^{\infty} b_r$ converges to a sum T, then $\sum_{r=1}^{\infty} c_r$ converges to a sum ST. A similar result has already been given in 5.2 for the sum (and difference) of two convergent series.

Example 12. The product of e^{2x} and e^{-x} may be written, using (65), as

$$e^{2x}e^{-x} = \left(1+2x+\frac{(2x)^2}{2!}+\frac{(2x)^3}{3!}+\ldots\right)\left(1-x+\frac{x^2}{2!}-\ldots\right) \tag{66}$$

$$= 1+x+\frac{x^2}{2!}+\frac{x^3}{3!}+\ldots = e^x. \tag{67}$$

5.8 Rearrangement of Series

Any series formed from $\sum\limits_{r=1}^{\infty} a_r$ by taking its terms in a different order is called a rearrangement of $\sum\limits_{r=1}^{\infty} a_r$. For example,

$$1+\frac{1}{3^2}-\frac{1}{2^2}+\frac{1}{5^2}+\frac{1}{7^2}-\frac{1}{4^2}\ldots \tag{68}$$

is a rearrangement of the absolutely convergent series

$$1-\frac{1}{2^2}+\frac{1}{3^2}-\frac{1}{4^2}+\frac{1}{5^2}-\frac{1}{6^2}+\frac{1}{7^2}-\ldots \tag{69}$$

such that two positive terms alternate with one negative term throughout the series. Similarly, two possible rearrangements of the conditionally convergent series

$$1-\tfrac{1}{2}+\tfrac{1}{3}-\tfrac{1}{4}\ldots \tag{70}$$

are
$$1+\tfrac{1}{3}-\tfrac{1}{2}+\tfrac{1}{5}+\tfrac{1}{7}-\tfrac{1}{4}\ldots \tag{71}$$

and
$$1+\tfrac{1}{3}+\tfrac{1}{5}-\tfrac{1}{2}-\tfrac{1}{4}+\tfrac{1}{7}\ldots. \tag{72}$$

In (71) two positive terms alternate with one negative term and in (72) three positive terms alternate with two negative terms; both of these are different from the original series (70) in which one positive term alternates with one negative term. Since in any rearrangement of an infinite series the pattern of N positive terms alternating with M negative terms can be chosen at will and can be continued throughout the series, it would be surprising if the sum of a rearranged series were equal to the sum of the original series. It can be proved, however, that provided we restrict ourselves either to series of positive terms or to series that are absolutely convergent, the terms may be rearranged in any way without affecting the sums of the series. This result is not true for series that are conditionally convergent, and any rearrangement of terms in a series of this type will usually lead to a series with a different sum. For example, (68) will have the same sum as (69), since (69) is absolutely convergent (p-series, $p = 2$). The series (70), however, is only conditionally convergent since the series formed from the absolute values of its terms is the divergent harmonic series. Hence we must expect that the sums of the two rearranged series (71) and (72) will be different from each other and different from the sum of (70). By way of justifying this we now show how to find the sums of (70) and (71) so verifying that they are different.

Consider (70) first: then by 5.1 (2) we have the sum S defined by

$$S = \lim_{n \to \infty} S_n = \lim_{n \to \infty} S_{2n} = \lim_{n \to \infty} \left(1 - \frac{1}{2} + \frac{1}{3} \dots - \frac{1}{2n} \right) \quad (73)$$

$$= \lim_{n \to \infty} \left\{ \left(1 + \frac{1}{2} + \frac{1}{3} + \dots + \frac{1}{2n} \right) - \left(1 + \frac{1}{2} + \frac{1}{3} \dots + \frac{1}{n} \right) \right\}. \quad (74)$$

Now from 5.3, Example 8, we have

$$\lim_{n \to \infty} \left\{ \left(1 + \frac{1}{2} + \frac{1}{3} + \dots + \frac{1}{n} \right) - \log_e (n+1) \right\} = \gamma,$$

where γ is Euler's constant. Hence

$$\left(1 + \frac{1}{2} + \frac{1}{3} + \dots + \frac{1}{n} \right) - \log_e (n+1) = \varepsilon_n, \quad (75)$$

where $\varepsilon_n \to \gamma$ as $n \to \infty$. Using (75) in (74) we get

$$S = \lim_{n \to \infty} \left[\{ \varepsilon_{2n} + \log_e (2n+1) \} - \{ \varepsilon_n + \log_e (n+1) \} \right] = \log_e 2, \quad (76)$$

since ε_{2n} and ε_n both tend to γ as $n \to \infty$.

Similarly the series (71) may be written as

$$S_{3n} = \left(1 + \frac{1}{3} - \frac{1}{2} \right) + \left(\frac{1}{5} + \frac{1}{7} - \frac{1}{4} \right) + \dots + \left(\frac{1}{4n-3} + \frac{1}{4n-1} - \frac{1}{2n} \right)$$

$$= \left(1+\frac{1}{3}+\frac{1}{5}+ \dots +\frac{1}{4n-1}\right)-\frac{1}{2}\left(1+\frac{1}{2}+\frac{1}{3}+ \dots +\frac{1}{n}\right)$$

$$= \left(1+\frac{1}{2}+\frac{1}{3}+\frac{1}{4}+ \dots +\frac{1}{4n}\right)-\frac{1}{2}\left(1+\frac{1}{2}+\frac{1}{3}+ \dots +\frac{1}{2n}\right)$$

$$-\frac{1}{2}\left(1+\frac{1}{2}+\frac{1}{3}+ \dots +\frac{1}{n}\right). \qquad (77)$$

Using (75), (77) becomes

$$\{\varepsilon_{4n}+\log_e (4n+1)\}-\tfrac{1}{2}\{\varepsilon_{2n}+\log_e (2n+1)\}-\tfrac{1}{2}\{\varepsilon_n +\log_e (n+1)\},$$

which tends to $\frac{3}{2}\log_e 2$ as $n \to \infty$, since ε_{4n}, ε_{2n} and ε_n all tend to γ as $n \to \infty$. Hence the sum of (71) differs from the sum of (70) by a factor $\frac{3}{2}$.

5.9 Power Series

An important type of series is the power series defined by

$$\sum_{r=0}^{\infty} a_r x^r = a_0+a_1 x+a_2 x^2 \dots , \qquad (78)$$

where a_0, a_1, a_2, ... are constants. The values of x for which (78) converges may be found using d'Alembert's ratio test given in 5.6. Hence for the series to be absolutely convergent we must have (from (57))

$$\lim_{r \to \infty}\left|\frac{a_{r+1}x^{r+1}}{a_r x^r}\right| = |x| \lim_{r \to \infty}\left|\frac{a_{r+1}}{a_r}\right| = k < 1. \qquad (79)$$

This condition may be more conveniently expressed as

$$|x| < R, \qquad (80)$$

where R, the radius of convergence, is defined by

$$R = \lim_{r \to \infty}\left|\frac{a_r}{a_{r+1}}\right|, \qquad (81)$$

provided the limit exists.

Writing (80) in full as

$$-R < x < R, \tag{82}$$

we see that the series converges absolutely provided x lies in the open interval (see Chapter 1, 1.2) $-R$ to R. This interval is called the interval (or range) of convergence. When $k = 1$, the ratio test gives no information and consequently the series may converge or diverge at $|x| = R$ (that is, at $x = \pm R$). However, as we shall see in the following examples, we may test the convergence of the series for these two particular values of x by direct substitution into the series.

Finally, by the ratio test, a power series obviously diverges for any value of x which lies outside the interval of convergence.

Example 13. The exponential series (see 6.2 (40))

$$1 + x + \frac{x^2}{2!} + \frac{x^3}{3!} + \ldots + \frac{x^r}{r!} + \ldots \tag{83}$$

is absolutely convergent for all x, since by (79)

$$|x| \lim_{r \to \infty} \left| \frac{a_{r+1}}{a_r} \right| = |x| \lim_{r \to \infty} \left| \frac{r!}{(r+1)!} \right| = 0 \tag{84}$$

irrespective of the value of x. Similarly, by (81), the radius of convergence is infinite and, by (82), the interval of convergence is therefore

$$-\infty < x < \infty. \tag{85}$$

Hence the series (83) represents the function e^x for all x.

Example 14. The series

$$x - \frac{x^3}{3} + \frac{x^5}{5} - \ldots + (-1)^r \frac{x^{2r+1}}{(2r+1)} + \ldots \tag{86}$$

is convergent for $|x| < 1$, since by (81)

$$R = \lim_{r \to \infty} \left| \frac{a_r}{a_{r+1}} \right| = \lim_{r \to \infty} \left| -\frac{2r+1}{2r-1} \right| = 1. \tag{87}$$

The interval of convergence is therefore

$$-1 < x < 1. \tag{88}$$

At the end points $x = \pm 1$, the series may converge or diverge. Putting $x = 1$ in (86), the series becomes

$$1 - \tfrac{1}{3} + \tfrac{1}{5} - \tfrac{1}{7} + \ldots, \tag{89}$$

which converges by 5.4 since the terms alternate in sign and continually decrease in magnitude. At $x = -1$, the series becomes again

$$-1 + \tfrac{1}{3} - \tfrac{1}{5} + \tfrac{1}{7} + \ldots. \tag{90}$$

Hence (86) converges if, and only if,

$$-1 \leqq x \leqq 1. \tag{91}$$

5.10 Operations with Power Series

(a) The sum, difference or product of two power series with common intervals of convergence leads to a third series which converges for the common interval of convergence of the first two series. (This result follows from the general properties of series given in 5.2 and 5.7.)

(b) The series obtained by term-by-term differentiation (or integration) of a given convergent power series is a power series with the same interval of convergence.

Consider the power series

$$S = \sum_{r=0}^{\infty} a_r x^r = a_0 + a_1 x + a_2 x^2 + \ldots \tag{92}$$

which converges if

$$|x| < \lim_{r \to \infty} \left| \frac{a_r}{a_{r+1}} \right| = R. \tag{93}$$

Then

$$\frac{dS}{dx} = \sum_{r=0}^{\infty} r a_r x^{r-1} = a_1 + 2a_2 x + 3a_3 x^2 + \ldots \tag{94}$$

86

converges if

$$|x| < \lim_{r \to \infty} \left| \frac{ra_r}{(r+1)a_{r+1}} \right| = \lim_{r \to \infty} \left| \frac{r}{r+1} \right| \lim_{r \to \infty} \left| \frac{a_r}{a_{r+1}} \right| = R. \tag{95}$$

Similarly,

$$\int S \, dx = \sum_{r=0}^{\infty} \frac{a_r x^{r+1}}{r+1} = a_0 x + \frac{a_1 x^2}{2} + \frac{a_2 x^3}{3} + \dots \tag{96}$$

converges if

$$|x| < \lim_{r \to \infty} \left| \frac{(r+2)a_r}{(r+1)a_{r+1}} \right| = \lim_{r \to \infty} \left| \frac{r+2}{r+1} \right| \lim_{r \to \infty} \left| \frac{a_r}{a_{r+1}} \right| = R. \tag{97}$$

Hence the differentiated and integrated series have the same intervals of convergence as the series from which they are derived. However, it does not follow that if the series converges at one (or both) of the end points ($x = \pm R$) of the interval of convergence that the differentiated or integrated series necessarily also converges at these points. As before, the convergence of these series at the two end points must be considered separately. Furthermore we may prove that differentiating or integrating a power series term-by-term within its interval of convergence is the same as differentiating or integrating the function it represents.

(c) If two power series converge for a common interval of convergence then one series may be substituted into the other to give a third series which converges in that common interval. For example, the series for $e^{e^{-x}}$ may be obtained by writing $y = e^{-x}$ and using the series

$$e^y = 1 + y + \frac{y^2}{2!} + \frac{y^3}{3!} + \dots . \tag{98}$$

Hence

$$e^{e^{-x}} = 1 + \left(1 - x + \frac{x^2}{2!} \dots\right) + \frac{1}{2!}\left(1 - x + \frac{x^2}{2!} - \dots\right)^2$$
$$+ \frac{1}{3!}\left(1 - x + \frac{x^2}{2!} - \dots\right)^3 + \dots . \tag{99}$$

5.11 Integration and Differentiation of Series

Although a power series may be differentiated and integrated term-by-term as many times as desired within its interval of convergence without introducing difficulties of convergence, this is not usually the case for an infinite series of arbitrary functions. For example, the series

$$2(\sin x - \tfrac{1}{2}\sin 2x + \tfrac{1}{3}\sin 3x - \ldots) = -2\sum_{r=1}^{\infty}(-1)^r\frac{\sin rx}{r} \quad (100)$$

can be shown by the methods of Chapter 15 (see 15.6) to represent the function x in the interval $-\pi < x < \pi$. However, term-by-term differentiation of this series gives

$$2(\cos x - \cos 2x + \cos 3x - \ldots) = -2\sum_{r=1}^{\infty}(-1)^r\cos rx, \quad (101)$$

which diverges for all x since $\lim_{r\to\infty}|\cos rx| \neq 0$. Clearly then the derivative of the sum of the terms in (100) is not equal to the sum of the derivatives of these terms. Nevertheless term-by-term integration of (100) gives the series

$$-2\left(\cos x - \frac{\cos 2x}{2^2} + \frac{\cos 3x}{3^2} - \ldots\right) + A = 2\sum_{r=1}^{\infty}(-1)^r\frac{\cos rx}{r^2} + A,$$

$$(102)$$

where A is an integration constant. With a proper choice of A this series can be shown to converge to $\dfrac{x^2}{2}$ in $-\pi < x < \pi$. Hence in this case the integral of the sum of terms in (100) is the same as the sum of the integrals of these terms.

The conditions under which we may interchange the operations of integration and summation, and differentiation and summation, without introducing convergence difficulties depend on the notion of the uniform convergence of a series of functions which is beyond the scope of this book. Consequently wherever we have need to interchange the order of such operations (see particularly Chapters 10, 12, 14 and 15) the reader may assume that the procedure is justified.

PROBLEMS 5

1. Examine the following series for convergence

   (a) $\displaystyle\sum_{r=1}^{\infty} \frac{1}{2r(r+1)}$,   (b) $\displaystyle\sum_{r=1}^{\infty} \frac{r!}{10^r}$,   (c) $1+\displaystyle\sum_{r=2}^{\infty} \frac{1}{\log_e r}$,

   (d) $\displaystyle\sum_{r=1}^{\infty} \frac{1}{(r+\frac{1}{2})^2}$,   (e) $\displaystyle\sum_{r=2}^{\infty} \frac{1}{r(\log_e r)^p}$,   (f) $\displaystyle\sum_{r=1}^{\infty} \frac{r+1}{r^2+2}$,

   (g) $\displaystyle\sum_{r=1}^{\infty} r^2 x^r, (x>0)$,   (h) $\displaystyle\sum_{r=1}^{\infty} \frac{r^r}{r!}$,   (i) $\displaystyle\sum_{r=1}^{\infty} \sin \frac{\pi r}{4}$,

   (j) $\displaystyle\sum_{r=1}^{\infty} \frac{1}{\sqrt{(r^2+r)}}$,   (k) $\displaystyle\sum_{r=1}^{\infty} \frac{r^p}{r!}$.

2. Examine the convergence of $\displaystyle\sum_{r=1}^{\infty} \frac{\cos r\theta}{r}$ for $\theta = 0, \pi/2$ and $2\pi/3$.

3. Show that if the conditionally convergent series

$$1-\frac{1}{\sqrt{2}}+\frac{1}{\sqrt{3}}-\frac{1}{\sqrt{4}}+\ldots$$

   is rearranged as the series

$$\left(1+\frac{1}{\sqrt{3}}-\frac{1}{\sqrt{2}}\right)+\left(\frac{1}{\sqrt{5}}+\frac{1}{\sqrt{7}}-\frac{1}{\sqrt{4}}\right)+\ldots,$$

   where two positive terms always alternate with one negative term, the series diverges.

4. Find the values of x for which the series

$$1-2(x-1)+3(x-1)^2+\ldots+r(-1)^{r-1}(x-1)^{r-1}+\ldots$$

   converges.

5. Prove that the binomial series

$$1+mx+\frac{m(m-1)}{2!}x^2+\frac{m(m-1)(m-2)}{3!}x^3+\ldots$$

   converges if $-1 < x < 1$.

6. Show that $\sum_{r=1}^{\infty} \dfrac{(-1)^r x^r}{r}$ is absolutely convergent if $-1 < x < 1$, and that it is conditionally convergent if $x = 1$, and divergent if $x = -1$.

7. Using the series for e^x, the hyperbolic functions defined in Chapter 1, 1.3 (h) may be written as

$$\sinh x = x + \frac{x^3}{3!} + \frac{x^5}{5!} + \dots$$

and
$$\cosh x = 1 + \frac{x^2}{2!} + \frac{x^4}{4!} + \dots .$$

Determine the interval of convergence for each of these functions and show that

$$2 \sinh x \cosh x = \left(2x + \frac{(2x)^3}{3!} + \frac{(2x)^5}{5!} + \dots \right)$$

for all x. Verify that this is the same as $\sinh 2x$.

8. Defining $\sin x$ by the series $\sum_{r=0}^{\infty} (-1)^r \dfrac{x^{2r+1}}{(2r+1)!}$, determine the interval of convergence and find, using the series for e^x, a power series for $e^{\sin x}$. State the values of x for which this series is valid.

9. Expand $f(x) = \sin^{-1}(2x\sqrt{(1-x^2)})$ in powers of x by first differentiating $f(x)$ and expanding the result, and then integrating term-by-term.

10. By expanding $(1 + x^2)^{-1}$ and integrating term-by-term, show that

$$\tan^{-1} x = x + \frac{x^3}{3} + \frac{x^5}{5} \dots,$$

and determine the values of x for which the series is convergent.

11. Show by integrating the series

$$S = \sum_{r=1}^{\infty} r x^{r-1}$$

term-by-term and summing, and then differentiating the sum, that $S = (1-x)^{-2}$. For what values of x is this valid?

12. The Bernoulli numbers $B_r(r = 0, 1, 2 \dots)$ are defined by the relation

$$\frac{x}{e^x - 1} = \sum_{r=0}^{\infty} B_r \frac{x^r}{r!}.$$

Show that $B_0 = 1$, $B_1 = -\frac{1}{2}$, $B_2 = \frac{1}{6}$, $B_4 = -\frac{1}{30}$, and

$$B_3 = B_5 = B_7 \dots = 0.$$

13. If $y^2 = x - 3x^2 + 6x^3 \dots$ for small values of x, show that $x = y^2 + 3y^4 + 12y^6 + \dots$.

14. Discuss the convergence of the following series, distinguishing between absolute and conditional convergence, and giving precise statements of any theorems on series to which appeal is made:

(i) $\sum_{r=1}^{\infty} (-\frac{1}{2})^r \frac{3 \cdot 6 \dots (3r)}{2 \cdot 5 \dots (3r-1)}$,

(ii) $\sum_{r=1}^{\infty} \frac{(-1)^r}{\sqrt{r}}$,

(iii) $\sum_{r=1}^{\infty} \frac{(-1)^r r^2}{3r^2 + 1}$.   (C.U.)

15. Cauchy's condensation test states that if $f(n)$ is a positive decreasing function of n, and a is any positive number > 1, then the two series

$$\sum_{n=1}^{\infty} f(n) \quad \text{and} \quad \sum_{n=1}^{\infty} a^n f(a^n)$$

are either both convergent or both divergent.

Use this test to investigate the convergence of

$$\sum_{n=1}^{\infty} \frac{1}{n (\log n)^p}.$$

CHAPTER 6

Taylor and Maclaurin Series

6.1 Taylor's Theorem

We now state an important theorem which enables functions to be expanded in power series in x in a given interval. (Examples of the series representation of a few functions have already been given in the last chapter; see 5.9 (83) and Problems 7, 8 and 10.)

Theorem 1. (Taylor's Theorem). If $f(x)$ is a continuous, single-valued function of x with continuous derivatives $f'(x), f''(x) \ldots$ up to and including $f^{(n)}(x)$ in a given interval $a \leqq x \leqq b$, and if $f^{(n+1)}(x)$ exists in $a < x < b$, then

$$f(x) = f(a) + \frac{(x-a)}{1!}f'(a) + \frac{(x-a)^2}{2!}f''(a) + \ldots$$

$$+ \ldots + \frac{(x-a)^n}{n!}f^{(n)}(a) + E_n(x), \qquad (1)$$

where

$$E_n(x) = \frac{(x-a)^{n+1}}{(n+1)!} f^{(n+1)}(\xi) \qquad (2)$$

and $a < \xi < x$.

The term E_n is a remainder term and represents the error involved in approximating to $f(x)$ by the polynomial

$$f(a) + \frac{(x-a)}{1!}f'(a) + \ldots + \frac{(x-a)^n}{n!}f^{(n)}(a). \qquad (3)$$

An alternative form of (1) may be obtained by changing x to $a+x$. Then

$$f(a+x) = f(a) + \frac{x}{1!}f'(a) + \frac{x^2}{2!}f''(a) + \ldots + \frac{x^n}{n}f^{(n)}(a) + E_n(x), \qquad (4)$$

where now, from (2),

$$E_n(x) = \frac{x^{n+1}}{(n+1)!} f^{(n+1)}(a+\theta x) \qquad (5)$$

and $0 < \theta < 1$.

A special case of (1) and (2) (or (4) and (5)) is when $a = 0$. Then

$$f(x) = f(0) + xf'(0) + \frac{x^2}{2!} f''(0) + \dots + \frac{x^n}{n!} f^{(n)}(0) + E_n(x), \qquad (6)$$

where

$$E_n(x) = \frac{x^{n+1}}{(n+1)!} f^{(n+1)}(\theta x) \qquad (7)$$

and $0 < \theta < 1$.

Theorem 2. If $\lim_{n \to \infty} E_n(x) = 0$, then $f(x)$ may be represented by the power series (see (1))

$$f(x) = f(a) + \frac{(x-a)}{1!} f'(a) + \frac{(x-a)^2}{2!} f''(a) + \dots = \sum_{r=0}^{\infty} \frac{(x-a)^r}{r!} f^{(r)}(a), \qquad (8)$$

or its equivalent form (see (4))

$$f(a+x) = f(a) + \frac{x}{1!} f'(a) + \frac{x^2}{2!} f''(a) + \dots = \sum_{r=0}^{\infty} \frac{x^r}{r!} f^{(r)}(a). \qquad (9)$$

These two series are Taylor's series for $f(x)$.

The special case $a = 0$ gives, with $\lim_{n \to \infty} E_n(x) = 0$,

$$f(x) = f(0) + xf'(0) + \frac{x^2}{2!} f''(0) + \dots = \sum_{r=0}^{\infty} \frac{x^r}{r!} f^{(r)}(0), \qquad (10)$$

which is known as Maclaurin's series.

Clearly both the Taylor series and the Maclaurin series only represent the function $f(x)$ in their intervals of convergence. The Taylor series is often referred to as a series expansion of $f(x)$ about the point $x = a$, and the Maclaurin series as an expansion about the point $x = 0$ (later we shall meet functions which have no Maclaurin series but which nevertheless can be expanded about some other point $x = a$, $a \neq 0$).

The form of Taylor's series may be verified in the following way. Let

$$f(x) = A_0 + A_1(x-a) + A_2(x-a)^2 + A_3(x-a)^3 + \dots, \qquad (11)$$

where A_0, A_1, A_2 ... are constants. Then differentiating term-by-term we have

$$f'(x) = A_1 + 2A_2(x-a) + 3A_3(x-a)^2 + \dots \qquad (12)$$

$$f''(x) = 2A_2 + 3 \cdot 2A_3(x-a) + 4 \cdot 3A_4(x-a)^2 + \dots \qquad (13)$$

$$f'''(x) = 3!A_3 + 4!A_4(x-a) + \dots, \qquad (14)$$

and in general,

$$f^{(n)}(x) = n!A_n + (n+1)!A_{n+1}(x-a) + \dots. \qquad (15)$$

Putting $x = a$ in (11)–(15) now gives

$$\left. \begin{array}{l} f(a) = A_0, \ f'(a) = A_1, \\ f''(a) = 2!A_2, \ f'''(a) = 3!A_3, \\ f^{(n)}(a) = n!A_n. \end{array} \right\} \qquad (16)$$

Hence using these values for the constants A_0, A_1, ... in (11) we obtain the Taylor series (8).

6.2 Standard Expansions

Before listing the Maclaurin series for some of the simple functions we illustrate the use of the Taylor and Maclaurin series by the following examples.

Example 1. Suppose we want to expand the function $f(x) = e^{3x}$ about $x = 0$ using Maclaurin's series.

Then, since $f'(x) = 3e^{3x}$, $f''(x) = 9e^{3x}, \ldots f^{(n)}(x) = 3^n e^{3x}$ and $f'(0) = 3$, $f''(0) = 9, \ldots f^{(n)}(0) = 3^n$, we have from (6)

$$e^{3x} = 1 + 3x + \frac{(3x)^2}{2!} + \frac{(3x)^3}{3!} + \ldots + \frac{(3x)^n}{n!} + E_n(x), \qquad (17)$$

where, by (7)

$$E_n(x) = \frac{(3x)^{n+1}}{(n+1)!} e^{3\theta x}, \quad (0 < \theta < 1). \qquad (18)$$

For any given finite value of x, say $x = c$, it is clear that

$$\lim_{n \to \infty} E_n(c) = 0. \qquad (19)$$

Hence by (10), e^{3x} may be represented by the infinite series

$$1 + 3x + \frac{(3x)^2}{2!} + \frac{(3x)^3}{3!} + \ldots = \sum_{r=0}^{\infty} \frac{(3x)^r}{r!}. \qquad (20)$$

Using the d'Alembert ratio test we find that (20) converges absolutely for all x. The possible error involved in approximating to e^{3x} (for a given value of x) by a finite number of terms of (20) may be found using (18) as follows.

Suppose $x = 0.02$ and $n = 3$. Then, since $0 < \theta < 1$, E_n must satisfy the inequality relation

$$\frac{(0.06)^4}{4!} < E_n < \frac{(0.06)^4}{4!} e^{0.06}, \qquad (21)$$

which gives (approximately)

$$5 \times 10^{-7} < E_n < 6 \times 10^{-7}. \qquad (22)$$

In other words, by taking only four terms ($n = 3$) of (20), the value of the resulting finite series for $x = 0.02$ differs from the exact value of $e^{3(0.02)}$ by a small number of the order of 5×10^{-7}.

On the other hand, with the same number of terms but with $x = \frac{1}{3}$, (18) gives

$$\frac{1}{4!} < E_n < \frac{e}{4!} \qquad (23)$$

or $$0{\cdot}042 < E_n < 0{\cdot}133. \qquad (24)$$

Taking four terms of the Maclaurin series therefore is not a good approximation to e^{3x} for $x = \frac{1}{3}$ and more terms should be taken if the error is to be reduced.

Example 2. As an example of Taylor's expansion we expand the function $f(x) = \cos x$ about the point $x = \frac{\pi}{3}$. Differentiating we have $f'(x) = -\sin x, f''(x) = -\cos x$, and in general

$$f^{(n+1)}(x) = \cos\left(x + \frac{n+1}{2}\pi\right). \qquad (25)$$

Hence

$$\left.\begin{aligned}
f\left(\frac{\pi}{3}\right) &= \cos\frac{\pi}{3} = \frac{1}{2}, \\[2mm]
f'\left(\frac{\pi}{3}\right) &= -\sin\frac{\pi}{3} = -\frac{\sqrt{3}}{2}, \\[2mm]
f''\left(\frac{\pi}{3}\right) &= -\cos\frac{\pi}{3} = -\frac{1}{2}, \\[2mm]
f'''\left(\frac{\pi}{3}\right) &= \sin\frac{\pi}{3} = \frac{\sqrt{3}}{2},
\end{aligned}\right\} \qquad (26)$$

and so on. Using these results in (1) we find

$$\cos x = \frac{1}{2} - \left(x - \frac{\pi}{3}\right)\frac{\sqrt{3}}{2} - \frac{\left(x - \frac{\pi}{3}\right)^2}{2!}\frac{1}{2} + \frac{\left(x - \frac{\pi}{3}\right)^3}{3!}\frac{\sqrt{3}}{2} + \dots$$

$$+ \frac{\left(x - \frac{\pi}{3}\right)^n}{n!}\cos\left(\frac{\pi}{3} + \frac{n}{2}\pi\right) + E_n(x), \qquad (27)$$

where, from (2) and (25),

$$E_n(x) = \frac{\left(x - \frac{\pi}{3}\right)^{n+1}}{(n+1)!}\cos\left(\xi + \frac{n+1}{2}\pi\right). \qquad (28)$$

Hence, since $\left| \cos\left(\xi + \dfrac{n+1}{2}\pi\right) \right| \leqq 1$, (28) may be written as

$$|E_n(x)| < \frac{\left|\left(x - \dfrac{\pi}{3}\right)^{n+1}\right|}{(n+1)!} \tag{29}$$

Again (as in Example 1) for any given value of x, say $x = c$, $E_n(c)$ may be made as small as we please by choosing sufficiently large values of n. Hence, since $\lim\limits_{n \to \infty} E_n = 0$, $\cos x$ may be represented by the infinite Taylor series (8) as

$$\cos x = \frac{1}{2} - \left(x - \frac{\pi}{3}\right)\frac{\sqrt{3}}{2} - \left(x - \frac{\pi}{3}\right)^2 \frac{1}{2 \cdot 2!} + \left(x - \frac{\pi}{3}\right)^3 \frac{\sqrt{3}}{2 \cdot 3!}$$

$$+ \dots + \frac{\left(x - \dfrac{\pi}{3}\right)}{r!} \cos\left(\frac{\pi}{3} + \frac{r\pi}{2}\right) + \dots , \tag{30}$$

which, by the ratio test, converges for all x.

This series is useful in evaluating the cosines of angles without the use of tables. For example, $\cos 61°$ may be evaluated by putting $x = \dfrac{61\pi}{180}$ radians in (30) which then gives

$$\cos 61° = \cos \frac{61\pi}{180} = \frac{1}{2} - \frac{\sqrt{3}}{2}\left(\frac{\pi}{180}\right) - \frac{1}{2 \cdot 2!}\left(\frac{\pi}{180}\right)^2$$

$$+ \frac{\sqrt{3}}{2} \cdot \frac{1}{3!}\left(\frac{\pi}{180}\right)^3 + \dots \tag{31}$$

The error involved by taking a finite number of terms of this series may easily be estimated from (29). For example, with two terms $(n = 1)$

$$\cos 61° \simeq \frac{1}{2} - \frac{\sqrt{3}}{2}\left(\frac{\pi}{180}\right) = 0\cdot 4849 \tag{32}$$

correct to four decimal places, with a possible error given by

$$\left| E_1\left(\frac{61\pi}{180}\right) \right| \leq \frac{1}{2!}\left(\frac{\pi}{180}\right)^2 = 0.0001 \tag{33}$$

to the same number of decimal places. The value of cos 61° obtained from tables is found to be 0·4848 (again corrected to four decimal places).

We now give the first few terms of the Maclaurin series for some elementary functions

(i) $(1+x)^\alpha = 1 + \alpha x + \dfrac{\alpha(\alpha-1)}{2!}x^2 + \dfrac{\alpha(\alpha-1)(\alpha-2)}{3!}x^3 + \ldots$ (34)

   for $|x| < 1$, where α is any real number,

(ii) $\sin x = x - \dfrac{x^3}{3!} + \dfrac{x^5}{5!} - \dfrac{x^7}{7!} + \ldots$ for all x, (35)

(iii) $\cos x = 1 - \dfrac{x^2}{2!} + \dfrac{x^4}{4!} - \dfrac{x^6}{6!} + \ldots$ for all x, (36)

(iv) $\tan x = x + \dfrac{x^3}{3} + \dfrac{2x^5}{15} + \dfrac{17x^7}{315} + \ldots$, for $-\dfrac{\pi}{2} < x < \dfrac{\pi}{2}$, (37)

(v) $\log_e (1+x) = x - \dfrac{x^2}{2} + \dfrac{x^3}{3} - \dfrac{x^4}{4} + \ldots$ for $-1 < x \leq 1$, (38)

(vi) $\dfrac{1}{2}\log_e\left(\dfrac{1+x}{1-x}\right) = x + \dfrac{x^3}{3} + \dfrac{x^5}{5} + \ldots$ for $-1 < x < 1$, (39)

(vii) $e^x = 1 + x + \dfrac{x^2}{2!} + \dfrac{x^3}{3!} + \ldots$ for all x, (40)

(viii) $\sinh x = \dfrac{e^x - e^{-x}}{2} = x + \dfrac{x^3}{3!} + \dfrac{x^5}{5!} + \ldots$ for all x, (41)

(ix) $\cosh x = \dfrac{e^x + e^{-x}}{2} = 1 + \dfrac{x^2}{2!} + \dfrac{x^4}{4!} + \dfrac{x^6}{6!} + \ldots$ for all x. (42)

The series for functions which are simple combinations of these elementary functions may be obtained using the properties of power series given in Chapter 5, 5.10. For example, substituting the series for $\sin x$ in the exponential series we have (5.10 (c))

$$e^{\sin x} = 1 + x + \frac{x^2}{2!} - \frac{3x^4}{4!} - \frac{8x^5}{5!} + \dots \tag{43}$$

for all x.

Similarly (using 5.10 (a)).

$$\cosh x \sin x = x + \frac{x^3}{3} - \frac{x^5}{30} \dots \tag{44}$$

for all x.

Finally, it should be noted that functions like $\dfrac{e^{-x}}{x}$, $\log_e x$ and $\cot x$ have no Maclaurin expansions since they are not defined at $x = 0$. Nevertheless we may expand such functions about some other point using Taylor's series. For example, expanding $\log_e x$ about $x = 1$ we find

$$\log_e x = (x-1) - \tfrac{1}{2}(x-1)^2 + \tfrac{1}{3}(x-1)^3 - \tfrac{1}{4}(x-1)^4 + \dots \tag{45}$$

for $0 < x \leq 2$.

In deriving the Maclaurin series for certain functions it is sometimes convenient to use Leibnitz's formula given in Chapter 3, 3.6, to obtain the higher differential coefficients. We illustrate this method by an example.

Example 3. If $y = \sin(m \sin^{-1} x)$, where m is a constant, then differentiating twice we find that y satisfies the differential equation

$$(1-x^2)\frac{d^2y}{dx^2} - x\frac{dy}{dx} + m^2 y = 0. \tag{46}$$

Using Leibnitz's formula we have (for $n > 0$)

$$(1-x^2)\frac{d^{n+2}y}{dx^{n+2}} - (2n+1)x\frac{d^{n+1}y}{dx^{n+1}} + (m^2 - n^2)\frac{d^n y}{dx^n} = 0, \tag{47}$$

which gives, with $x = 0$,

$$\frac{d^{n+2}y}{dx^{n+2}} = (n^2 - m^2)\frac{d^n y}{dx^n}. \tag{48}$$

Hence (48) is a relation between the values of all the differential coefficients of y evaluated at $x = 0$; this is exactly what is required in developing the Maclaurin series for y. Since $y(0) = 0$, $y'(0) = m$ it can be easily verified using (48) that

$$y = mx + \frac{m(1-m^2)x^3}{3!} + \frac{m(1-m^2)(9-m^2)x^5}{5!} + \dots . \qquad (49)$$

6.3 Evaluation of Limits

Suppose we have two functions $f(x)$ and $g(x)$ which are zero when $x = a$. Then although the ratio $\dfrac{f(a)}{g(a)}$ is an undefined quantity $\left(\dfrac{0}{0}\right)$, nevertheless the limit of $\dfrac{f(x)}{g(x)}$ as $x \to a$ may exist. An example of this type of ratio has already been met in Chapter 2, 2.4 where it was shown by a geometrical argument that

$$\lim_{x \to 0} \left(\frac{\sin x}{x} \right) = 1. \qquad (50)$$

We now show how to proceed analytically with limits of this type. Consider the ratio of $f(x)$ and $g(x)$ and let both functions be expanded about the point $x = a$ using Taylor's theorem. Then

$$\frac{f(x)}{g(x)} = \frac{f(a) + (x-a)f'(a) + \dfrac{(x-a)^2}{2!}f''(a) + \dots}{g(a) + (x-a)g'(a) + \dfrac{(x-a)^2}{2!}g''(a) + \dots} \qquad (51)$$

Now by assumption $f(a) = g(a) = 0$. Hence

$$\frac{f(x)}{g(x)} = \frac{f'(a) + \dfrac{(x-a)}{2!}f''(a) + \dots}{g'(a) + \dfrac{(x-a)}{2!}g''(a) + \dots} \qquad (52)$$

and consequently

$$\lim_{x \to a} \frac{f(x)}{g(x)} = \frac{f'(a)}{g'(a)}, \qquad (53)$$

provided $g'(a)$ is non-zero. Equation (53) states that the limit of the ratio of two functions as $x \to a$ where both functions are zero at $x = a$ is given by the ratio of the derivatives of the functions each evaluated at $x = a$. If, however, $f'(a) = g'(a) = 0$ then the same procedure must be applied to the ratio $\dfrac{f'(x)}{g'(x)}$. Consequently if $f(a) = g(a) = 0$ and $f'(a) = g'(a) = 0$, we have

$$\lim_{x \to a} \frac{f(x)}{g(x)} = \frac{f''(a)}{g''(a)}, \tag{54}$$

provided $g''(a)$ is non-zero. Provided the limit exists it is usually possible to find a value of n such that

$$\lim_{x \to a} \frac{f(x)}{g(x)} = \frac{f^{(n)}(a)}{g^{(n)}(a)}. \tag{55}$$

This method of evaluating limits is sometimes more conveniently expressed by writing (53) as

$$\lim_{x \to a} \frac{f(x)}{g(x)} = \lim_{x \to a} \frac{f'(x)}{g'(x)}, \tag{56}$$

which is usually known as l'Hospital's rule.

We illustrate these results by the following examples.

Example 4. Using (56)

$$\lim_{x \to 1} \left\{ \frac{\log_e x}{x^2 - 1} \right\} = \lim_{x \to 1} \left\{ \frac{1/x}{2x} \right\} = \frac{1}{2}. \tag{57}$$

Example 5. To evaluate

$$\lim_{x \to 0} \ (\cos x)^{1/x} \tag{58}$$

we put $y = (\cos x)^{1/x}$ and consider the behaviour of

$$\log_e y = \frac{\log_e \cos x}{x}. \tag{59}$$

101

Then by (56)

$$\lim_{x \to 0} \log_e y = \lim_{x \to 0} \left\{ \frac{\log_e \cos x}{x} \right\} = \lim_{x \to 0} \left\{ \frac{-\tan x}{1} \right\} = 0. \quad (60)$$

Hence, since as $x \to 0$, $\log_e y \to 0$, we have

$$y = (\cos x)^{1/x} \to 1. \quad (61)$$

Example 6. This example illustrates the repeated use of l'Hospital's rule. For (by (56))

$$\lim_{x \to 0} \left(\frac{\tan x - x}{x - \sin x} \right) = \lim_{x \to 0} \left(\frac{\sec^2 x - 1}{1 - \cos x} \right). \quad (62)$$

But $\dfrac{\sec^2 x - 1}{1 - \cos x}$ is of the form $\dfrac{0}{0}$ when $x = 0$. Hence we apply l'Hospital's rule again which gives

$$\lim_{x \to 0} \left(\frac{\sec^2 x - 1}{1 - \cos x} \right) = \lim_{x \to 0} \left(\frac{2 \sec^2 x \tan x}{\sin x} \right)$$

$$= \lim_{x \to 0} (2 \sec^3 x) = 2. \quad (63)$$

The second application of l'Hospital's rule could have been avoided by rewriting the right-hand side of (62) as

$$\lim_{x \to 0} \left(\frac{\sec^2 x - 1}{1 - \cos x} \right) = \lim_{x \to 0} \left(\frac{(1 - \cos^2 x) \sec^2 x}{1 - \cos x} \right)$$

$$= \lim_{x \to 0} \{ (1 + \cos x) \sec^2 x \}$$

$$= \lim_{x \to 0} \sec^2 x + \lim_{x \to 0} \sec x = 2, \quad (64)$$

(using Theorem 1, Chapter 3).

Example 7. If $f(x)$ and $g(x)$ both tend to infinity as $x \to a$, we may still apply l'Hospital's rule by writing

$$\lim_{x \to a} \frac{f(x)}{g(x)} = \lim_{x \to a} \left\{ \frac{1/g(x)}{1/f(x)} \right\}, \quad (65)$$

where the ratio $\left\{\dfrac{1/g(x)}{1/f(x)}\right\}$ is of the form $\dfrac{0}{0}$ at $x = a$.

Similarly if $\dfrac{f(x)}{g(x)}$ becomes either $\dfrac{0}{0}$ or $\dfrac{\infty}{\infty}$ as $x \to \infty$ we may write (putting $x = 1/y$)

$$\lim_{x\to\infty}\frac{f(x)}{g(x)} = \lim_{y\to 0}\frac{f\left(\dfrac{1}{y}\right)}{g\left(\dfrac{1}{y}\right)} = \lim_{y\to 0}\left\{\frac{-\dfrac{1}{y^2}f'\left(\dfrac{1}{y}\right)}{-\dfrac{1}{y^2}g'\left(\dfrac{1}{y}\right)}\right\} = \lim_{y\to 0}\frac{f'\left(\dfrac{1}{y}\right)}{g'\left(\dfrac{1}{y}\right)}$$

$$= \lim_{x\to\infty}\left\{\frac{f'(x)}{g'(x)}\right\}. \qquad (66)$$

Hence l'Hospital's rule applies when $a \equiv \infty$.
For example,

$$\lim_{x\to\infty}(x^3 e^{-x^2}) = \lim_{x\to\infty}\left\{\frac{x^3}{e^{x^2}}\right\} = \lim_{x\to\infty}\left\{\frac{3x^2}{2xe^{x^2}}\right\}$$

$$= \frac{3}{2}\lim_{x\to\infty}\left\{\frac{x}{e^{x^2}}\right\} = \frac{3}{2}\lim_{x\to\infty}\left\{\frac{1}{2xe^{x^2}}\right\} = 0. \qquad (67)$$

Similarly, if $n > 0$, it follows that

$$\lim_{x\to\infty}(x^n e^{-x^2}) = 0, \qquad (68)$$

and (by putting $x = 1/y$) that

$$\lim_{y\to 0}\left\{\frac{e^{-1/y^2}}{y^n}\right\} = 0. \qquad (69)$$

Example 8. The use of l'Hospital's rule may often be avoided by using series expansions. For example,

$$\lim_{x\to 0}\left(\frac{\sin x}{x}\right) = \lim_{x\to 0}\left(\frac{x - \dfrac{x^3}{3!} + \dfrac{x^5}{5!} - \cdots}{x}\right) = \lim_{x\to 0}\left(1 - \frac{x^2}{3!} + \frac{x^4}{5!} - \cdots\right) = 1,$$
$$(70)$$

as found earlier.

PROBLEMS 6

1. Obtain the first three non-vanishing terms of the Maclaurin series for the following functions

    (a) $\sqrt{(1+x)}$,   (b) $\tan^{-1} x$,   (c) $\log_e\left(\dfrac{1}{1-x}\right)$,

    (d) $\sec x$,   (e) $\cos(\sin x)$,   (f) $e^x \sin^{-1} x$.

2. Prove that

$$\sin x = x - \frac{x^3}{3!} + \frac{x^5}{5!} - \frac{x^7}{7!}\cos(\theta x),$$

and that

$$e^{-x^2} = 1 - x^2 + \frac{x^4}{2} - \frac{x^6}{6}e^{-\theta^2 x^2},$$

where $0 < \theta < 1$.

3. Expand $\sqrt{x}$ about $x = 1$ using Taylor's series, and show that this expansion is valid only for $0 \leqq x \leqq 2$.

4. Expand $\{\log_e(1+x)\}^2$ in powers of x as far as x^4. Hence, or otherwise, determine

    (i) whether $\cos 2x + \{\log_e(1+x)\}^2$ has a maximum, minimum or point of inflection at $x = 0$.

    (ii) whether $\dfrac{\{\log_e(1+x)\}^2}{x(1-\cos x)}$ has a finite limit at $x = 0$ and if so,

    what is the value of this limit?   (C.U.)

5. Evaluate the following limits

    (a) $\lim\limits_{x\to 0} \dfrac{\tan x - \sin x}{x^3}$,   (b) $\lim\limits_{x\to 0} \dfrac{\sin x + x}{x + x^2}$,

    (c) $\lim\limits_{x\to 0} x^2 \log_e x$,   (d) $\lim\limits_{x\to -1} \dfrac{\sin \pi x}{1 + x}$,

    (e) $\lim\limits_{x\to \infty} \dfrac{2x \cos x}{x + 1}$,   (f) $\lim\limits_{x\to \pi/2} (\sin x)^{\tan x}$,

    (g) $\lim\limits_{x\to 0} (x^x)$.

6. If $y = e^{\sin^{-1} x}$ prove that

$$(1-x^2)\frac{d^2y}{dx^2} - x\frac{dy}{dx} - y = 0$$

and that

$$(1-x^2)\frac{d^{n+2}y}{dx^{n+2}} - (2n+1)\,x\,\frac{d^{n+1}y}{dx^{n+1}} - (n^2+1)\frac{d^n y}{dx^n} = 0.$$

Hence verify the Maclaurin expansion

$$e^{\sin^{-1} x} = 1 + x + \frac{x^2}{2} + \frac{x^3}{3} + \frac{5}{24}x^4 + \dots .$$

7. By expanding the integrand in a power series in λ, show that

$$\int_0^{\pi} \cos(\lambda \sin x)\,dx = \pi\left(1 - \frac{\lambda^2}{4} + \frac{\lambda^4}{64} + \dots \right).$$

8. Obtain the value of $\sin 31°$ by expanding $\sin x$ to four terms about the point $x = \frac{\pi}{6}$.

9. Show, using equation (69) 6.3, that $f(x) = e^{-1/x^2}$ has no Maclaurin expansion but is always equal to the remainder term for all n and all x. Draw a rough graph of the function.

10. It is given that $y(x)$ is a function of x satisfying the differential equation

$$\frac{d^2y}{dx^2} = xy.$$

By differentiating n times show that for $n \geq 1$,

$$\frac{d^{n+2}y}{dx^{n+2}} = n\,\frac{d^{n-1}y}{dx^{n-1}}$$

at $x = 0$. Hence derive from Maclaurin's expansion an expression for y as an ascending power series in x such that

$$y = 1, \frac{dy}{dx} = 0 \quad \text{at} \quad x = 0. \qquad\qquad \text{(C.U.)}$$

11. Use Taylor's Theorem to show that, when h is small,

(a) $f'(a) = \dfrac{f(a+h)-f(a-h)}{2h}$ with an error of order $\frac{1}{6}h^2 f'''(a)$,

(b) $f''(a) = \dfrac{f(a+h)-2f(a)+f(a-h)}{h^2}$ with an error of order

$$\frac{1}{12}h^2 f''''(a).$$

Taking $f(x) = \sin x$ and $h = \pi/12$, find from (a) and (b) the approximate values of the first and second differential co-efficients of $\sin x$ at $x = \pi/4$, and compare them with tabulated values.

CHAPTER 7

Complex Variable

7.1 Complex Numbers

Besides the real numbers, more general numbers may be invented. In particular, numbers called complex numbers form an extremely valuable generalisation of real numbers and their use enables many results in real variable theory to be obtained in a concise way. It is usual to denote a typical complex number by z, where z is defined in terms of an ordered pair of real numbers (x, y) and the *imaginary number* $i = \sqrt{(-1)}$ such that

$$z = (x, y) = x + iy. \qquad (1)$$

The real numbers x and y are called the real part and the imaginary part of z respectively. We see that, by inventing the imaginary number i to represent $\sqrt{(-1)}$, equations which have no solutions in the real number system, such as

$$x^2 + 1 = 0 \qquad (2)$$

and

$$x^2 - 2x + 2 = 0, \qquad (3)$$

now have solutions in terms of complex numbers (i.e. for (2), $x = \pm i$, and for (3), $x = 1 \pm i$). In fact in the complex number system a general nth degree equation has exactly n roots (we shall meet this point again in Chapter 20). A simple consequence of the definition of i is that powers of i may be expressed in terms of ± 1 and i itself. For example,

$$i^2 = -1, \quad i^3 = i^2 i = -i, \quad i^4 = (i^2)^2 = 1, \qquad (4)$$

and

$$i^{-1} = \frac{1}{i} = \frac{i}{i^2} = -i, \quad i^{-2} = -1, \quad i^{-3} = \frac{1}{i^3} = \frac{1}{-i} = i. \qquad (5)$$

Now since a complex number is defined in terms of an ordered pair of real numbers (x, y) it follows that every complex number can be represented by just one point in a plane. Conversely every point in the plane may be associated with just one complex number. In other words there exists a (1-1) correspondence between the infinite set of complex numbers and the points of a plane. Consider now the plane defined by the rectangular Cartesian coordinate system xOy (see Fig. 7.1). Then if lengths along the x-axis represent the real part of z, and lengths along the y-axis the imaginary part, the point P uniquely represents the complex number z. For example, the complex number $z = 2 + 3i$ corresponds to the point obtained by going two units along the x-axis (from O) and then 3 units along a line parallel to the y-axis. It is usual to refer loosely to the x-axis as the real axis and to the y-axis as the imaginary axis, and to speak of the whole diagram as the Argand diagram.

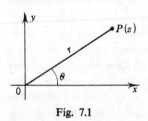

Fig. 7.1

If instead of x and y we adopt polar coordinates r and θ given by

$$x = r \cos \theta, \, y = r \sin \theta \tag{6}$$

we have, from (1),

$$z = x + iy = r(\cos \theta + i \sin \theta). \tag{7}$$

The number r (represented by the length of OP in the Argand diagram (see Fig. 7.1)) is called the modulus of z and is written as $|z|$ or mod z. It is always taken as positive and is zero only when $z = 0$. The angle θ is called the argument of z (written as arg z) and is the angle between the line OP and the positive direction of the x-axis. However, θ is not unique since the angles $\theta + 2k\pi$ (k zero or any integer) are also arguments for the same complex number. We therefore define the principal value of the argument of a complex number as that value of θ which satisfies the inequality

$$-\pi < \theta \leq \pi. \tag{8}$$

(It is convention that the principal value of the argument of a complex number is implied when arg z is written with a capital A, thus —Arg z.) From the Argand diagram (Fig. 7.1) we now find

$$|z| = r = \sqrt{(x^2 + y^2)} \tag{9}$$

and
$$\sin\theta = \frac{y}{\sqrt{(x^2+y^2)}}, \quad \cos\theta = \frac{x}{\sqrt{(x^2+y^2)}} \tag{10}$$

for all θ. For example, if $z = 2+3i$, then $z = r = \sqrt{(2^2+3^2)} = \sqrt{13}$, and $\sin\theta = \dfrac{3}{\sqrt{13}}$, $\cos\theta = \dfrac{2}{\sqrt{13}}$. An important complex number associated with any complex number z is its complex conjugate $\bar{z}$. If $z = x+iy$, then $\bar{z}$ is defined as

$$\bar{z} = x - iy, \tag{11}$$

from which we see that

$$|z| = |\bar{z}|. \tag{12}$$

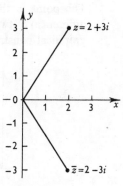

In other words in the Argand diagram $\bar{z}$ is the mirror image of z in the real axis. For example, the complex conjugate of $z=2+3i$ is $\bar{z} = 2-3i$ (see Fig. 7.2).

Fig. 7.2

7.2 Operations with Complex Numbers
If $z_1 = x_1+iy_1$ and $z_2 = x_2+iy_2$ are two complex numbers then

(a) the sum (or difference) of z_1 and z_2 is defined as the complex number whose real part is the sum (or difference) of the real parts of z_1 and z_2, and whose imaginary part is the sum (or difference) of the imaginary parts of z_1 and z_2. Expressing this definition mathematically we have

$$z_1 \pm z_2 = (x_1+iy_1) \pm (x_2+iy_2) = (x_1 \pm x_2)+i(y_1 \pm y_2). \tag{13}$$

Using (13) it can now easily be seen that complex numbers (like real numbers) satisfy the commutative and associative laws of addition in that

$$z_1+z_2 = z_2+z_1 \tag{14}$$

and
$$z_1+(z_2+z_3) = (z_1+z_2)+z_3, \tag{15}$$

where z_3 is any complex number.

109

It should be noted that if $z = x + iy$ is any complex number, then

$$z + \bar{z} = 2x \qquad (16)$$

and

$$z - \bar{z} = 2iy. \qquad (17)$$

The point in the Argand diagram representing the sum (or difference) of two complex numbers may be found by a simple graphical construction. Suppose in Fig. 7.3, P represents z_1

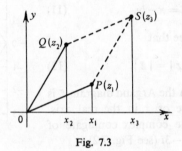

Fig. 7.3

and Q represents z_2. Then completing the parallelogram $OPSQ$, the point S clearly represents the complex number $z_1 + z_2$ since its real and imaginary parts are respectively $x_1 + x_2$ and $y_1 + y_2$. The construction of the point representing the difference of z_1 and z_2 is left to the reader.

(b) The product of z_1 and z_2 is defined as

$$z_1 z_2 = (x_1 + iy_1)(x_2 + iy_2) = (x_1 x_2 - y_1 y_2) + i(x_1 y_2 + x_2 y_1), (18)$$

the last expression in (18) being obtained from the middle expression by multiplying the terms out and using $i^2 = -1$. Here again, like real numbers, the commutative and associative laws of multiplication

$$z_1 z_2 = z_2 z_1 \qquad (19)$$

and

$$z_1(z_2 z_3) = (z_1 z_2)z_3 \qquad (20)$$

are satisfied (using the definition (18)).

The product of a complex number $z = x + iy$ and its complex conjugate $\bar{z}$ is (from (18))

$$z\bar{z} = (x+iy)(x-iy) = x^2 + y^2 = r^2 = |z|^2 \qquad (21)$$

and, since $|z|$ is always positive,

$$|z| = \sqrt{(z\bar{z})}. \qquad (22)$$

Similarly

$$|z^n| = (z^n\bar{z}^n)^{1/2} = (z\bar{z})^{n/2} = |z|^n, \qquad (23)$$

where n is any real number.

Likewise for any two complex numbers

$$|z_1 z_2|^2 = z_1 z_2 \bar{z}_1 \bar{z}_2 = z_1 \bar{z}_1 \cdot z_2 \bar{z}_2 = |z_1|^2 |z_2|^2 \qquad (24)$$

and, since all moduli are positive,

$$|z_1 z_2| = |z_1| |z_2|. \qquad (25)$$

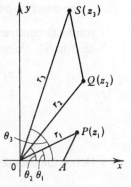

Fig. 7.4

As with addition and subtraction a graphical construction may also be set up for the point in the Argand diagram representing the product $z_1 z_2$. Suppose in Fig. 7.4, P and Q represent the points z_1 and z_2 respectively, and OA represents one unit length along the real axis. Then using the (r, θ) form of a complex number given in (7)

$$z_1 z_2 = r_1 r_2 (\cos\theta_1 + i\sin\theta_1)(\cos\theta_2 + i\sin\theta_2)$$
$$= r_1 r_2 \{\cos(\theta_1 + \theta_2) + i\sin(\theta_1 + \theta_2)\}. \qquad (26)$$

The complex number representing $z_1 z_2$ therefore has a modulus $r_1 r_2$ and an argument $\theta_1 + \theta_2$. Suppose now a point S is drawn such that $\triangle SOQ$ is similar to $\triangle POA$. Then if S is denoted by z_3, we have

$$\frac{r_3}{r_2} = \frac{r_1}{1}, \qquad (27)$$

and

$$\theta_3 - \theta_2 = \theta_1. \qquad (28)$$

Hence $r_3 = r_1r_2$ and $\theta_3 = \theta_1 + \theta_2$, and consequently $S(=z_3)$ represents the product z_1z_2.

(c) The division of z_2 by z_1 may be defined by writing

$$\frac{z_2}{z_1} = \frac{x_2 + iy_2}{x_1 + iy_1} = \frac{(x_2 + iy_2)(x_1 - iy_1)}{(x_1 + iy_1)(x_1 - iy_1)}$$

$$= \frac{x_1x_2 + y_1y_2}{x_1{}^2 + y_1{}^2} + i\frac{x_1y_2 - y_1x_2}{x_1{}^2 + y_1{}^2} \qquad (29)$$

provided $x_1{}^2 + y_1{}^2 = |z_1|^2 \neq 0$. Clearly division by zero is an undefined operation in the complex number system in the same way that division by zero is undefined in the real number system.

The complex number representing $\frac{z_2}{z_1}$ may be obtained by a graphical construction in the Argand diagram similar to that used in (b). Writing z_1 and z_2 in (r, θ) form we have

$$\frac{z_2}{z_1} = \frac{r_2(\cos\theta_2 + i\sin\theta_2)}{r_1(\cos\theta_1 + i\sin\theta_1)}$$

$$= \frac{r_2}{r_1}\{\cos(\theta_2 - \theta_1) + i\sin(\theta_2 - \theta_1)\}. \qquad (30)$$

Hence the complex number representing $\frac{z_2}{z_1}$ has a modulus $\frac{r_2}{r_1}$ and an argument $\theta_2 - \theta_1$.

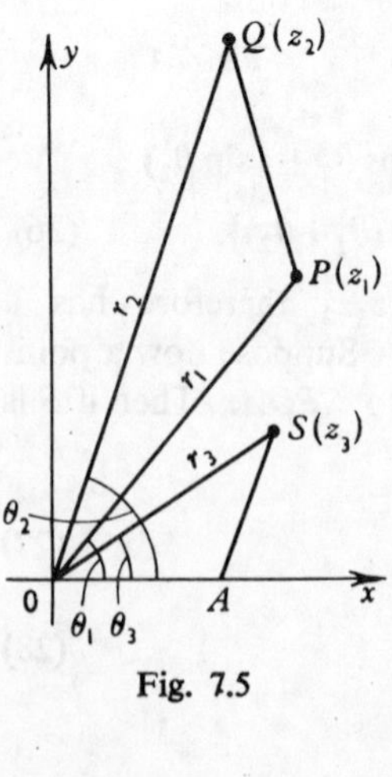

Fig. 7.5

If now in Fig. 7.5, the point $S(=z_3)$ is drawn such that $\triangle SOA$ is similar to $\triangle QOP$ (where again OA has unit length) then

$$\frac{r_3}{1} = \frac{r_2}{r_1} \qquad (31)$$

and $$\theta_2 - \theta_1 = \theta_3. \qquad (32)$$

Consequently S is the point representing $\frac{z_2}{z_1}$.

112

We illustrate the rules (a), (b) and (c) by the following example.

Example 1. If $z_1 = -3+5i$ and $z_2 = 2-3i$, then

$$z_1+z_2 = (-3+2)+i(5-3) = -1+2i, \tag{33}$$

$$z_1-z_2 = (-3-2)+i(5+3) = -5+8i, \tag{34}$$

$$z_1z_2 = (-3+5i)(2-3i) = (-3.2+5.3)+i(5.2+3.3) = 9+19i, \tag{35}$$

and $\quad \dfrac{z_1}{z_2} = \dfrac{-3+5i}{2-3i} = \dfrac{(-3+5i)(2+3i)}{(2-3i)(2+3i)} = \dfrac{-21+i}{13}. \tag{36}$

Further, since $|z_1| = \sqrt{(3^2+5^2)} = \sqrt{34}, |z_2| = \sqrt{(2^2+3^2)} = \sqrt{13}$ and $|z_1z_2| = \sqrt{(9^2+19^2)} = \sqrt{442}$, we find

$$|z_1z_2| = |z_1||z_2|, \tag{37}$$

which verifies (25).

7.3 The Exponential and Circular Functions

We have already seen in Chapter 5, 5.9, that the power series

$$\sum_{r=0}^{\infty} a_r x^r = a_0+a_1x+a_2x^2+ \ldots \tag{38}$$

(where the a_r are real constants and x is real) converges if $|x| < R$, the radius of convergence R being defined by $\lim\limits_{r \to \infty} \left| \dfrac{a_r}{a_{r+1}} \right|$. If in (38) x is replaced by a complex number z, the power series will again converge if $|z| < R$ which clearly contains the special case of real z. Hence we may define functions of z in terms of power series in z such that when z is real they reduce to the real variable definitions. In particular we define (by analogy with the real variable Maclaurin series given in Chapter 6, 6.2).

$$e^z = 1+z+\frac{z^2}{2!}+\frac{z^3}{3!}+\frac{z^4}{4!}+ \ldots \tag{39}$$

$$\sin z = z-\frac{z^3}{3!}+\frac{z^5}{5!}-\frac{z^7}{7!}+ \ldots \tag{40}$$

$$\cos z = 1 - \frac{z^2}{2!} + \frac{z^4}{4!} - \frac{z^6}{6!} + \dots . \tag{41}$$

These series converge for $0 \leq |z| < \infty$.

Using (39)–(41) we now find the important result

$$e^{i\theta} = \cos\theta + i\sin\theta, \tag{42}$$

where θ is any real number. This result enables the (r, θ) form of a complex number given by (2) to be expressed as

$$z = r(\cos\theta + i\sin\theta) = re^{i\theta}, \tag{43}$$

or (since θ is not unique (see 7.1)) more generally as

$$z = re^{i(\theta + 2\pi k)}, \tag{44}$$

where k is any integer.

Changing the sign of i in (42) we have

$$e^{-i\theta} = \cos\theta - i\sin\theta, \tag{45}$$

which when added to or subtracted from (42) itself leads to

$$\cos\theta = \frac{e^{i\theta} + e^{-i\theta}}{2} \tag{46}$$

and

$$\sin\theta = \frac{e^{i\theta} - e^{-i\theta}}{2i}. \tag{47}$$

These two results may in fact be taken as definitions of the circular functions $\cos\theta$ and $\sin\theta$ and are useful in showing the relationship between $\cos\theta$ and $\sin\theta$ and the hyperbolic functions $\cosh\theta$ and $\sinh\theta$ (see Chapter 8, 8.3). The following examples illustrate the use of the $re^{i\theta}$ form of z.

Example 2. To write $z = (4+3i)e^{i\pi/3}$ in the form $u + iv$, where u and v are real numbers, we put

$$e^{i\pi/3} = \cos\frac{\pi}{3} + i\sin\frac{\pi}{3} = \frac{1}{2} + i\frac{\sqrt{3}}{2}. \tag{48}$$

Hence

$$z = \frac{1}{2}(1+i\sqrt{3})(4+3i) = \left(\frac{4-3\sqrt{3}}{2}\right) + i\left(\frac{3+4\sqrt{3}}{2}\right), \quad (49)$$

which is of the required form.

Example 3. Since

$$e^{i\theta} = \cos\theta + i\sin\theta,$$

we have

$$e^{i\pi/2} = \cos\frac{\pi}{2} + i\sin\frac{\pi}{2} = i, \quad (50)$$

$$e^{i\pi} = \cos\pi + i\sin\pi = -1, \quad (51)$$

$$e^{3i\pi/2} = \cos\frac{3\pi}{2} + i\sin\frac{3\pi}{2} = -i, \quad (52)$$

and

$$e^{2i\pi} = \cos 2\pi + i\sin 2\pi = 1. \quad (53)$$

In general if n is an integer

$$e^{n\pi i} = \cos n\pi + i\sin n\pi = (-1)^n, \quad (54)$$

$$e^{2n\pi i} = \cos 2n\pi + i\sin 2n\pi = 1, \quad (55)$$

and

$$e^{(2n+1)\pi i/2} = \cos\left(\frac{2n+1}{2}\right)\pi + i\sin\left(\frac{2n+1}{2}\right)\pi = i(-1)^n. \quad (56)$$

If now $z_1 = re^{i\theta}$ is a given complex number represented by the point P in the Argand diagram (see Fig. 7.6) and the line OP is rotated through an angle α then the point Q represents a complex number z_2, where

$$z_2 = re^{i(\theta+\alpha)} = re^{i\theta}e^{i\alpha} = z_1 e^{i\alpha}. \quad (57)$$

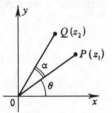

Fig. 7.6

For example, using (50), a rotation of z_1 through an angle $\alpha = \pi/2$ leads to a complex number $z_2 = iz_1$. Similarly, if $\alpha = \pi$, using (51), $z_2 = -z_1$.

115

Example 4. From the general (r, θ) form of a complex number

$$z = re^{i(\theta + 2\pi k)}, \quad (k \text{ zero or any integer}), \tag{58}$$

we may define the logarithmic function of z as

$$\log_e z = \log_e (re^{i(\theta + 2\pi k)}) = \log_e r + i(\theta + 2\pi k). \tag{59}$$

Using the relations between (r, θ) and (x, y) given by (9) and (10), (59) may be written in the alternative form

$$\log_e z = \log_e \sqrt{(x^2 + y^2)} + i \left(\tan^{-1} \frac{y}{x} + 2\pi k \right), \tag{60}$$

which gives the real and imaginary parts of $\log_e z$ in terms of x and y; $\log_e z$ is clearly a many-valued function. The principal value of $\log_e z$ is written as $\mathrm{Log}_e z$ and corresponds to taking the principal value of the argument in (59) (i.e. $k = 0$).

For example, $\mathrm{Log}\,(-1) = i\pi$.

7.4 de Moivre's Theorem and Applications
If z_1 and z_2 are two complex numbers given by

$$z_1 = r_1(\cos \theta_1 + i \sin \theta_1) \tag{61}$$

$$z_2 = r_2(\cos \theta_2 + i \sin \theta_2) \tag{62}$$

then by 7.2, (26)

$$z_1 z_2 = r_1 r_2 \{ \cos (\theta_1 + \theta_2) + i \sin (\theta_1 + \theta_2) \}. \tag{63}$$

Similarly, if we have a set of n complex numbers

$$\left. \begin{aligned} z_1 &= r_1(\cos \theta_1 + i \sin \theta_1), \\ z_2 &= r_2(\cos \theta_2 + i \sin \theta_2), \\ &\;\;\cdot \qquad \cdot \qquad \cdot \\ &\;\;\cdot \qquad \cdot \qquad \cdot \\ &\;\;\cdot \qquad \cdot \qquad \cdot \\ z_n &= r_n(\cos \theta_n + i \sin \theta_n), \end{aligned} \right\} \tag{64}$$

then

$$z_1 z_2 \ldots z_n = (r_1 r_2 \ldots r_n)\{\cos (\theta_1 + \theta_2 + \ldots + \theta_n)$$
$$+ i \sin (\theta_1 + \theta_2 + \ldots + \theta_n)\}. \quad (65)$$

Putting $r_1 = r_2 = r_3 = \ldots = r_n = 1$ and $\theta_1 = \theta_2 = \theta_3 = \ldots = \theta_n = \theta$, (65) gives

$$z^n = (\cos \theta + i \sin \theta)^n = \cos n\theta + i \sin n\theta, \quad (66)$$

where n is any positive integer. This result is known as de Moivre's Theorem and is also valid both when n is a negative integer, and a fractional number of the form p/q, where p and q are integers. For example, suppose n is a negative integer, say, $-m$, where m is a positive integer. Then

$$(\cos \theta + i \sin \theta)^{-m} = \frac{1}{(\cos \theta + i \sin \theta)^m} = \frac{1}{\cos m\theta + i \sin m\theta}$$

$$= (\cos m\theta + i \sin m\theta)^{-1} = \cos m\theta - i \sin m\theta$$

$$= \cos (-m)\theta + i \sin (-m) \, \theta, \quad (67)$$

which proves the theorem. A similar argument applies when n is fractional. Two applications of de Moivre's Theorem are now given.

(a) Expansion of $\cos^n \theta$, $\sin^n \theta$, $\cos n\theta$ *and* $\sin n\theta$

If $z = \cos \theta + i \sin \theta$, then $1/z = \cos \theta - i \sin \theta$.
Consequently

$$z + \frac{1}{z} = 2 \cos \theta \quad (68)$$

and

$$z - \frac{1}{z} = 2i \sin \theta. \quad (69)$$

In the same way, since

$$z^n = (\cos \theta + i \sin \theta)^n = \cos n\theta + i \sin n\theta \quad (70)$$

and

$$\frac{1}{z^n} = (\cos \theta + i \sin \theta)^{-n} = \cos n\theta - i \sin n\theta, \quad (71)$$

117

we have $$z^n + \frac{1}{z^n} = 2 \cos n\theta, \tag{72}$$

and $$z^n - \frac{1}{z^n} = 2i \sin n\theta. \tag{73}$$

These results are useful in expanding powers of $\cos \theta$ and $\sin \theta$ in terms of multiple angles as shown in the following example.

Example 5. To express $\cos^6 \theta$ in multiple angles we use (68) and write

$$2^6 \cos^6 \theta = \left(z + \frac{1}{z}\right)^6$$

$$= \left(z^6 + \frac{1}{z^6}\right) + 6\left(z^4 + \frac{1}{z^4}\right) + 15\left(z^2 + \frac{1}{z^2}\right) + 20. \tag{74}$$

By (72) the bracketed terms in (74) may all be expressed as cosines of multiple angles giving

$$2^6 \cos^6 \theta = 2 \cos 6\theta + 12 \cos 4\theta + 30 \cos 2\theta + 20, \tag{75}$$

or

$$\cos^6 \theta = \tfrac{1}{32}\{\cos 6\theta + 6 \cos 4\theta + 15 \cos 2\theta + 10\}. \tag{76}$$

Similarly, using (69) we have

$$2^5 i \sin^5 \theta = \left(z - \frac{1}{z}\right)^5 = \left(z^5 - \frac{1}{z^5}\right) - 5\left(z^3 - \frac{1}{z^3}\right) + 10\left(z - \frac{1}{z}\right), \tag{77}$$

which gives, by (73),

$$2^5 \sin^5 \theta = 2 \sin 5\theta - 10 \sin 3\theta + 20 \sin \theta. \tag{78}$$

Hence $$\sin^5 \theta = \tfrac{1}{16}\{\sin 5\theta - 5 \sin 3\theta + 10 \sin \theta\}. \tag{79}$$

The converse problem of expressing the cosines and sines of multiple angles in powers of $\cos \theta$ and $\sin \theta$ may also be solved using de Moivre's Theorem, as shown by the following example.

118

Example 6. To express $\cos 6\theta$ and $\sin 6\theta$ in terms of powers of $\cos \theta$ and $\sin \theta$ we write

$$(\cos 6\theta + i \sin 6\theta) = (\cos \theta + i \sin \theta)^6$$
$$= \cos^6 \theta + 6i \cos^5 \theta \sin \theta - 15 \cos^4 \theta \sin^2\theta$$
$$+ 20i^3 \cos^3 \theta \sin^3 \theta + 15 \cos^2 \theta \sin^4 \theta$$
$$+ 6i \cos \theta \sin^5 \theta - \sin^6 \theta. \tag{80}$$

The real and imaginary parts of this expression then give respectively

$$\cos 6\theta = \cos^6\theta - 15\cos^4 \theta \sin^2\theta + 15\cos^2 \theta \sin^4\theta - \sin^6 \theta \tag{81}$$

and

$$\sin 6\theta = 6\cos^5\theta \sin \theta - 20 \cos^3\theta \sin^3\theta + 6 \cos \theta \sin^5\theta. \tag{82}$$

(b) Solution of Equations

The use of de Moivre's Theorem in finding the complex roots of numbers and the equivalent problem of the solution of equations in z is best shown by the following examples.

Example 7. To find the n complex roots of the equation

$$z^n - 1 = 0, \tag{83}$$

where n is a positive integer, we use the general complex form of unity

$$1 = e^{2\pi k i} = \cos 2\pi k + i \sin 2\pi k, \tag{84}$$

where k is any integer (or zero). de Moivre's Theorem now gives

$$z = 1^{1/n} = (\cos 2\pi k + i \sin 2\pi k)^{1/n} = \cos \frac{2\pi k}{n} + i \sin\frac{2\pi k}{n}, \tag{85}$$

which, letting $k = 0, 1, 2, \ldots (n-1)$, has n distinct values $z_1, z_2, z_3 \ldots z_n$. These are the roots of (83) (or, in other words, the n roots of unity). For example, if $n = 6$, the six roots of (83) are the six values of

$$z = \cos \frac{2\pi k}{6} + i \sin \frac{2\pi k}{6} \tag{86}$$

obtained by putting $k = 0, 1, 2, 3, 4, 5.$

Hence

$$
\left.\begin{array}{ll}
k = 0, \; z_1 = \cos 0 + i \sin 0 & = \qquad 1, \\[2mm]
k = 1, \; z_2 = \cos \dfrac{\pi}{3} + i \sin \dfrac{\pi}{3} & = \quad \dfrac{1}{2} + i\dfrac{\sqrt{3}}{2}, \\[2mm]
k = 2, \; z_3 = \cos \dfrac{2\pi}{3} + i \sin \dfrac{2\pi}{3} & = -\dfrac{1}{2} + i\dfrac{\sqrt{3}}{2}, \\[2mm]
k = 3, \; z_4 = \cos \pi + i \sin \pi & = \qquad -1, \\[2mm]
k = 4, \; z_5 = \cos \dfrac{4\pi}{3} + i \sin \dfrac{4\pi}{3} & = -\dfrac{1}{2} - i\dfrac{\sqrt{3}}{2}, \\[2mm]
k = 5, \; z_6 = \cos \dfrac{5\pi}{3} + i \sin \dfrac{5\pi}{3} & = \quad \dfrac{1}{2} - i\dfrac{\sqrt{3}}{2}.
\end{array}\right\} \tag{87}
$$

No new roots are obtained by giving k any other values. For example, the root corresponding to $k = 6$ just reproduces the root corresponding to $k = 0$; similarly $k = 7$ reproduces $k = 1$, $k = -1$ reproduces $k = 5$, and so on.

It is important to notice that in taking roots of complex numbers the general form

$$
z = re^{i(\theta + 2\pi k)}, \quad k \text{ any integer (or zero)}, \tag{88}
$$

which allows for the many-valuedness of the argument, must always be used. For example, only one root of unity would have been obtained if we had written (84) as

$$
1 = \cos 0 + i \sin 0. \tag{89}
$$

Although (89) is certainly correct, it is not general enough when taking roots of complex numbers.

Example 8. The solutions of the equation

$$
z^4 - 1 = i\sqrt{3} \tag{90}
$$

are obtained by writing

$$
z^4 = 1 + i\sqrt{3} = 2\left(\frac{1}{2} + i\frac{\sqrt{3}}{2}\right)
$$

$$
= 2\left\{\cos\left(\frac{\pi}{3} + 2\pi k\right) + i \sin\left(\frac{\pi}{3} + 2\pi k\right)\right\}. \tag{91}
$$

Hence using de Moivre's Theorem,

$$z = 2^{1/4}\left\{\cos\left(\frac{\pi}{12}+\frac{\pi k}{2}\right)+i\sin\left(\frac{\pi}{12}+\frac{\pi k}{2}\right)\right\}, \tag{92}$$

where $k = 0, 1, 2, 3$. The four roots are therefore

$$\left.\begin{aligned}
&k=0,\ z_1 = 2^{1/4}\left(\cos\frac{\pi}{12}+i\sin\frac{\pi}{12}\right),\\
&k=1,\ z_2 = 2^{1/4}\left(\cos\frac{7\pi}{12}+i\sin\frac{7\pi}{12}\right),\\
&k=2,\ z_3 = 2^{1/4}\left(\cos\frac{13\pi}{12}+i\sin\frac{13\pi}{12}\right) = -z_1,\\
&k=3,\ z_4 = 2^{1/4}\left(\cos\frac{19\pi}{12}+i\sin\frac{19\pi}{12}\right) = -z_2.
\end{aligned}\right\} \tag{93}$$

All other k-values (integral) reproduce these four roots.

7.5 Other Applications of Complex Numbers

(a) Summation of Series

Example 9. To sum the finite series

$$C = 1+a\cos\theta+a^2\cos 2\theta+ \ldots +a^n\cos n\theta, \tag{94}$$

where a is a constant, we form the series

$$S = a\sin\theta+a^2\sin 2\theta+ \ldots +a^n\sin n\theta \tag{95}$$

and consider the complex series

$$C+iS = 1+ae^{i\theta}+a^2e^{2i\theta}+ \ldots +a^n e^{ni\theta}. \tag{96}$$

Since (96) is a geometric series with a common ratio $k = ae^{i\theta}$, we have (see Chapter 5)

$$C+iS = \frac{1-(ae^{i\theta})^{n+1}}{1-ae^{i\theta}}. \tag{97}$$

Consequently C and S are respectively the real and imaginary parts of the right-hand expression in (97). After some simplification we finally find that

$$C = \frac{1 - a \cos \theta + a^{n+2} \cos n\theta - a^{n+1} \cos (n+1)\theta}{1 - 2a \cos \theta + a^2} \qquad (98)$$

and

$$S = \frac{a \sin \theta + a^{n+2} \sin n\theta - a^{n+1} \sin (n+1)\theta}{1 - 2a \cos \theta + a^2}. \qquad (99)$$

Example 10. The series

$$C = 1 + {}^nC_1 \cos \theta + {}^nC_2 \cos 2\theta + \ldots + {}^nC_n \cos n\theta \qquad (100)$$

may be summed by writing

$$S = {}^nC_1 \sin \theta + {}^nC_2 \sin 2\theta + \ldots + {}^nC_n \sin n\theta \qquad (101)$$

so that

$$C + iS = 1 + \sum_{r=1}^{n} {}^nC_r e^{ir\theta} = (1 + e^{i\theta})^n. \qquad (102)$$

Now

$$(1 + e^{i\theta})^n = (1 + \cos \theta + i \sin \theta)^n$$

$$= \left(2 \cos^2 \frac{\theta}{2} + 2i \sin \frac{\theta}{2} \cos \frac{\theta}{2} \right)^n = 2^n \cos^n \frac{\theta}{2} \left(\cos \frac{\theta}{2} + i \sin \frac{\theta}{2} \right)^n$$

$$= \left(2 \cos \frac{\theta}{2} \right)^n \left(\cos \frac{n\theta}{2} + i \sin \frac{n\theta}{2} \right) \qquad (103)$$

(by de Moivre's Theorem).

Hence taking the real part of (103) we find

$$C = 2 \left(\cos \frac{\theta}{2} \right)^n \cos \frac{n\theta}{2}. \qquad (104)$$

Similarly the imaginary part of (103) gives

$$S = \left(2 \cos \frac{\theta}{2} \right)^n \sin \frac{n\theta}{2}. \qquad (105)$$

The sums of many series involving sines and cosines may be obtained using the methods of these last two examples.

(b) Evaluation of Integrals

A similar approach to that used above in summing series may be applied to the evaluation of certain integrals.

Example 11. Suppose we wish to evaluate

$$C = \int_0^t e^{ax} \cos bx \, dx, \tag{106}$$

where a, b and t are constants. Then putting

$$S = \int_0^t e^{ax} \sin bx \, dx \tag{107}$$

and forming $C + iS$, we have

$$C + iS = \int_0^t e^{ax}(\cos bx + i \sin bx) \, dx = \int_0^t e^{(a+ib)x} \, dx$$

$$= \frac{e^{(a+ib)t} - 1}{a + ib}. \tag{108}$$

Simplifying the right-hand side of (108) and taking the real and imaginary parts we have respectively

$$C = \frac{e^{at}(a \cos bt + b \sin bt) - a}{a^2 + b^2} \tag{109}$$

and

$$S = \frac{e^{at}(a \sin bt - b \cos bt) + b}{a^2 + b^2}. \tag{110}$$

These results could have been obtained by integrating (106) and (107) directly by parts, but in some cases it is found more convenient to proceed as shown in this example.

(c) Transformations

It is sometimes useful in more complicated applications of complex variable theory to associate with a complex number $z = x + iy$

123

another complex number $w = u + iv$ defined by a relation $w = f(z)$, where f is a known function. In this way a given point in the Argand diagram of z (the z-plane) usually becomes a different point in the Argand diagram of w (the w-plane), and, more important, a given region in the z-plane usually transforms into a different shaped region in the w-plane. When dealing, therefore, with the variation of some quantity over a complicated region of the z-plane it is often convenient to make a transformation to the w-plane such that the region concerned becomes a simpler one to deal with (for instance, a circle or a rectangle). This is represented diagramatically in Fig. 7.7, the point $P(z)$ being transformed into the point $Q(w)$. If in fact

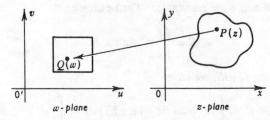

Fig. 7.7

to every point in the z-plane there corresponds just one point in the w-plane, and to every point in the w-plane there corresponds just one point in the z-plane, the transformation is said to be one-to-one. We now consider two examples of simple transformations.

Example 12. If $w = u + iv$ and $z = x + iy$ are related by

$$w = \frac{1}{z},\tag{111}$$

then

$$u + iv = \frac{1}{x + iy} = \frac{x - iy}{x^2 + y^2}.\tag{112}$$

Hence

$$u = \frac{x}{x^2 + y^2}\tag{113}$$

and

$$v = \frac{-y}{x^2 + y^2}.\tag{114}$$

Putting $y = mx$ in (113) and (114) we have

$$u = \frac{1}{x(1+m^2)}, \quad v = -\frac{m}{x(1+m^2)} \tag{115}$$

which, on eliminating x, give the equation

$$v = -mu. \tag{116}$$

Consequently straight lines through the origin of the z-plane transform into straight lines through the origin of the w-plane but with different gradients (see Fig. 7.8).

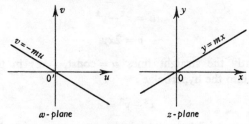

Fig. 7.8

Similarly, circles $x^2 + y^2 = a^2$ in the z-plane give

$$u = \frac{x}{a^2}, \quad v = -\frac{y}{a^2} \tag{117}$$

which define the circles

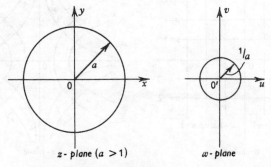

Fig. 7.9

125

$$u^2 + v^2 = 1/a^2 \tag{118}$$

in the w-plane. Hence circles with their centres at the origin of the z-plane are transformed into circles with centres at the origin of the w-plane but with inverse radii (see Fig. 7.9).

Example 13. If w and z are related by the transformation

$$w = z^2 \tag{119}$$

then

$$u + iv = (x + iy)^2 = (x^2 - y^2) + 2ixy. \tag{120}$$

Hence

$$u = x^2 - y^2 \tag{121}$$

and

$$v = 2xy. \tag{122}$$

Consequently the straight lines $u = $ const, $(=c)$ in the w-plane transform into the hyperbolae

$$x^2 - y^2 = c \tag{123}$$

in the z-plane, whilst the straight lines $v = $ const, $(=c)$ transform into the rectangular hyperbolae

$$xy = c/2 . \tag{124}$$

(The curves for $c = -2$, -1, 1 and 2 are shown in Fig. 7.10, equivalent regions being shaded.)

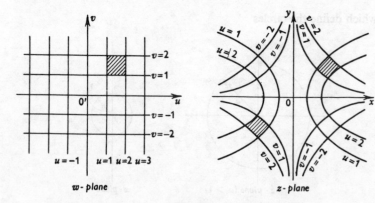

w-plane   z-plane

Fig. 7.10

We now see that in the case of the transformation $w = 1/z$ (Example 12) straight lines are transformed into straight lines, whereas in the case of $w = z^2$ (Example 13) this is not so. This is due to a fundamental difference in the forms of the two transformations. In fact $w = 1/z$ is a special case of an important transformation called the bilinear transformation defined by

$$w = \frac{az+b}{cz+d},\tag{125}$$

where a, b, c, d are constants such that $ad \neq bc$. It can be proved that this transformation is such that straight lines and circles in the z-plane are transformed into either straight lines or circles in the w-plane. The transformation $w = z^2$ is clearly not of this type.

7.6. Functions of a Complex Variable

In the last section we were dealing with special cases of the general relationship

$$w = f(z),\tag{126}$$

where $w = u + iv$ and $z = x + iy$ are both complex numbers. As z varies, $f(z)$ is said to be a function of the complex variable z, and concepts such as limits, continuity and differentiability arise for these functions as they did (see Chapter 3, 3.3) for functions of a real variable.

Firstly, in much of what we shall need we meet the idea of single-valuedness. A function $w = f(z)$ is said to be single-valued if for each value of z there is precisely one value of w. For example, $w = z^2$ is a single-valued function of z whereas $w = z^{1/2}$ is not.

Secondly, we say that $f(z)$ tends to a limit w_0 as $z \to z_0$—that is,

$$\lim_{z \to z_0} f(z) = w_0,\tag{127}$$

if for any real number $\varepsilon > 0$ there exists a real number η such that

$$|f(z) - w_0| < \varepsilon \quad \text{when} \quad |z - z_0| < \eta.\tag{128}$$

This definition clearly generalises the definition of the limit of a real function given in Chapter 3, equation (5). However, whereas in real variable theory the limit point can be approached along only two possible paths (that is, along the x-axis in the positive and negative senses), in complex theory (128) shows that z must lie within a circular region centred at z_0 and of radius η, and that z may approach z_0 along any path in this region which joins the two points.

Example 14. Find

$$\lim_{z \to i} \left(\frac{z^2+1}{z-i} \right). \tag{129}$$

Clearly

$$\lim_{z \to i} \left(\frac{z^2+1}{z-i} \right) = \lim_{z \to i} \left[\frac{(z+i)(z-i)}{z-i} \right] = \lim_{z \to i} (z+i) \tag{130}$$

$$= 2i. \tag{131}$$

We show that (128) is indeed satisfied here by asking if an η can be found for which

$$\left| \frac{z^2+1}{z-i} - 2i \right| < \varepsilon. \tag{132}$$

Now (132) becomes

$$\left| \frac{z^2+1-2iz-2}{z-1} \right| = \left| \frac{(z-i)^2}{(z-i)} \right| = |z-i| < \varepsilon, \tag{133}$$

whence (128) is satisfied by taking $\eta = \varepsilon$.

We note that although the function $(z^2+1)/(z-i)$ is not defined at $z = i$, nevertheless the limit does exist (as with, for example, the function $\frac{\sin x}{x}$ as $x \to 0$ in real variable theory).

A function $f(z)$ (assumed single-valued) is said to be continuous at $z = z_0$ if

(a) $\lim_{z \to z_0} f(z)$ exists,

(b) the function is defined for $z = z_0$

and

(c) if $\lim_{z \to z_0} f(z) = f(z_0)$,

where again z may approach z_0 along any path in the z-plane. A complex function is said to be continuous in some region R of the

z-plane if it is continuous at all points of that region. Finally we mention that, as with real variable theory, it may be proved that if $f(z)$ and $g(z)$ are both continuous at $z = z_0$, then $f+g$ and fg are continuous at $z = z_0$. Likewise f/g is continuous at $z = z_0$ provided $g(z_0) \neq 0$. Clearly z^n, where n is a positive integer, is a continuous function for all z since z is a continuous function. Hence the polynomial

$$f(z) = a_0 + a_1 z + a_2 z^2 + \ldots + a_n z^n \qquad (134)$$

(where the a_i are complex numbers) is everywhere continuous. Similarly, by the preceding results, the rational function

$$\frac{a_0 + a_1 z + a_2 z^2 + \ldots + a_n z^n}{b_0 + b_1 z + b_2 z^2 + \ldots + b_m z^m}, \qquad (135)$$

where the b_i are complex numbers also, is continuous everywhere except at the points at which the denominator vanishes.

7.7 Differentiation of Functions of a Complex Variable

Unlike the concepts of limiting values and continuity discussed in the last section, that of differentiability is not quite so straightforward.

We define (again by analogy with the real variable theory) the derivative of a function $w = f(z)$ at a point $z = z_0$ by the limit

$$\left(\frac{dw}{dz}\right)_{z=z_0} = f'(z_0) = \lim_{z \to z_0} \frac{f(z) - f(z_0)}{z - z_0}. \qquad (136)$$

Since, as we have seen in 7.6, the notion of a limit in complex variable theory implies that z may approach z_0 from any direction in the z-plane, the existence of the derivative requires that (136) exists and always has the same value whatever path is chosen. This is a fairly severe requirement and imposes certain conditions on the nature of a differentiable complex function—conditions which do not arise in real variable theory. To show this in detail we now calculate the limit $f'(z_0)$ in (136) in two distinct ways. First let $z - z_0 = \Delta z$. Then

$$f'(z_0) = \lim_{\Delta z \to 0} \left\{ \frac{f(z_0 + \Delta z) - f(z_0)}{\Delta z} \right\}. \qquad (137)$$

Now suppose z approaches z_0 along a line through z_0 parallel to the real (x) axis. Then, since $\Delta z = \Delta x + i\,\Delta y$, it follows that $\Delta z = \Delta x$ for this particular path.

Consequently, using $f(z) = u(x,y) + iv(x,y)$, we have

$$f'(z_0) = \lim_{\Delta x \to 0} \left\{ \frac{u(x_0 + \Delta x, y_0) + iv(x_0 + \Delta x, y_0) - u(x_0, y_0) - iv(x_0, y_0)}{\Delta x} \right\}$$
$$(138)$$

$$= \lim_{\Delta x \to 0} \left\{ \frac{u(x_0 + \Delta x, y_0) - u(x_0, y_0)}{\Delta x} \right\}$$

$$+ i \lim_{\Delta x \to 0} \left\{ \frac{v(x_0 + \Delta x, y_0) - v(x_0, y_0)}{\Delta x} \right\} \quad (139)$$

$$= \left(\frac{\partial u}{\partial x} \right)_{x = x_0} + i \left(\frac{\partial v}{\partial x} \right)_{x = x_0} \quad (140)$$

Now we choose a second path along which we evaluate (137)—namely, a line through z_0 parallel to the imaginary (y) axis. For this path $\Delta z = i\,\Delta y$, and we have

$$f'(z_0) = \lim_{\Delta y \to 0} \left\{ \frac{u(x_0, y_0 + \Delta y) + iv(x_0, y_0 + \Delta y) - u(x_0, y_0) - iv(x_0, y_0)}{i\,\Delta y} \right\}$$
$$(141)$$

$$= \lim_{\Delta y \to 0} \left\{ \frac{u(x_0, y_0 + \Delta y) - u(x_0, y_0)}{i\,\Delta y} \right\}$$

$$+ i \lim_{\Delta y \to 0} \left\{ \frac{v(x_0, y_0 + \Delta y) - v(x_0, y_0)}{i\,\Delta y} \right\} \quad (142)$$

$$= \frac{1}{i} \left(\frac{\partial u}{\partial y} \right)_{y = y_0} + \left(\frac{\partial v}{\partial y} \right)_{y = y_0}. \quad (143)$$

Since, however, $f'(z_0)$ is to be uniquely defined at z_0 whatever path is chosen, (140) and (143) must agree. Hence, equating the real and imaginary parts of these two equations we finally find

$$\frac{\partial u}{\partial x} = \frac{\partial v}{\partial y}, \quad \frac{\partial u}{\partial y} = -\frac{\partial v}{\partial x} \quad (144)$$

at $z_0 = x_0 + iy_0$. These relations are known as the Cauchy–Riemann equations and are necessary conditions for $f(z)$ to be differentiable. Any function $w = f(z) = u(x,y) + iv(x,y)$ which does not satisfy these conditions is not a differentiable function.

Example 15. Consider $f(z) = \bar{z} = x - iy$. Then

$$u(x,y) = x, \quad v(x,y) = -y \tag{145}$$

whence

$$\frac{\partial u}{\partial x} = 1, \quad \frac{\partial v}{\partial y} = -1, \quad \frac{\partial u}{\partial y} = 0, \quad \frac{\partial v}{\partial x} = 0. \tag{146}$$

Hence the Cauchy–Riemann equations are not satisfied and consequently $\bar{z}$ is not a differentiable function.

Example 16. Consider $f(z) = |z|$.
Then

$$u(x,y) = \sqrt{(x^2+y^2)}, \quad v(x,y) = 0, \tag{147}$$

and again it is clear that the Cauchy–Riemann equations are not satisfied. Hence $|z|$ is not a differentiable function.

We have at this stage only shown that the Cauchy–Riemann equations are necessary for differentiability. However, it may be proved (and this is the real importance of the Cauchy–Riemann equations) that provided the partial derivatives $\dfrac{\partial u}{\partial x}, \dfrac{\partial v}{\partial x}, \dfrac{\partial u}{\partial y}, \dfrac{\partial v}{\partial y}$ are continuous in some domain containing the point z_0, and if the Cauchy–Riemann equations are satisfied, then the function is differentiable at that point. In other words, under these conditions, the Cauchy–Riemann equations are both necessary and *sufficient* conditions for a function to be differentiable.

Example 17. Consider $f(z) = e^z$.
Then

$$e^z = e^x e^{iy} = e^x(\cos y + i\sin y) \tag{148}$$

so

$$u(x,y) = e^x \cos y, \quad v(x,y) = e^x \sin y. \tag{149}$$

The partial derivatives are continuous and are easily seen to satisfy

131

the Cauchy–Riemann equations. Hence e^z is a differentiable function. Its derivative may be obtained from either (140) or (143) giving

$$\frac{d(e^z)}{dz} = f'(z) = \frac{\partial u}{\partial x} + i\frac{\partial v}{\partial y} \tag{150}$$

$$= e^x \cos y + ie^x \sin y = e^z. \tag{151}$$

Finally, we emphasize again that functions which are differentiable form a special, rather restricted, but vitally important class of functions. We say that a function of a complex variable is *analytic* at a point $z = z_0$ if it is defined and has a derivative at z_0, and is analytic in a domain if it is analytic at every point of that domain.

PROBLEMS 7

1. If $z_1 = 3+2i$ and $z_2 = -1+i$, find z_1+z_2, z_1-z_2, z_1z_2 and z_1/z_2. Evaluate $\bar{z}_1 z_1$ and $\bar{z}_2 z_2$.

2. Find the real and imaginary parts of

(a) $\dfrac{2+3i}{3+2i}$, (b) $\log_e\left\{\dfrac{1}{2}(\sqrt{3}+i)\right\}$, (c) $(1+i)^{iy}$,

(d) $\dfrac{1}{i^5}$, (e) $\left(-\dfrac{1}{2}+i\dfrac{\sqrt{3}}{2}\right)^2$, (f) $\tan(x+iy)$.

3. Prove that

$$\frac{1+\cos\alpha+i\sin\alpha}{1-\cos\alpha+i\sin\alpha} = \cot\frac{\alpha}{2} \cdot e^{i[\alpha-(\pi/2)]}. \tag{L.U.}$$

4. Solve the equations

(a) $(z+1)^7+(z-1)^7 = 0$,

(b) $z^4+1 = i\sqrt{3}$,

(c) $\log_e(x+iy) = 2+\dfrac{\pi i}{3}$.

5. By considering the roots of

$$z^{2n+1}+1 = 0 \quad (n \text{ integral})$$

show that

$$\sum_{k=-n}^{n} \cos\left(\frac{2k+1}{2n+1}\right)\pi = 0. \qquad \text{(C.U.)}$$

6. Prove that, if $z = \cos\theta + i\sin\theta$ and n is any positive integer,

$$z^n - \frac{1}{z^n} = 2i\sin n\theta.$$

Show that

$$z^{2n} - 1 = (z^2 - 1) \prod_{r=1}^{n-1}\left\{z^2 - 2z\cos\frac{r\pi}{n} + 1\right\}.$$

By substituting $z = \cos\theta + i\sin\theta$, deduce that

$$\frac{\sin n\theta}{\sin\theta} = 2^{n-1} \prod_{r=1}^{n-1}\left\{\cos\theta - \cos\frac{r\pi}{n}\right\}. \qquad \text{(L.U.)}$$

7. From $\dfrac{1}{1-z} = 1 + z + z^2 + z^3 \ldots$, $(|z| < 1)$, prove that

$$\sum_{n=0}^{\infty} r^n \cos n\theta = \frac{1 - r\cos\theta}{1 - 2r\cos\theta + r^2}$$

and

$$\sum_{n=0}^{\infty} r^n \sin n\theta = \frac{r\sin\theta}{1 - 2r\cos\theta + r^2}, \quad (r < 1).$$

8. Show that

$$e^{\cos x}\cos(\sin x) = 1 + \frac{\cos x}{1!} + \frac{\cos 2x}{2!} + \frac{\cos 3x}{3!} + \ldots$$

$$e^{\cos x}\sin(\sin x) = \frac{\sin x}{1!} + \frac{\sin 2x}{2!} + \frac{\sin 3x}{3!} + \ldots.$$

Hence show that

$$\int_0^\pi e^{\cos x}\cos(\sin x)\,dx = \pi.$$

9. Using complex numbers, sum the following series

   (a) $\displaystyle\sum_{r=1}^{\infty} \frac{\sin r\theta}{r!}$, (b) $\displaystyle\sum_{r=0}^{\infty} \frac{\cos r\theta}{2^r}$, (c) $\displaystyle\sum_{r=1}^{n} \sin r\theta$.

10. Find the curves represented by the equations

   (a) $|z+1| = |z-i|$,

   (b) $\arg\left(\dfrac{z+1}{z-1}\right) = \alpha$, $\alpha = \text{const}$,

   (c) $z = \sin(\alpha + it)$, α const, t varying.

11. Show that

   (a) $\cos 5\theta = 16 \cos^5 \theta - 20 \cos^3 \theta + 5 \cos \theta$,

   (b) $\cos^4 \theta = \frac{1}{8}(\cos 4\theta + 4 \cos 2\theta + 3)$.

12. Find the transforms of the circles $|z| = 1$, and $|z| = \frac{1}{2}$ under the bilinear transformation

$$w = \frac{z-2}{z-1},$$

and determine the domain of the w-plane that corresponds to the region between these two circles. (L.U.)

13. Find all the values of Log $(1+i)$, Log i, $(1+i)^i$ and i^i.

14. Determine all possible values of $\sin^{-1} 2$ and $\tan^{-1}(2i)$.

15. The points z_1, z_2, z_3 form an equilateral triangle in the Argand diagram. $z_1 = 4+6i, z_2 = (1-i)z_1$.
   Show that z_3 must have one of two values and determine these values.
   What are the vertices of the regular hexagon of which z_1 is the centre and z_2 one vertex? (C.U.)

16. If $w = 1/z$, show that the lines $u = \text{constant}$, $v = \text{constant}$ in the w-plane are orthogonal circles in the z-plane which have their centres on the coordinate axes and pass through the origin.

17. $ABCD$ is the region whose two sides AB, CD are the parabolae

134

$y^2 = 4b(b+x)$, $y^2 = 4c(c-x)$, and whose other two sides are the x-axis and the parabola $y^2 = 4a(a+x)$, where a, b and c are constants. Show that the transformation $w = z^{1/2}$ maps $ABCD$ on to a rectangle in the w-plane.

18. Find the constant λ such that $e^{\lambda x} \cos 3y$ is the real part of an analytic function of z. Find the corresponding imaginary part.

19. Determine which of the following functions satisfy the Cauchy–Riemann equations:

    (a) $f(z) = z^3 - iz^2 + 1$.
    (b) $f(z) = z \sin z + |z|^2$.
    (c) $f(z) = \sinh \bar{z}$.

20. Find the points, if any, at which the following functions are not analytic:

    (a) $\dfrac{1}{(z+1)^3}$,   (b) $\dfrac{z}{z+3}$,   (c) $\dfrac{\cos z}{z^2+1}$.

CHAPTER 8

Hyperbolic Functions

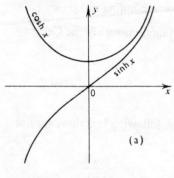

(a)

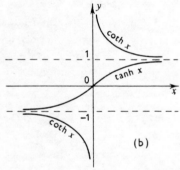

(b)

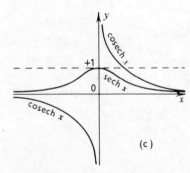

(c)

Fig. 8.1

8.1 Definitions

The two basic hyperbolic functions sinh x and cosh x have already been defined in Chapter 1, 1.3 (h), but for completeness we restate these here, together with some additional hyperbolic functions which can be derived from them.

If

$$\sinh x = \frac{e^x - e^{-x}}{2} \qquad (1)$$

and

$$\cosh x = \frac{e^x + e^{-x}}{2} \qquad (2)$$

then we define (by analogy with the circular functions)

$$\tanh x = \frac{\sinh x}{\cosh x} = \frac{e^x - e^{-x}}{e^x + e^{-x}}, \qquad (3)$$

$$\coth x = \frac{1}{\tanh x} = \frac{e^x + e^{-x}}{e^x - e^{-x}}, \qquad (4)$$

$$\operatorname{sech} x = \frac{1}{\cosh x} = \frac{2}{e^x + e^{-x}}, \qquad (5)$$

and

$$\operatorname{cosech} x = \frac{1}{\sinh x} = \frac{2}{e^x - e^{-x}}, \qquad (6)$$

Rough graphs of these six functions are given in Figs. 8 (a), (b) and (c).

136

We remark here that hyperbolic functions of a complex argument z may be defined by simply replacing x by z in (1)–(6).

It follows directly from (1) and (2) that

$$\cosh^2 x - \sinh^2 x = 1 \tag{7}$$

and that

$$\sinh x \cosh x = \tfrac{1}{4}(e^{2x} - e^{-2x}) = \tfrac{1}{2}\sinh 2x. \tag{8}$$

We now see that the reason for calling these functions hyperbolic is that the locus defined by the parametric equations

$$\left. \begin{array}{l} x = a \cosh t \\ y = a \sinh t \end{array} \right\} \tag{9}$$

is (in virtue of (7)) the hyperbola

$$x^2 - y^2 = a^2. \tag{10}$$

(In the case of the (circular) functions $\sin t$ and $\cos t$ the parametric equations $x = a \cos t$, $y = a \sin t$ define the circle $x^2 + y^2 = a^2$, since $\cos^2 t + \sin^2 t = 1$.)

The following identities may be obtained from the definitions of the hyperbolic functions given in (1)–(6)

(a) $\operatorname{sech}^2 x + \tanh^2 x = 1,$ (11)

(b) $\coth^2 x - \operatorname{cosech}^2 x = 1,$ (12)

(c) $\sinh(x \pm y) = \sinh x \cosh y \pm \cosh x \sinh y,$ (13)

(d) $\cosh(x \pm y) = \cosh x \cosh y \pm \sinh x \sinh y,$ (14)

(e) $\tanh(x \pm y) = \dfrac{\tanh x \mp \tanh y}{1 \pm \tanh x \tanh y},$ (15)

(f) $\sinh 2x = 2 \sinh x \cosh x,$ (see (8)), (16)

(g) $\cosh 2x = \cosh^2 x + \sinh^2 x,$ (17)

(h) $\tanh 2x = \dfrac{2 \tanh x}{1 + \tanh^2 x}.$ (18)

8.2 Derivatives and Indefinite Integrals

The differential coefficients of hyperbolic functions may be obtained directly from their exponential forms. For example,

$$\frac{d}{dx}(\cosh x) = \frac{d}{dx}\left(\frac{e^x + e^{-x}}{2}\right) = \frac{e^x - e^{-x}}{2} = \sinh x, \qquad (19)$$

and

$$\frac{d}{dx}(\sinh x) = \frac{d}{dx}\left(\frac{e^x - e^{-x}}{2}\right) = \frac{e^x + e^{-x}}{2} = \cosh x. \qquad (20)$$

Similarly we find

$$\frac{d}{dx}(\tanh x) = \operatorname{sech}^2 x \qquad (21)$$

and

$$\frac{d}{dx}(\coth x) = -\operatorname{cosech}^2 x. \qquad (22)$$

From the definition of indefinite integration as the inverse operation to differentiation (see Chapter 4) it follows from (19)–(22) that

$$\int \sinh x \, dx = \cosh x + c, \qquad (23)$$

$$\int \cosh x \, dx = \sinh x + c, \qquad (24)$$

$$\int \operatorname{sech}^2 x \, dx = \tanh x + c, \qquad (25)$$

$$\int \operatorname{cosech}^2 x \, dx = -\coth x + c, \qquad (26)$$

where in each case c is a constant of integration. A short list of indefinite integrals involving hyperbolic functions is given in Table 2.

Integrals involving hyperbolic functions may sometimes be evaluated (as shown in the following example) by using a ' t ' substitution similar to that given for circular functions (see Chapter 4, 4.4).

Example 1. To evaluate $\int \dfrac{dx}{3 + 4 \cosh x}$ we make the substitution

$$t = \tanh \frac{x}{2}. \tag{27}$$

Then

$$\sinh x = \frac{2t}{1 - t^2}, \quad \cosh x = \frac{1 + t^2}{1 - t^2}, \quad dx = \frac{2\,dt}{1 - t^2}. \tag{28}$$

Hence, using (28),

$$\int \frac{dx}{3 + 4 \cosh x} = 2 \int \frac{dt}{7 + t^2} = \frac{2}{\sqrt{7}} \tan^{-1} \left(\frac{1}{\sqrt{7}} \tanh \frac{x}{2} \right) + c, \tag{29}$$

where c is a constant of integration.

TABLE 2

$f(x)$	$F(x) = \int f(x)\,dx$	$f(x)$	$F(x) = \int f(x)\,dx$
$\sinh x$	$\cosh x$	$\dfrac{1}{\sqrt{(x^2 + a^2)}}$	$\sinh^{-1} \dfrac{x}{a}$
$\cosh x$	$\sinh x$	$\dfrac{1}{\sqrt{(x^2 - a^2)}}$	$\cosh^{-1} \dfrac{x}{a}, \quad (x > a)$
$\tanh x$	$\log_e \cosh x$	$\dfrac{1}{a^2 - x^2}$	$\dfrac{1}{a} \tanh^{-1} \dfrac{x}{a}, \quad (\lvert x \rvert < a)$
$\coth x$	$\log_e \lvert \sinh x \rvert$	$\dfrac{1}{x^2 - a^2}$	$-\dfrac{1}{a} \coth^{-1} \dfrac{x}{a}, (\lvert x \rvert > a)$
$\operatorname{sech}^2 x$	$\tanh x$	$\operatorname{sech} x$	$\sin^{-1}(\tanh x)$
$\operatorname{cosech}^2 x$	$-\coth x$		

8.3 Relation between Hyperbolic and Circular Functions
From the definitions

$$\sinh x = \frac{e^x - e^{-x}}{2}, \quad \cosh x = \frac{e^x + e^{-x}}{2} \tag{30}$$

139

given in 8.1, and the expressions

$$\sin x = \frac{e^{ix} - e^{-ix}}{2i}, \quad \cos x = \frac{e^{ix} + e^{-ix}}{2} \tag{31}$$

derived in Chapter 7, 7.3, we find

$$\sinh ix = i \sin x, \tag{32}$$

$$\cosh ix = \cos x, \tag{33}$$

and consequently

$$\tanh ix = i \tan x. \tag{34}$$

Alternatively (32)–(34) may be written (by replacing ix by x) as

$$i \sinh x = \sin ix, \tag{35}$$

$$\cosh x = \cos ix, \tag{36}$$

and

$$i \tanh x = \tan ix. \tag{37}$$

These relations are useful in finding the real and imaginary parts of hyperbolic functions of a complex variable z as shown by the following example.

Example 2. If $z = x + iy$, then by (14)

$$\cosh z = \cosh (x + iy) = \cosh x \cosh iy + \sinh x \sinh iy. \tag{38}$$

Using (32) and (33), (38) becomes

$$\cosh z = \cosh x \cos y + i \sinh x \sin y, \tag{39}$$

from which the real and imaginary parts are seen to be $\cosh x \cos y$ and $\sinh x \sin y$ respectively.

8.4 Inverse Hyperbolic Functions

Just as we define inverse circular functions ($\sin^{-1} x$, $\cos^{-1} x$, etc.), so we may define inverse hyperbolic functions. For example, if $y = \cosh x$ then $x = \cosh^{-1} y$ is called the inverse hyperbolic cosine of y. Inverse hyperbolic functions may be expressed in terms of the logarithmic function as shown below.

(a) If $y = \sinh^{-1} x$, then $\sinh y = x$, and consequently (from (7)) $\cosh y = \sqrt{(x^2 + 1)}$.

Hence, since

$$e^y = \cosh y + \sinh y, \tag{40}$$

we have

$$e^y = x + \sqrt{(x^2 + 1)} \tag{41}$$

or, alternatively,

$$y = \sinh^{-1} x = \log_e (x + \sqrt{(x^2 + 1)}) \tag{42}$$

for all x. The graph of $\sinh^{-1} x$ is shown in Fig. 8.2.

(b) If $y = \cosh^{-1} x$, then $x = \cosh y$, and consequently

$$\sinh y = \pm \sqrt{(x^2 - 1)}. \tag{43}$$

Hence, using (40),

$$e^y = x \pm \sqrt{(x^2 - 1)}, \tag{44}$$

which gives

$$y = \cosh^{-1} x = \log_e \{x \pm \sqrt{(x^2 - 1)}\} \tag{45}$$

provided $x \geqq 1$. The ambiguity of sign arises in (43) since $\sinh y$ may be negative. Consequently $\cosh^{-1} x$ is a double-valued function in that for every value of x (> 1) there are two values of y. However, as shown in Fig. 8.2, these y-values only differ in sign since $\cosh^{-1} x$ is symmetrical about the x-axis. It is usual to define the principal value of $y = \cosh^{-1} x$ as that part corresponding to $y \geqq 0$.

(c) If $y = \tanh^{-1} x$, then $x = \tanh y$, and hence, from (11),

$$\operatorname{sech} y = \sqrt{(1 - x^2)}. \tag{46}$$

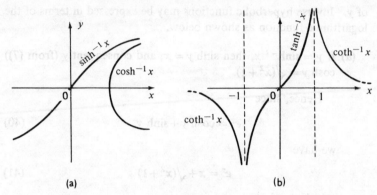

(a)   (b)

Fig. 8.2

Again, using (40),

$$e^y = \cosh y + \sinh y = \frac{1}{\sqrt{(1-x^2)}} + \frac{x}{\sqrt{(1-x^2)}} = \sqrt{\left(\frac{1+x}{1-x}\right)}, \quad (47)$$

and hence

$$y = \tanh^{-1} x = \tfrac{1}{2} \log_e \left(\frac{1+x}{1-x}\right), \quad (48)$$

for $|x| < 1$.

The derivatives of the inverse hyperbolic functions may be obtained either from their formal definitions or from their logarithmic forms. For example, from (42),

$$\frac{d}{dx}(\sinh^{-1} x) = \frac{d}{dx}\{\log_e [x + \sqrt{(x^2+1)}]\} = \frac{1}{\sqrt{(x^2+1)}}. \quad (49)$$

PROBLEMS 8

1. Show that

(a) $\operatorname{cosech}^{-1} x = \log_e \left\{\dfrac{1}{x} + \sqrt{\left(1 + \dfrac{1}{x^2}\right)}\right\}$, for $|x| > 0$,

(b) $\operatorname{sech}^{-1} x = \log_e \left\{\dfrac{1}{x} \pm \sqrt{\left(\dfrac{1}{x^2} - 1\right)}\right\}$, for $0 < x \leqq 1$,

(c) $\coth^{-1} x = \frac{1}{2} \log_e \left(\frac{x+1}{x-1} \right)$, for $|x| > 1$,

and obtain the differential coefficients of these functions.

2. Express $\sinh(x+iy)$ in the form $u+iv$, where x, y, u, v are all real, and show that

$$| \sinh(x+iy) |^2 = \frac{1}{2}(\cosh 2x - \cos 2y).$$

If $\sinh(x+iy) = e^{i\pi/3}$ prove that either

$$x = \log_e \left(\frac{\sqrt{6}+\sqrt{2}}{2} \right), \quad y = 2n\pi + \frac{\pi}{4},$$

or

$$x = \log_e \left(\frac{\sqrt{6}-\sqrt{2}}{2} \right), \quad y = 2n\pi + \frac{3\pi}{4},$$

n being any integer. (L.U.)

3. Verify the following inequalities

(a) $\operatorname{sech} x < \operatorname{cosech} x < \coth x$, for $x > 0$,

(b) $\dfrac{x}{\sqrt{(x^2+1)}} < \sinh^{-1} x < x$, for $x > 0$,

(c) $\cosh^{-1} x > \left(\dfrac{x-1}{x+1} \right)^{\frac{1}{2}}$, for $x > 1$.

4. Differentiate the functions

(a) $\tan^{-1}(\sinh x)$, (b) $\log_e \tanh x$, (c) $\sinh^{-1}(\sin \sqrt{x})$,

(d) $\tanh^{-1}[\sqrt{(2-x)}]$, (e) $(\cosh x)^{\sin x}$.

5. Obtain the following series using Maclaurin's expansion

(a) $\tanh x = x - \dfrac{x^3}{3} + \dfrac{2x^5}{15} + \ldots$

(b) $\tanh^{-1} x = x + \dfrac{x^3}{3} + \dfrac{x^5}{5} + \ldots$

(c) $\operatorname{sech} x = 1 - \dfrac{x^2}{2} + \dfrac{5x^4}{24} + \ ....$

6. Evaluate using l'Hospital's rule, or otherwise,

(a) $\lim\limits_{x \to 0} \left\{ \dfrac{x - \sinh x}{x - \sin x} \right\}$,

(b) $\lim\limits_{x \to 0} \left\{ \dfrac{x - \sin^{-1} x}{\log_e \cos x} \right\}$,

(c) $\lim\limits_{x \to \infty} e^{(\sinh^{-1} x / \cosh^{-1} x)}$.

7. Show that $y = (\sinh^{-1} x)^2$ satisfies the equation

$$(1 + x^2) \dfrac{d^2 y}{dx^2} + x \dfrac{dy}{dx} = 2.$$

Hence, or otherwise, show that

$$y = x^2 - \dfrac{2^2}{3 \cdot 4} x^4 + \dfrac{2^2 \cdot 4^2}{3 \cdot 4 \cdot 5 \cdot 6} x^6 - \ ....$$ (L.U.)

8. Evaluate the integrals

(a) $\displaystyle\int \sinh x \sin x \, dx,$

(b) $\displaystyle\int \operatorname{sech} x \, dx,$

(c) $\displaystyle\int_0^\infty e^{-3x} \cosh^2 x \, dx,$

(d) $\displaystyle\int_0^1 x^2 \sinh x \, dx,$

(e) $\displaystyle\int_0^a (\sinh^{-1} x)\sqrt{(1 + x^2)} \, dx.$

9. If $u + iv = \log_e \left(\dfrac{x + iy + a}{x + iy - a} \right)$, prove that

(a) $x^2 + y^2 - 2ax \coth u + a^2 = 0,$

(b) $x = a \dfrac{\sinh u}{\cosh u - \cos v},$

(c) $|x + iy|^2 = a^2 \dfrac{\cosh u + \cos v}{\cosh u - \cos v}.$ (L.U.)

10. If $z = x + iy$ solve the equation $\cosh z = 1 + \sinh z.$

CHAPTER 9

Partial Differentiation

9.1 Functions of Several Independent Variables

The concepts of continuity and differentiability of functions of one real independent variable have already been discussed in Chapter 3 and in this section we extend these ideas to functions of two or more real independent variables x, y, u, v ... (or x_1, x_2, x_3 ...). We first discuss, however, some general properties of functions of this type.

Consider, for example, a function of two variables x and y defined by

$$f(x, y) = x^2 - 2y^2. \tag{1}$$

Then the value of $f(x, y)$ is determined by (1) for every number pair (x, y). For instance, if $(x, y) = (0, 0)$ we have $f(0, 0) = 0$ and if $(x, y) = (1, 0)$, $f(1, 0) = 1$.

In general we may represent every pair of numbers (x, y) by a point P in the (x, y) plane of a rectangular Cartesian coordinate system and denote the corresponding value of $f(x, y)$ by the length of the line PP' drawn parallel to the z-axis (see Fig. 9.1). The locus of all points such as P' is then a surface in the (x, y, z) space which represents the function $f(x, y)$. However, this simple geometrical picture is impossible to visualise when dealing with functions of three or more independent variables.

Returning now for simplicity to functions of two independent variables, we notice that many functions are only defined within a certain region of the (x, y) plane. (This is analogous to the one-variable case where $f(x)$ is defined in a certain interval of x.) For example, the real function

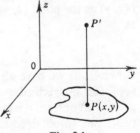

Fig. 9.1

$$f(x, y) = \sqrt{(a^2 - x^2 - y^2)} \tag{2}$$

is only defined within and on the boundary of the circle $x^2 + y^2 = a^2$;

145

outside this region it takes on imaginary values. Similarly the function

$$f(x, y) = \tan \frac{y}{x} \tag{3}$$

is undefined along the line $x = 0$. The function given in (1), however, is defined for all values of x and y. It is usual to denote the region of definition of a function of several independent variables by the letter R.

If now a function $f(x, y)$ has just one real value for every (x, y) value within its region of definition R, we say that it is a single-valued function of x and y within R. Similarly, when two or more values of the function are obtained for a given (x, y) value we call the function two-valued or many-valued. For instance, the function defined by (1) is single-valued over the region R given by $-\infty < x < \infty$, $-\infty < y < \infty$, whereas the function defined by (2) is two-valued over the region R given by $x^2 + y^2 < a^2$ (since both signs of the square root may be taken) and single-valued (equal to zero) on the boundary of the circle $x^2 + y^2 = a^2$.

Another important concept already defined in Chapter 3, 3.2 for functions of one independent variable is that of continuity. When discussing the continuity of functions of two or more independent variables similar considerations apply. Suppose $f(x, y)$ is a real single-valued function of x and y. Then if $f(x, y)$ approaches a value l as x approaches a and y approaches b, l is said to be a limit of $f(x, y)$ as the point (x, y) approaches the point (a, b) and is written as

$$\lim_{(x, y) \to (a, b)} f(x, y) = l. \tag{4}$$

However, as we have already seen in the one-variable case, x may approach a specified point $x = a$ from either the negative side $(-\infty \to a)$ or from the positive side $(+\infty \to a)$, and the values of the two limits so obtained may be different. The same is true of (4); the way in which $(x, y) \to (a, b)$ may determine the value of l. However, there is now much more freedom than in the one-variable case since (see Fig. 9.2) the point $Q(x, y)$ may approach the point $P(a, b)$ along any of the infinity of curves, say c, which lie in the (x, y) plane and

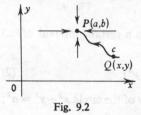

Fig. 9.2

146

which pass through P. If, however, the limit exists independently of the way in which Q approaches P and is such that

$$\lim_{(x, y) \to (a, b)} f(x, y) = f(a, b), \tag{5}$$

(assuming that $f(a, b)$ exists), then $f(x, y)$ is said to be a continuous function of x and y at the point (a, b). Likewise, if a function $f(x, y)$ is continuous at every point of a region R of the (x, y) plane it is said to be continuous over that region.

9.2 First Partial Derivatives

Suppose $f(x, y)$ is a real single-valued function of two independent variables x and y. Then the partial derivative of $f(x, y)$ with respect to x is defined as

$$\left(\frac{\partial f}{\partial x}\right)_y = \lim_{\delta x \to 0}\left\{\frac{f(x+\delta x, y) - f(x, y)}{\delta x}\right\}. \tag{6}$$

Similarly the partial derivative of $f(x, y)$ with respect to y is defined as

$$\left(\frac{\partial f}{\partial y}\right)_x = \lim_{\delta y \to 0}\left\{\frac{f(x, y+\delta y) - f(x, y)}{\delta y}\right\}. \tag{7}$$

In other words the partial derivative of $f(x, y)$ with respect to x may be thought of as the ordinary derivative of $f(x, y)$ with respect to x obtained by treating y as a constant. Similarly, the partial derivative of $f(x, y)$ with respect to y may be found by treating x as a constant and evaluating the ordinary derivative of $f(x, y)$ with respect to y. The variable which is to be held constant in the differentiation is denoted by a subscript as shown in (6) and (7). Alternative notations, however, exist for partial derivatives and one of the more useful and compact of these is to denote $\left(\dfrac{\partial f}{\partial x}\right)_y$ by f_x, and $\left(\dfrac{\partial f}{\partial y}\right)_x$ by f_y. The subscripts appearing on the f now denote the variables with respect to which $f(x, y)$ is to be differentiated.

The following examples illustrate the evaluation of first partial derivatives.

Example 1. If
$$f(x, y) = x^2 - 2y^2 \tag{8}$$
(see (1)), then
$$f_x = \left(\frac{\partial f}{\partial x}\right)_y = \lim_{\delta x \to 0} \left\{ \frac{[(x+\delta x)^2 - 2y^2] - (x^2 - 2y^2)}{\delta x} \right\} \tag{9}$$

$$= \lim_{\delta x \to 0} \left(\frac{2x\,\delta x + (\delta x)^2}{\delta x} \right) = 2x. \tag{10}$$

Similarly
$$f_y = \left(\frac{\partial f}{\partial y}\right)_x = \lim_{\delta y \to 0} \left\{ \frac{[x^2 - 2(y+\delta y)^2] - (x^2 - 2y^2)}{\delta y} \right\} \tag{11}$$

$$= \lim_{\delta y \to 0} \left\{ \frac{-4y\,\delta y - 2(\delta y)^2}{\delta y} \right\} = -4y. \tag{12}$$

Example 2. The last example illustrated the technique of partial differentiation from first principles (i.e. by the evaluation of a limit). We now differentiate partially by keeping certain variables constant as required. For example, if

$$f(x, y) = \sin^2 x \cos y + \frac{x}{y^2}, \tag{13}$$

then keeping y constant we find

$$f_x = \left(\frac{\partial f}{\partial x}\right)_y = 2 \sin x \cos x \cos y + \frac{1}{y^2}. \tag{14}$$

Similarly, keeping x constant,

$$f_y = \left(\frac{\partial f}{\partial y}\right)_x = -\sin^2 x \sin y - \frac{2x}{y^3}. \tag{15}$$

Example 3. To obtain the partial derivatives of a function of n independent variables any $n-1$ of these variables must be held constant and the differentiation carried out with respect to the remaining variable. There are therefore n first partial derivatives of such a function. For example, if

$$f(x, y, z) = e^{2x} \cos xy \tag{16}$$

148

then

$$f_x = \left(\frac{\partial f}{\partial x}\right)_{y, z} = -ye^{2z} \sin xy, \tag{17}$$

$$f_y = \left(\frac{\partial f}{\partial y}\right)_{x, z} = -xe^{2z} \sin xy, \tag{18}$$

and

$$f_z = \left(\frac{\partial f}{\partial z}\right)_{x, y} = 2e^{2z} \cos xy. \tag{19}$$

9.3 Function of a Function

It is a well-known property of functions of one independent variable that if f is a function of a variable u, and u is a function of a variable x, then

$$\frac{df}{dx} = \frac{df}{du} \cdot \frac{du}{dx}. \tag{20}$$

This result may be immediately extended to the case when f is a function of two or more independent variables. Suppose $f = f(u)$ and $u = u(x, y)$. Then, by the definition of a partial derivative,

$$f_x = \left(\frac{\partial f}{\partial x}\right)_y = \frac{df}{du}\left(\frac{\partial u}{\partial x}\right)_y, \tag{21}$$

$$f_y = \left(\frac{\partial f}{\partial y}\right)_x = \frac{df}{du}\left(\frac{\partial u}{\partial y}\right)_x. \tag{22}$$

Example 4. If

$$f(x, y) = \tan^{-1}\frac{y}{x} \tag{23}$$

then putting $u = y/x$ we have

$$f_x = \left(\frac{\partial f}{\partial x}\right)_y = \frac{d}{du}(\tan^{-1} u)\left(\frac{\partial u}{\partial x}\right)_y = -\frac{y}{x^2 + y^2} \tag{24}$$

and

$$f_y = \left(\frac{\partial f}{\partial y}\right)_x = \frac{d}{du}(\tan^{-1} u)\left(\frac{\partial u}{\partial y}\right)_x = \frac{x}{x^2 + y^2}. \tag{25}$$

Example 5. If $f(u) = \sin u$ and $u = \sqrt{(x^2 + y^2)}$ then

$$f_x = \left(\frac{\partial f}{\partial x}\right)_y = (\cos u)\,\frac{x}{\sqrt{(x^2+y^2)}} = \frac{x\cos\sqrt{(x^2+y^2)}}{\sqrt{(x^2+y^2)}}, \qquad (26)$$

and

$$f_y = \left(\frac{\partial f}{\partial y}\right)_x = (\cos u)\,\frac{y}{\sqrt{(x^2+y^2)}} = \frac{y\cos\sqrt{(x^2+y^2)}}{\sqrt{(x^2+y^2)}}. \qquad (27)$$

9.4 Higher Partial Derivatives

Provided the first partial derivatives of a function are differentiable we may differentiate them partially to obtain the second partial derivatives. The four second partial derivatives of $f(x, y)$ are therefore

$$f_{xx} = \frac{\partial^2 f}{\partial x^2} = \frac{\partial}{\partial x}\,f_x = \frac{\partial}{\partial x}\left(\frac{\partial f}{\partial x}\right)_y, \qquad (28)$$

$$f_{yy} = \frac{\partial^2 f}{\partial y^2} = \frac{\partial}{\partial y}\,f_y = \frac{\partial}{\partial y}\left(\frac{\partial f}{\partial y}\right)_x, \qquad (29)$$

$$f_{xy} = \frac{\partial^2 f}{\partial x\,\partial y} = \frac{\partial}{\partial x}\,f_y = \frac{\partial}{\partial x}\left(\frac{\partial f}{\partial y}\right)_x, \qquad (30)$$

and

$$f_{yx} = \frac{\partial^2 f}{\partial y\,\partial x} = \frac{\partial}{\partial y}\,f_x = \frac{\partial}{\partial y}\left(\frac{\partial f}{\partial x}\right)_y. \qquad (31)$$

Higher partial derivatives than the second may be obtained in a similar way.

Example 6. We have already seen in Example 4 that if

$$f(x, y) = \tan^{-1}\frac{y}{x} \qquad (32)$$

then

$$\left(\frac{\partial f}{\partial x}\right)_y = -\frac{y}{x^2+y^2}, \quad \left(\frac{\partial f}{\partial y}\right)_x = \frac{x}{x^2+y^2}. \qquad (33)$$

150

Hence, differentiating these first derivatives partially, we obtain

$$f_{xx} = \frac{\partial^2 f}{\partial x^2} = \frac{\partial}{\partial x}\left(-\frac{y}{x^2+y^2}\right) = \frac{2xy}{(x^2+y^2)^2} \tag{34}$$

and

$$f_{yy} = \frac{\partial^2 f}{\partial y^2} = \frac{\partial}{\partial y}\left(\frac{x}{x^2+y^2}\right) = -\frac{2xy}{(x^2+y^2)^2}. \tag{35}$$

Also

$$f_{xy} = \frac{\partial^2 f}{\partial x\partial y} = \frac{\partial}{\partial x}\left(\frac{x}{x^2+y^2}\right) = \frac{y^2-x^2}{(x^2+y^2)^2} \tag{36}$$

and

$$f_{yx} = \frac{\partial^2 f}{\partial y\partial x} = \frac{\partial}{\partial y}\left(-\frac{y}{x^2+y^2}\right) = \frac{y^2-x^2}{(x^2+y^2)^2}. \tag{37}$$

Since (36) and (37) are equal we have

$$\frac{\partial^2 f}{\partial x\partial y} = \frac{\partial^2 f}{\partial y\partial x}, \tag{38}$$

which shows that the operators $\frac{\partial}{\partial x}$ and $\frac{\partial}{\partial y}$ are commutative. We shall return to this point in the next section. Finally we note that if (34) and (35) are added then $f(x, y)$ satisfies the partial differential equation (Laplace's equation in two variables)

$$\frac{\partial^2 f}{\partial x^2} + \frac{\partial^2 f}{\partial y^2} = 0. \tag{39}$$

In general, any function satisfying this equation is called a harmonic function (see Chapter 24).

9.5 Commutative Property of Partial Differentiation

In 9.4 Example 6 we have shown that the second partial derivatives f_{xy} and f_{yx} of the function $f(x, y) = \tan^{-1}\frac{y}{x}$ are equal. This is in fact the case for most functions as can be verified by choosing a few functions at random. It can be proved that a sufficient (but not necessary) condition that $f_{xy} = f_{yx}$ at some point (a, b) is that both f_{xy} and f_{yx} are continuous at (a, b), and in all that follows it will be assumed that this condition is satisfied.

9.6 Total Derivatives

Suppose $f(x, y)$ is a continuous function defined in a region R of the xy-plane, and that both $\left(\dfrac{\partial f}{\partial x}\right)_y$ and $\left(\dfrac{\partial f}{\partial y}\right)_x$ are continuous in this region. We now consider the change in the value of the function brought about by allowing small changes in x and y.

If δf is the change in $f(x, y)$ due to changes δx and δy in x and y then

$$\delta f = f(x+\delta x, y+\delta y) - f(x, y) \tag{40}$$

$$= f(x+\delta x, y+\delta y) - f(x, y+\delta y) + f(x, y+\delta y) - f(x, y). \tag{41}$$

Now by definition (see 9.2 (6) and (7))

$$\frac{\partial}{\partial x} f(x, y+\delta y) = \lim_{\delta x \to 0} \left\{ \frac{f(x+\delta x, y+\delta y) - f(x, y+\delta y)}{\delta x} \right\} \tag{42}$$

and

$$\frac{\partial}{\partial y} f(x, y) = \lim_{\delta y \to 0} \left\{ \frac{f(x, y+\delta y) - f(x, y)}{\delta y} \right\}. \tag{43}$$

Consequently

$$f(x+\delta x, y+\delta y) - f(x, y+\delta y) = \left[\frac{\partial}{\partial x} f(x, y+\delta y) + \alpha \right] \delta x, \tag{44}$$

and

$$f(x, y+\delta y) - f(x, y) = \left[\frac{\partial}{\partial y} f(x, y) + \beta \right] \delta y, \tag{45}$$

where α and β satisfy the conditions

$$\lim_{\delta x \to 0} \alpha = 0 \quad \text{and} \quad \lim_{\delta y \to 0} \beta = 0. \tag{46}$$

Using (44) and (45) in (41) we now find

$$\delta f = \left[\frac{\partial}{\partial x} f(x, y+\delta y)+\alpha\right] \delta x + \left[\frac{\partial}{\partial y} f(x, y)+\beta\right] \delta y. \qquad (47)$$

Furthermore, since all first derivatives are continuous by assumption, the first term of (47) may be written as

$$\frac{\partial}{\partial x} f(x, y+\delta y) = \frac{\partial f(x, y)}{\partial x}+\gamma, \qquad (48)$$

where γ satisfies the condition

$$\lim_{\delta y \to 0} \gamma = 0. \qquad (49)$$

Hence, using (48), (47) now becomes

$$\delta f = \frac{\partial f(x, y)}{\partial x} \delta x + \frac{\partial f(x, y)}{\partial y} \delta y + (\alpha+\gamma) \delta x + \beta \, \delta y. \qquad (50)$$

The expression

$$\delta f \simeq \frac{\partial f(x, y)}{\partial x} \delta x + \frac{\partial f(x, y)}{\partial y} \delta y \qquad (51)$$

obtained by neglecting the small terms $(\alpha+\gamma) \delta x$ and $\beta \, \delta y$ in (50) represents, to the first order in δx and δy, the change in $f(x, y)$ due to changes δx and δy in x and y respectively.

It is easily seen that the first term of (51) represents the change in $f(x, y)$ due to a change δx in x keeping y constant; similarly the second term is the change in $f(x, y)$ due to a change δy in y keeping x constant. The total differential is nothing more than the sum of these two effects. In the case of a function of n independent variables $f(x_1, x_2, \dots x_n)$ we have

$$\delta f \simeq \frac{\partial f}{\partial x_1} \delta x_1 + \frac{\partial f}{\partial x_2} \delta x_2 + \dots + \frac{\partial f}{\partial x_n} \delta x_n = \sum_{r=1}^{n} \frac{\partial f}{\partial x_r} \delta x_r. \qquad (52)$$

The following example illustrates the use of these results.

Example 7. To find the change in

$$f(x, y) = xe^{xy} \tag{53}$$

when the values of x and y are slightly changed from 1 and 0 to $1 + \delta x$ and δy respectively. We first use (51) to obtain

$$\delta f \simeq (xye^{xy} + e^{xy}) \, \delta x + x^2 e^{xy} \, \delta y. \tag{54}$$

Hence putting $x = 1$, $y = 0$ in (54) we have

$$\delta f \simeq \delta x + \delta y. \tag{55}$$

For example, if $\delta x = 0 \cdot 10$ and $\delta y = 0 \cdot 05$, then $\delta f \simeq 0 \cdot 15$.

We now return to the exact expression for δf given in (50). Suppose $u = f(x, y)$ and that both x and y are differentiable functions of a variable t so that

$$x = x(t), \quad y = y(t) \tag{56}$$

and

$$u = u(t). \tag{57}$$

Hence dividing (50) by δt and proceeding to the limit $\delta t \to 0$ (which implies $\delta x \to 0$, $\delta y \to 0$ and consequently α, β, $\gamma \to 0$) we have

$$\frac{du}{dt} = \frac{\partial f}{\partial x} \cdot \frac{dx}{dt} + \frac{\partial f}{\partial y} \cdot \frac{dy}{dt}. \tag{58}$$

This expression is called the total derivative of $u(t)$ with respect to t. It is easily seen that if

$$u = f(x_1, x_2, x_3 \ldots x_n), \tag{59}$$

where $x_1, x_2, x_3 \ldots x_n$ are all differentiable functions of a variable t, then $u = u(t)$ and

$$\frac{du}{dt} = \frac{\partial f}{\partial x_1} \cdot \frac{dx_1}{dt} + \frac{\partial f}{\partial x_2} \cdot \frac{dx_2}{dt} + \ldots + \frac{\partial f}{\partial x_n} \cdot \frac{dx_n}{dt} = \sum_{r=1}^{n} \frac{\partial f}{\partial x_r} \cdot \frac{dx_r}{dt}. \tag{60}$$

Example 8. Suppose

$$u = f(x, y) = x^2 + y^2 \tag{61}$$

and

$$x = \sinh t, \quad y = t^2. \tag{62}$$

Then by direct substitution we have

$$u(t) = \sinh^2 t + t^4 \tag{63}$$

and consequently

$$\frac{du}{dt} = 2 \sinh t \cosh t + 4t^3. \tag{64}$$

We now obtain this result using the expression for the total derivative. Since

$$\frac{\partial f}{\partial x} = 2x, \quad \frac{\partial f}{\partial y} = 2y, \tag{65}$$

$$\frac{dx}{dt} = \cosh t, \quad \frac{dy}{dt} = 2t, \tag{66}$$

(58) gives

$$\frac{du}{dt} = 2x \cosh t + 4yt \tag{67}$$

$$= 2 \sinh t \cosh t + 4t^3, \tag{68}$$

as before.

9.7 Implicit Differentiation

A special case of the total derivative (58) arises when y is itself a function of x (i.e. $t = x$). Consequently u is a function of x only and

$$\frac{du}{dx} = \frac{\partial f}{\partial x} + \frac{\partial f}{\partial y} \cdot \frac{dy}{dx}. \tag{69}$$

Example 9. Suppose

$$u = f(x, y) = \tan^{-1} \frac{x}{y} \tag{70}$$

and

$$y = \sin x. \tag{71}$$

Then by (69) we have

$$\frac{du}{dx} = \frac{y}{x^2 + y^2} - \frac{x}{x^2 + y^2} \cos x \tag{72}$$

$$= \frac{\sin x - x \cos x}{x^2 + \sin^2 x}. \tag{73}$$

This result could have been obtained by the slightly more laborious method of substituting (71) into (70) and then differentiating with respect to x in the usual way.

When y is defined as a function of x by the equation

$$u = f(x, y) = 0 \tag{74}$$

y is called an implicit function of x. Since u is identically zero its total derivative must vanish, and consequently from (69)

$$\frac{dy}{dx} = -\left(\frac{\partial f}{\partial x}\right)_y \Big/ \left(\frac{\partial f}{\partial y}\right)_x. \tag{75}$$

Example 10. The gradient of the tangent at any point (x, y) of the conic

$$f(x, y) = ax^2 + 2hxy + by^2 + 2gx + 2fy + c = 0, \tag{76}$$

(where a, h, b, g, f and c are constants) is, by (75),

$$\frac{dy}{dx} = -\frac{2ax + 2hy + 2g}{2by + 2hx + 2f}. \tag{77}$$

Example 11. The pair of equations

$$F(x, y, z) = 0, \quad G(x, y, z) = 0, \tag{78}$$

where F and G are differentiable functions of x, y and z define, for example, y and z as functions of x. Hence, since the total derivatives of $F(x, y, z)$ and $G(x, y, z)$ are identically zero, we have

$$\frac{\partial F}{\partial x} + \frac{\partial F}{\partial y} \cdot \frac{dy}{dx} + \frac{\partial F}{\partial z} \cdot \frac{dz}{dx} = 0 \qquad (79)$$

and

$$\frac{\partial G}{\partial x} + \frac{\partial G}{\partial y} \cdot \frac{dy}{dx} + \frac{\partial G}{\partial z} \cdot \frac{dz}{dx} = 0, \qquad (80)$$

whence

$$\frac{dy}{dx} = -\left(\frac{\partial F}{\partial x} \cdot \frac{\partial G}{\partial z} - \frac{\partial F}{\partial z} \cdot \frac{\partial G}{\partial x}\right)\left(\frac{\partial F}{\partial y} \cdot \frac{\partial G}{\partial z} - \frac{\partial F}{\partial z} \cdot \frac{\partial G}{\partial y}\right)^{-1} \qquad (81)$$

and

$$\frac{dz}{dx} = \left(\frac{\partial F}{\partial x} \cdot \frac{\partial G}{\partial y} - \frac{\partial F}{\partial y} \cdot \frac{\partial G}{\partial x}\right)\left(\frac{\partial F}{\partial y} \cdot \frac{\partial G}{\partial z} - \frac{\partial F}{\partial z} \cdot \frac{\partial G}{\partial y}\right)^{-1}. \qquad (82)$$

For example, if

$$F(x, y, z) = x^2 + y^2 + z^2,$$
$$G(x, y, z) = x^2 - y^2 + 2z^2, \qquad (83)$$

then

$$\frac{\partial F}{\partial x} = 2x, \quad \frac{\partial F}{\partial y} = 2y, \quad \frac{\partial F}{\partial z} = 2z, \qquad (84)$$

$$\frac{\partial G}{\partial x} = 2x, \quad \frac{\partial G}{\partial y} = -2y, \quad \frac{\partial G}{\partial z} = 4z, \qquad (85)$$

and hence, by (81) and (82),

$$\frac{dy}{dx} = -\frac{x}{3y}, \quad \frac{dz}{dx} = -\frac{2x}{3z}. \qquad (86)$$

9.8 Higher Total Derivatives

We have already seen that if $u = f(x, y)$ and x and y are differentiable functions of t then

$$\frac{du}{dt} = \frac{\partial f}{\partial x} \cdot \frac{dx}{dt} + \frac{\partial f}{\partial y} \cdot \frac{dy}{dt}. \qquad (87)$$

To find $\dfrac{d^2u}{dt^2}$ we note from (87) that the operator $\dfrac{d}{dt}$ can be written as

$$\frac{d}{dt} \equiv \frac{dx}{dt} \cdot \frac{\partial}{\partial x} + \frac{dy}{dt} \cdot \frac{\partial}{\partial y}. \tag{88}$$

Hence

$$\frac{d^2u}{dt^2} = \frac{d}{dt}\left(\frac{du}{dt}\right) = \left(\frac{dx}{dt} \cdot \frac{\partial}{\partial x} + \frac{dy}{dt} \cdot \frac{\partial}{\partial y}\right)\left(\frac{\partial f}{\partial x} \cdot \frac{dx}{dt} + \frac{\partial f}{\partial y} \cdot \frac{dy}{dt}\right) \tag{89}$$

$$= \frac{\partial^2 f}{\partial x^2}\left(\frac{dx}{dt}\right)^2 + 2\frac{\partial^2 f}{\partial x \partial y}\left(\frac{dx}{dt}\right)\left(\frac{dy}{dt}\right) + \frac{\partial^2 f}{\partial y^2}\left(\frac{dy}{dt}\right)^2$$

$$+ \frac{\partial f}{\partial x} \cdot \frac{d^2 x}{dt^2} + \frac{\partial f}{\partial y} \cdot \frac{d^2 y}{dt^2}, \tag{90}$$

where we have assumed that $f_{xy} = f_{yx}$. Higher total derivatives may be obtained in a similar way.

A special case of (90) which will be needed later is when

$$\frac{dx}{dt} = h, \quad \frac{dy}{dt} = k, \tag{91}$$

where h and k are constants. We then have

$$\frac{d^2u}{dt^2} = h^2 \frac{\partial^2 f}{\partial x^2} + 2hk \frac{\partial^2 f}{\partial x \partial y} + k^2 \frac{\partial^2 f}{\partial y^2}, \tag{92}$$

which, if we define the differential operator $*D$ by

$$*D = h\frac{\partial}{\partial x} + k\frac{\partial}{\partial y}, \tag{93}$$

may be written symbolically as

$$\frac{d^2u}{dt^2} = \left(h\frac{\partial}{\partial x} + k\frac{\partial}{\partial y}\right)^2 f = *D^2 f. \tag{94}$$

Similarly we find

$$\frac{d^3u}{dt^3} = h^3 \frac{\partial^3 f}{\partial x^3} + 3h^2 k \frac{\partial^3 f}{\partial x^2 \partial y} + 3hk^2 \frac{\partial^3 f}{\partial x \partial y^2} + k^3 \frac{\partial^3 f}{\partial y^3} \tag{95}$$

$$= \left(h\frac{\partial}{\partial x} + k\frac{\partial}{\partial y} \right)^3 f = *D^3 f, \tag{96}$$

assuming the commutative property of partial differentiation. In general,

$$\frac{d^n u}{dt^n} = \left(h\frac{\partial}{\partial x} + k\frac{\partial}{\partial y} \right)^n f = *D^n f, \tag{97}$$

where the operator $\left(h\dfrac{\partial}{\partial x} + k\dfrac{\partial}{\partial y} \right)^n$ is to be expanded by means of the binomial theorem.

9.9 Homogeneous Functions

A function $f(x, y)$ is said to be homogeneous of degree m if

$$f(kx, ky) = k^m f(x, y), \tag{98}$$

where k is a constant. A similar definition applies to a function of any number of independent variables. For example,

$$f(x, y) = x^3 + 4xy^2 - 3y^3 \tag{99}$$

is homogeneous of degree 3 since

$$(kx)^3 + 4(kx)(ky)^2 - 3(ky)^3 = k^3[x^3 + 4xy^2 - 3y^3]. \tag{100}$$

Similarly

$$f(x, y) = \frac{x^2 + y^2}{4xy} + \frac{y}{x}\sin\left(\frac{x}{y}\right) \tag{101}$$

is homogeneous of degree 0 since

$$\frac{(kx)^2 + (ky)^2}{4k^2 xy} + \frac{ky}{kx}\sin\left(\frac{kx}{ky}\right) = k^0\left\{ \frac{x^2 + y^2}{4xy} + \frac{y}{x}\sin\frac{x}{y} \right\}. \tag{102}$$

9.10 Euler's Theorem

Theorem 1. If $u = f(x_1, x_2, \dots x_n)$ is a homogeneous differentiable function of degree m in the independent variables $x_1, x_2, \dots x_n$, then

$$x_1\frac{\partial f}{\partial x_1} + x_2\frac{\partial f}{\partial x_2} + \dots + x_n\frac{\partial f}{\partial x_n} = mf. \tag{103}$$

159

To prove this theorem we define a new set of variables $y_1, y_2, \ldots y_n$ by the relations

$$x_1 = y_1 k, \quad x_2 = y_2 k, \quad \ldots x_n = y_n k, \tag{104}$$

where k is a constant. Then since u is homogeneous of degree m

$$u = f(y_1 k, y_2 k, \ldots y_n k) = k^m f(y_1, y_2, \ldots y_n). \tag{105}$$

Differentiating (105) with respect to k we find

$$\frac{du}{dk} = \frac{\partial f}{\partial x_1}\frac{dx_1}{dk} + \frac{\partial f}{\partial x_2}\frac{dx_2}{dk} + \ldots + \frac{\partial f}{\partial x_n}\frac{dx_n}{dk} = mk^{m-1}f(y_1, y_2 \ldots y_n) \tag{106}$$

or

$$\frac{du}{dk} = y_1 \frac{\partial f}{\partial x_1} + y_2 \frac{\partial f}{\partial x_2} + \ldots + y_n \frac{\partial f}{\partial x_n} = mk^{m-1}f. \tag{107}$$

Hence multiplying the last two expressions of (107) by k we have

$$x_1 \frac{\partial f}{\partial x_1} + x_2 \frac{\partial f}{\partial x_2} + \ldots + x_n \frac{\partial f}{\partial x_n} = mf, \tag{108}$$

which proves the theorem.

Example 12. The function

$$f(x, y) = x^3 + 4xy^2 - 3y^3 \tag{109}$$

is homogeneous of degree 3 and hence, by Euler's Theorem,

$$x \frac{\partial f}{\partial x} + y \frac{\partial f}{\partial y} = 3f. \tag{110}$$

This is easily verified since

$$\frac{\partial f}{\partial x} = 3x^2 + 4y^2, \quad \frac{\partial f}{\partial y} = 8xy - 9y^2. \tag{111}$$

Hence

$$x \frac{\partial f}{\partial x} + y \frac{\partial f}{\partial y} = x(3x^2 + 4y^2) + y(8xy - 9y^2)$$

$$= 3(x^3 + 4xy^2 - 3y^3) = 3f. \tag{112}$$

9.11 Change of Variables

We have seen earlier on in this chapter that if $u = f(x, y)$ is a continuous and differentiable function of the independent variables x, y and if x and y are differentiable functions of a variable t then

$$\frac{du}{dt} = \frac{\partial f}{\partial x} \cdot \frac{dx}{dt} + \frac{\partial f}{\partial y} \cdot \frac{dy}{dt}. \tag{113}$$

Suppose now that x and y are functions not just of one variable but of two, say s and t, such that

$$x = x(s, t), \quad y = y(s, t). \tag{114}$$

Clearly since u is a function of x and y it is also a function of s and t and necessarily has the two partial derivatives $\left(\dfrac{\partial u}{\partial s}\right)_t$ and $\left(\dfrac{\partial u}{\partial t}\right)_s$. Hence keeping t a constant and differentiating with respect to s, we have (following (113))

$$\left(\frac{\partial u}{\partial s}\right)_t = \frac{\partial f}{\partial x}\left(\frac{\partial x}{\partial s}\right)_t + \frac{\partial f}{\partial y}\left(\frac{\partial y}{\partial s}\right)_t. \tag{115}$$

Similarly, keeping s a constant and differentiating with respect to t

$$\left(\frac{\partial u}{\partial t}\right)_s = \frac{\partial f}{\partial x}\left(\frac{\partial x}{\partial t}\right)_s + \frac{\partial f}{\partial y}\left(\frac{\partial y}{\partial t}\right)_s. \tag{116}$$

Example 13. Given that $u = f(x, y)$ and

$$x = s^2 - t^2, \quad y = 2st, \tag{117}$$

prove that

$$s\,\frac{\partial u}{\partial s} - t\,\frac{\partial u}{\partial t} = 2(s^2 + t^2)\frac{\partial f}{\partial x}. \tag{118}$$

From (115), (116) and (117) we have

$$\frac{\partial u}{\partial s} = \frac{\partial f}{\partial x}\frac{\partial x}{\partial s} + \frac{\partial f}{\partial y}\frac{\partial y}{\partial s} = 2s\,\frac{\partial f}{\partial x} + 2t\,\frac{\partial f}{\partial y}, \tag{119}$$

$$\frac{\partial u}{\partial t} = \frac{\partial f}{\partial x}\frac{\partial x}{\partial t} + \frac{\partial f}{\partial y}\frac{\partial y}{\partial t} = -2t\frac{\partial f}{\partial x} + 2s\frac{\partial f}{\partial y}. \tag{120}$$

Hence multiplying (119) by s and (120) by t and subtracting we obtain (118) as required.

Example 14. Given $u = f(x, y)$ and

$$x = r\cos\theta, \quad y = r\sin\theta, \tag{121}$$

prove that

$$r\frac{\partial u}{\partial r} = x\frac{\partial f}{\partial x} + y\frac{\partial f}{\partial y} \tag{122}$$

and

$$\frac{\partial u}{\partial\theta} = x\frac{\partial f}{\partial y} - y\frac{\partial f}{\partial x}. \tag{123}$$

These results are easily obtained since from (115) we have

$$\frac{\partial u}{\partial r} = \frac{\partial f}{\partial x}\cos\theta + \frac{\partial f}{\partial y}\sin\theta, \tag{124}$$

which, on multiplying through by r, gives (122). Similarly from (116)

$$\frac{\partial u}{\partial\theta} = \frac{\partial f}{\partial x}(-r\sin\theta) + \frac{\partial f}{\partial y}r\cos\theta \tag{125}$$

$$= x\frac{\partial f}{\partial y} - y\frac{\partial f}{\partial x}, \tag{126}$$

which is (123).

Example 15. If x and y are rectangular Cartesian coordinates and if $u = f(x, y)$ satisfies Laplace's equation

$$\frac{\partial^2 f}{\partial x^2} + \frac{\partial^2 f}{\partial y^2} = 0, \tag{127}$$

obtain the form of this equation in polar coordinates (r, θ), where $x = r\cos\theta$, $y = r\sin\theta$.

162

From (115) and (116) we have

$$\left(\frac{\partial u}{\partial r}\right)_{\theta} = \frac{\partial f}{\partial x}\left(\frac{\partial x}{\partial r}\right)_{\theta} + \frac{\partial f}{\partial y}\left(\frac{\partial y}{\partial r}\right)_{\theta} \tag{129}$$

$$= \cos \theta \frac{\partial f}{\partial x} + \sin \theta \frac{\partial f}{\partial y}, \tag{130}$$

and

$$\left(\frac{\partial u}{\partial \theta}\right)_{r} = \frac{\partial f}{\partial x}\left(\frac{\partial x}{\partial \theta}\right)_{r} + \frac{\partial f}{\partial y}\left(\frac{\partial y}{\partial \theta}\right)_{r} \tag{131}$$

$$= -r \sin \theta \frac{\partial f}{\partial x} + r \cos \theta \frac{\partial f}{\partial y}. \tag{132}$$

Solving (130) and (132) for $\frac{\partial f}{\partial x}$ and $\frac{\partial f}{\partial y}$ we find

$$\frac{\partial f}{\partial x} = \cos \theta \frac{\partial u}{\partial r} - \frac{\sin \theta}{r} \frac{\partial u}{\partial \theta} \tag{133}$$

$$\frac{\partial f}{\partial y} = \sin \theta \frac{\partial u}{\partial r} + \frac{\cos \theta}{r} \frac{\partial u}{\partial \theta}. \tag{134}$$

Hence the operators $\frac{\partial}{\partial x}$ and $\frac{\partial}{\partial y}$ in polar coordinates are

$$\frac{\partial}{\partial x} = \cos \theta \frac{\partial}{\partial r} - \frac{\sin \theta}{r} \frac{\partial}{\partial \theta} \tag{135}$$

$$\frac{\partial}{\partial y} = \sin \theta \frac{\partial}{\partial r} + \frac{\cos \theta}{r} \frac{\partial}{\partial \theta}. \tag{136}$$

Consequently

$$\frac{\partial^2 f}{\partial x^2} = \frac{\partial}{\partial x}\left(\frac{\partial f}{\partial x}\right) = \cos \theta \frac{\partial}{\partial r}\left(\cos \theta \frac{\partial u}{\partial r} - \frac{\sin \theta}{r} \frac{\partial u}{\partial \theta}\right)$$

$$- \frac{\sin \theta}{r} \frac{\partial}{\partial \theta}\left(\cos \theta \frac{\partial u}{\partial r} - \frac{\sin \theta}{r} \frac{\partial u}{\partial \theta}\right) \tag{137}$$

$$= \cos^2 \theta \frac{\partial^2 u}{\partial r^2} - \frac{2 \sin \theta \cos \theta}{r} \frac{\partial^2 u}{\partial r \partial \theta} + \frac{\sin^2 \theta}{r^2} \frac{\partial^2 u}{\partial \theta^2}$$

$$+ \frac{\sin^2 \theta}{r} \frac{\partial u}{\partial r} + \frac{2 \sin \theta \cos \theta}{r^2} \frac{\partial u}{\partial \theta}. \tag{138}$$

Similarly

$$\frac{\partial^2 f}{\partial y^2} = \frac{\partial}{\partial y}\left(\frac{\partial f}{\partial y}\right) = \sin \theta \frac{\partial}{\partial r}\left(\sin \theta \frac{\partial u}{\partial r} + \frac{\cos \theta}{r} \frac{\partial u}{\partial \theta}\right)$$

$$+ \frac{\cos \theta}{r} \frac{\partial}{\partial \theta}\left(\sin \theta \frac{\partial u}{\partial r} + \frac{\cos \theta}{r} \frac{\partial u}{\partial \theta}\right) \tag{139}$$

$$= \sin^2 \theta \frac{\partial^2 u}{\partial r^2} + \frac{2 \sin \theta \cos \theta}{r} \frac{\partial^2 u}{\partial r \partial \theta} + \frac{\cos^2 \theta}{r^2} \frac{\partial^2 u}{\partial \theta^2}$$

$$+ \frac{\cos^2 \theta}{r} \frac{\partial u}{\partial r} - \frac{2 \sin \theta \cos \theta}{r^2} \frac{\partial u}{\partial \theta}. \tag{140}$$

Finally, adding (138) and (140), we have Laplace's equation in two dimensions

$$\frac{\partial^2 f}{\partial x^2} + \frac{\partial^2 f}{\partial y^2} = \frac{\partial^2 u}{\partial r^2} + \frac{1}{r} \frac{\partial u}{\partial r} + \frac{1}{r^2} \frac{\partial^2 u}{\partial \theta^2} = 0. \tag{141}$$

For the form of Laplace's equation in three dimensions in both spherical and cylindrical polar coordinates see Chapter 19, 19.8.

9.12 Taylor's Theorem for Functions of Two Independent Variables

Theorem 2. (Taylor's Theorem). If $f(x, y)$ is defined in a region R of the xy-plane and all its partial derivatives of orders up to and including the $(n+1)$th are continuous in R, then for any point (a, b) in this region

$$f(a+h, b+k) = f(a, b) + {}^*Df(a, b) + \frac{1}{2!}{}^*D^2f(a, b) + \ldots$$

$$\ldots + \frac{1}{n!}{}^*D^nf(a, b) + E_n, \tag{142}$$

where $*D$ is the differential operator defined by (92)–(97) as

$$*D = h\frac{\partial}{\partial x} + k\frac{\partial}{\partial y}, \tag{143}$$

and

$$*D^r f(a, b) \quad \text{means} \quad \left(h\frac{\partial}{\partial x} + k\frac{\partial}{\partial y}\right)^r f(x, y) \tag{144}$$

evaluated at the point (a, b). The Lagrange error term E_n is given by

$$E_n = \frac{1}{(n+1)!}*D^{n+1}f(a+\theta h, b+\theta k) \tag{145}$$

where $0 < \theta < 1$.

To prove this theorem we let

$$x = a + ht, \quad y = b + kt, \tag{146}$$

where a, b, h, k are constants and t is a variable. Then putting

$$f(x, y) = f(a + ht, b + kt) = u(t), \tag{147}$$

where $u(t)$ is a continuous function of t, we have by (97)

$$\frac{d^n u}{dt^n} = *D^n f. \tag{148}$$

Since by assumption all partial derivatives of $f(x, y)$ up to and including the $(n+1)$th order are continuous in R so also are the ordinary derivatives of u with respect to t. Hence $u(t)$ may be expanded by Maclaurin's series (see Chapter 6, 6.1 (6) and (7)) as

$$u(t) = u(0) + tu'(0) + \frac{t^2}{2!}u''(0) + \ldots + \frac{t^n}{n!}u^{(n)}(0) + E_n, \tag{149}$$

where

$$E_n = \frac{t^{n+1}}{(n+1)!}u^{(n+1)}(\theta t), \quad 0 < \theta < 1. \tag{150}$$

Hence using (146) and (147) we have

$$f(a+ht,\, b+kt) = f(a,\, b) + t*Df(a,\, b) + \frac{t^2}{2!}*D^2f(a,\, b) + \dots$$

$$\dots + \frac{t^n}{n!}*D^nf(a,\, b) + E_n, \qquad (151)$$

where now

$$E_n = \frac{t^{n+1}}{(n+1)!}*D^{n+1}f(a+h\theta t,\, b+k\theta t), \quad 0 < \theta < 1. \qquad (152)$$

Putting $t = 1$ in (151) and (152) we finally obtain Taylor's expansion (142) with the error term (145).

Theorem 3. If

$$\lim_{n \to \infty} E_n = 0, \qquad (153)$$

then

$$f(a+h,\, b+k) = f(a,\, b) + *Df(a,\, b) + \frac{*D^2}{2!}f(a,\, b) + \dots$$

$$= \sum_{r=0}^{\infty} \frac{1}{r!}*D^rf(a,\, b). \qquad (154)$$

In all that follows we shall assume that (153) is satisfied.

An alternative form of Taylor's series (154) may be obtained by putting

$$h = x-a, \quad k = y-b. \qquad (155)$$

Then

$$f(x,\, y) = f(a,\, b) + [(x-a)f_x(a,\, b) + (y-b)f_y(a,\, b)]$$

$$+ \frac{1}{2!}\{(x-a)^2f_{xx}(a,\, b) + 2(x-a)(y-b)f_{xy}(a,\, b)$$

$$+ (y-b)^2f_{yy}(a,\, b)\} + \dots, \qquad (156)$$

which is Taylor's expansion of $f(x, y)$ about the point (a, b). When there is no dependence on y, (156) reduces to Taylor's series for a function of one variable (6.1 (8)).

Example 16. Expand the function

$$f(x, y) = \sin xy \qquad (157)$$

about the point $\left(1, \dfrac{\pi}{3}\right)$ neglecting terms of degree three and higher

Here

$$f\left(1, \frac{\pi}{3}\right) = \frac{\sqrt{3}}{2},$$

$$f_x(x, y) = y \cos xy, \quad f_x\left(1, \frac{\pi}{3}\right) = \frac{\pi}{6},$$

$$f_y(x, y) = x \cos xy, \quad f_y\left(1, \frac{\pi}{3}\right) = \frac{1}{2},$$

$$f_{xx}(x, y) = -y^2 \sin xy, \quad f_{xx}\left(1, \frac{\pi}{3}\right) = -\frac{\pi^2\sqrt{3}}{18}, \qquad (158)$$

$$f_{xy}(x, y) = -xy \sin xy + \cos xy, \quad f_{xy}\left(1, \frac{\pi}{3}\right) = -\frac{\pi\sqrt{3}}{6} + \frac{1}{2},$$

$$f_{yy}(x, y) = -x^2 \sin xy, \quad f_{yy}\left(1, \frac{\pi}{3}\right) = -\frac{\sqrt{3}}{2}.$$

Hence substituting these results in (156) we have

$$\sin xy = \frac{\sqrt{3}}{2} + \left\{(x-1)\left(\frac{\pi}{6}\right) + \left(y - \frac{\pi}{3}\right)\left(\frac{1}{2}\right)\right\} + \frac{1}{2!}\left\{(x-1)^2\left(-\frac{\pi^2\sqrt{3}}{18}\right)\right.$$

$$\left. + 2(x-1)\left(y - \frac{\pi}{3}\right)\left(-\frac{\pi\sqrt{3}}{6} + \frac{1}{2}\right) + \left(y - \frac{\pi}{3}\right)^2\left(-\frac{\sqrt{3}}{2}\right)\right\} \qquad (159)$$

$$+ \text{terms of degree 3 and higher.}$$

9.13 Maxima and Minima of Functions of Two Variables

A function $f(x, y)$ is said to have a maximum value at a point $(x, y) = (a, b)$ if

$$f(a+h, b+k) - f(a, b) < 0, \tag{160}$$

where h and k are small arbitrary quantities.

Similarly $f(x, y)$ is said to have a minimum at $(x, y) = (a, b)$ if

$$f(a+h, b+k) - f(a, b) > 0. \tag{161}$$

These results may be interpreted geometrically (see Fig. 9.3) by noticing (in the manner of 9.1) that the surface $z = f(x, y)$ is higher

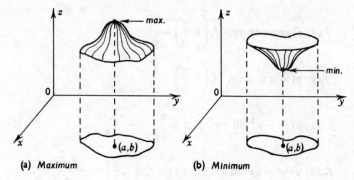

(a) Maximum   (b) Minimum

Fig. 9.3

at $(x, y) = (a, b)$ than at any neighbouring point when (160) is satisfied (thus corresponding to a maximum), and is lower at (a, b) than at any neighbouring point when (161) is satisfied (thus corresponding to a minimum).

Now if a maximum or minimum occurs at (a, b) the curves lying in the two planes $x = a$ and $y = b$ must also have maxima or minima at (a, b) (see Fig. 9.4). Consequently the tangents T_1 and T_2 to these curves at (a, b) must be parallel to the Ox and Oy axes respectively. This requires

$$\frac{\partial f}{\partial x} = 0, \quad \frac{\partial f}{\partial y} = 0 \tag{162}$$

at all maxima and minima. The solution of these equations gives the coordinates of points of possible maxima and minima, and also of points called saddle points which will be defined later. In general we speak of the solution of (162) as giving the stationary or critical

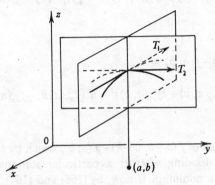

Fig. 9.4

points of $f(x, y)$. To decide whether a particular stationary point is a maximum, minimum or neither we now use the Taylor expansion of $f(x, y)$ in the form given by (142), namely

$$f(a+h, b+k) = f(a, b) + {}^*Df(a, b) + \frac{1}{2!}{}^*D^2f(a, b) + \dots , \quad (163)$$

where $\quad {}^*D = \left(h\frac{\partial}{\partial x} + k\frac{\partial}{\partial y} \right).$

If (a, b) is a stationary point then (162) gives

$${}^*Df(a, b) = 0. \quad (164)$$

Hence, neglecting terms of order h^3, k^3 and higher, we have

$$f(a+h, b+k) - f(a, b) = \tfrac{1}{2}\{h^2 f_{xx}(a, b) + 2hk f_{xy}(a, b) + k^2 f_{yy}(a, b)\}, \quad (165)$$

at a stationary point, where, for example, $f_{xx}(a, b)$ means $\dfrac{\partial^2 f(x, y)}{\partial x^2}$

169

evaluated at (a, b). We now see that (165) may be rewritten as either

$$f(a+h, b+k) - f(a, b) =$$

$$\frac{1}{2f_{xx}(a, b)}\{[hf_{xx}(a, b) + kf_{xy}(a, b)]^2 - k^2[f_{xy}^2(a, b) - f_{xx}(a, b)f_{yy}(a, b)]\},$$

$$(166)$$

or

$$f(a+h, b+k) - f(a, b) =$$

$$\frac{1}{2f_{yy}(a, b)}\{[hf_{xy}(a, b) + kf_{yy}(a, b)]^2 - h^2[f_{xy}^2(a, b) - f_{xx}(a, b)f_{yy}(a, b)]\}.$$

$$(167)$$

Clearly the sign of $f(a+h, b+k) - f(a, b)$, which by (160) and (161) is crucial in deciding whether a particular stationary point is a maximum or a minimum, is now, by (166) and (167), dependent on the values of h and k. However, if

$$\Delta \equiv f_{xy}^2(a, b) - f_{xx}(a, b)f_{yy}(a, b) < 0 \qquad (168)$$

then the terms in curly brackets in (166) and (167) are positive for all h and k.

Consequently, with $\Delta < 0$, the sign of $f(a+h, b+k) - f(a, b)$ depends entirely on the signs of $f_{xx}(a, b)$ and $f_{yy}(a, b)$. From (160) and (161) we deduce therefore that (a, b) is a maximum if

$$\Delta < 0, \quad f_{xx}(a, b) < 0, \qquad (169)$$

and a minimum if

$$\Delta < 0, \quad f_{xx}(a, b) > 0, \qquad (170)$$

We note that $\Delta < 0$, and $f_{xx}(a, b) \gtrless 0$ imply $f_{yy}(a, b) \gtrless 0$.

When $\Delta > 0$ the signs of the curly brackets in (166) and (167) depend on the values of h and k. In this case the stationary point (a, b) is called a saddle point. Such a point is neither a maximum nor a minimum, but is such that the point P is a maximum for the curve C_1 and a minimum for the curve C_2 (see Fig. 9.5).

When $\Delta = 0$ a more refined test is required to determine the nature of a given stationary point.

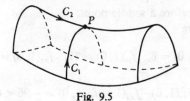

Fig. 9.5

Example 17. Consider the function

$$f(x, y) = x^4 + 4x^2y^2 - 2x^2 + 2y^2 - 1. \tag{171}$$

The conditions $\frac{\partial f}{\partial x} = 0, \frac{\partial f}{\partial y} = 0$ give the two equations

$$4x(x^2 + 2y^2 - 1) = 0, \tag{172}$$

and

$$4y(1 + 2x^2) = 0, \tag{173}$$

respectively.

Hence solving (172) and (173) we have

$$x = 0, \pm 1, \\ y = 0, \tag{174}$$

giving the stationary points of (171) as $(0, 0)$, $(1, 0)$ and $(-1, 0)$. We now test each of these points separately for a maximum, minimum or saddle point. To do this we first differentiate (171) twice to get

$$\left. \begin{array}{l} f_{xx} = 12x^2 + 8y^2 - 4, \\ f_{yy} = 8x^2 + 4, \\ f_{xy} = 16xy. \end{array} \right\} \tag{175}$$

Point $(0, 0)$. Using (175) we now have

$$f_{xx}(0, 0) = -4, \quad f_{yy}(0, 0) = 4, f_{xy}(0, 0) = 0, \tag{176}$$

whence

$$\Delta = f_{xy}^2(0, 0) - f_{xx}(0, 0)f_{yy}(0, 0) = 16 > 0. \tag{177}$$

171

This point is therefore a saddle point.

Point $(1, 0)$. Here

$$f_{xx}(1, 0) = 8, \quad f_{yy}(1, 0) = 12, \quad f_{xy}(1, 0) = 0 \qquad (178)$$

and

$$\Delta = f_{xy}^2(1, 0) - f_{xx}(1, 0)f_{yy}(1, 0) = -96 < 0. \qquad (179)$$

By (170) this point is therefore a minimum.

Point $(-1, 0)$. Since the values of $f_{xx}(-1, 0)$, $f_{yy}(-1, 0)$ and $f_{xy}(-1, 0)$ are identical with those given in (178), this point is also a minimum.

The function $f(x, y)$ defined in (171) therefore has two minima (at $(1, 0)$, $(-1, 0)$), and one saddle point (at $(0, 0)$). Finally it is easily found that $f(x, y) = -2$ at both minima, and $f(x, y) = -1$ at the saddle point. The reader should now attempt to sketch the surface $z = f(x, y)$ defined by (171).

Example 18. To find the maximim value of

$$f(x, y, z) = x^2y^2z^2 \qquad (180)$$

subject to the condition

$$x^2 + y^2 + z^2 = c^2, \qquad (181)$$

where c is a constant. Problems of this type where some constraint is applied (which effectively means that not all the variables are independent) are best dealt with by the method of Lagrange multipliers (see next section). However, in this particular example we can easily reduce the problem to one in two-independent variables by eliminating z to get

$$f(x, y) = x^2y^2(c^2 - x^2 - y^2) \qquad (182)$$

and proceeding as in Example 17.

The stationary points are easily found to be $(c, 0, 0)$, $(0, c, 0)$, $(0, 0, c)$ and $\left(\pm\dfrac{c}{\sqrt{3}}, \pm\dfrac{c}{\sqrt{3}}, \pm\dfrac{c}{\sqrt{3}} \right)$, where all possible combinations of signs are allowed. This second point is, in fact, eight symmetrically placed points in the form of a cube centred at the origin of the $Oxyz$-coordinate system. It is at these eight points that the function (180) takes on its maximum value $\left(=\dfrac{c^6}{27} \right)$

9.14 Lagrange Multipliers

In the last example a problem of maximising a function of three independent variables subject to a constraint was successfully dealt with by eliminating one of the variables. However, this approach may not always be possible. For example, if instead of the constraint (181), we had the relation $e^{-yz}\sin^2(x+z)+1=0$, it would not be possible to solve explicitly for z. An alternative method of dealing with maxima and minima problems which are subject to constraints and which overcomes this difficulty was developed by Lagrange. We indicate this method here for the case of functions of two variables only, but the technique may be extended to any number of variables.

Suppose $f(x,y)$ is to be examined for stationary points subject to the contraint

$$g(x,y)=0. \tag{183}$$

Now for $f(x,y)$ to be stationary we must have the total differential

$$df = \frac{\partial f}{\partial x}dx + \frac{\partial f}{\partial y}dy = 0. \tag{184}$$

This would normally lead to the usual equations

$$\frac{\partial f}{\partial x}=0, \quad \frac{\partial f}{\partial y}=0 \tag{185}$$

for the stationary points. However, dx and dy are not now independent but are related via the total differential of $g(x,y)$

$$dg = \frac{\partial g}{\partial x}dx + \frac{\partial g}{\partial y}dy = 0 \tag{186}$$

(using (183)).

Hence multiplying (186) by a parameter λ and adding to (184), we have

$$d(f+\lambda g) = \left(\frac{\partial f}{\partial x}+\lambda\frac{\partial g}{\partial x}\right)dx + \left(\frac{\partial f}{\partial y}+\lambda\frac{\partial g}{\partial y}\right)dy = 0. \tag{187}$$

We now choose λ such that

$$\frac{\partial f}{\partial x}+\lambda\frac{\partial g}{\partial x}=0, \tag{188}$$

whence it follows from (187) that

$$\frac{\partial f}{\partial y} + \lambda \frac{\partial g}{\partial y} = 0. \tag{189}$$

Equations (188), (189) and (183) are together sufficient to determine the stationary points and the value of the multiplier λ.

Example 19. To find the maximum distance from the origin $(0, 0)$ to the curve

$$3x^2 + 3y^2 + 4xy - 2 = 0. \tag{190}$$

Here we have to find the maximum value of the distance l, where

$$l^2 = f(x, y) = x^2 + y^2 \tag{191}$$

subject to the constraint

$$g(x, y) = 3x^2 + 3y^2 + 4xy - 2 = 0. \tag{192}$$

Now the Lagrange equations (188) and (189) give

$$2x + \lambda(6x + 4y) = 0, \tag{193}$$

$$2y + \lambda(6y + 4x) = 0, \tag{194}$$

which must be solved together with (192). From (193) and (194) we find $4\lambda(y^2 - x^2) = 0$ whence $y = \pm x$. With $y = x$, (192) gives $10x^2 - 2 = 0$ or $x = \pm\frac{1}{\sqrt{5}}$; with $y = -x$, (192) gives $2x^2 - 2 = 0$ or $x = \pm 1$. The stationary points are therefore

$$\left(\frac{1}{\sqrt{5}}, \frac{1}{\sqrt{5}}\right), \quad \left(-\frac{1}{\sqrt{5}}, -\frac{1}{\sqrt{5}}\right), \quad (1, -1), \quad (-1, 1).$$

From (191) we find that for the first two points $l^2 = \frac{2}{5}$, whilst for the last two $l^2 = 2$. Hence the maximum distance from the origin to the curve is $l = \sqrt{2}$.

PROBLEMS 9

1. Show that $f_{xy} = f_{yx}$ for the following functions

    (a) $f(x, y) = x^2 - xy + y^2$,   (b) $f(x, y) = x \sin(y - x)$,

    (c) $f(x, y) = e^y \log_e(x + y)$,   (d) $f(x, y) = \dfrac{xy}{x^2 + y^2}$.

2. Show that $V = (Ar^n + Br^{-n}) \cos(n\theta + \varepsilon)$, where A, B, n and ε are arbitrary constants, satisfies the equation

$$\frac{\partial^2 V}{\partial r^2} + \frac{1}{r}\frac{\partial V}{\partial r} + \frac{1}{r^2}\frac{\partial^2 V}{\partial \theta^2} = 0.$$

3. Find $\dfrac{\partial^3 f(x, y, z)}{\partial x\, \partial y\, \partial z}$ $(= f_{xyz})$ when (a) $f(x, y, z) = e^{xyz}$,

   (b) $f(x, y, z) = \dfrac{xy}{2x+z}$, and verify in each case that

$$f_{xyz} = f_{yzx} = f_{zxy}.$$

4. If $V = f(x - ct) + g(x + ct)$, where f and g are arbitrary functions and c is a constant, prove that

$$\frac{\partial^2 V}{\partial x^2} - \frac{1}{c^2}\frac{\partial^2 V}{\partial t^2} = 0.$$

5. Find $\dfrac{du}{dt}$ in two ways given that $u = x^n y^n$, and $x = \cos at$ $y = \sin bt$, where a, b and n are constants.

6. Find $\dfrac{du}{dx}$ in two ways given that $u = x^2 y + \dfrac{1}{y}$ and $y = \log_e x$.

7. If $V = x \log_e (x^2 + y^2) - 2y \tan^{-1}\left(\dfrac{y}{x}\right)$, find $\dfrac{\partial V}{\partial x}$ and $\dfrac{\partial V}{\partial y}$, and

   verify that

$$x\frac{\partial V}{\partial x} + y\frac{\partial V}{\partial y} = V + 2x. \qquad \text{(L.U.)}$$

8. Given that $V_1(x, y, z)$ and $V_2(x, y, z)$ are solutions of Laplace's equation

$$\frac{\partial^2 V}{\partial x^2} + \frac{\partial^2 V}{\partial y^2} + \frac{\partial^2 V}{\partial z^2} = 0,$$

show that $W(x, y, z) = V_1(x, y, z) + r^2 V_2(x, y, z)$ satisfies the equation

$$\left(\frac{\partial^2}{\partial x^2} + \frac{\partial^2}{\partial y^2} + \frac{\partial^2}{\partial z^2}\right)\left(\frac{\partial^2 W}{\partial x^2} + \frac{\partial^2 W}{\partial y^2} + \frac{\partial^2 W}{\partial z^2}\right) = 0,$$

where $r^2 = x^2 + y^2 + z^2$.

9. Evaluate $x\dfrac{\partial f}{\partial x} + y\dfrac{\partial f}{\partial y}$ when

   (a) $f = xy - \dfrac{1}{x+y}$,   (b) $f = \log_e \dfrac{y}{x}$,   (c) $f = f(xy)$.

10. If $f(x, y, z)$ is a homogeneous function of degree n, prove that

$$x^2 \frac{\partial^2 f}{\partial x^2} + y^2 \frac{\partial^2 f}{\partial y^2} + z^2 \frac{\partial^2 f}{\partial z^2} + 2xy \frac{\partial^2 f}{\partial x \partial y} + 2yz \frac{\partial^2 f}{\partial y \partial z} + 2zx \frac{\partial^2 f}{\partial z \partial x}$$

$$= n(n-1)f.$$

11. (a) If $x = a\cos\theta + b\sin\phi$, $y = b\cos\theta - a\sin\phi$, where a and b are constants, prove that

$$b\sin\theta\left(\frac{\partial\theta}{\partial x}\right)_y + a\cos\phi\left(\frac{\partial\phi}{\partial x}\right)_y = 0.$$

   (b) If $V = xf(u)$ and $u = y/x$, show that

$$x^2 \frac{\partial^2 V}{\partial x^2} + 2xy\frac{\partial^2 V}{\partial x \partial y} + y^2 \frac{\partial^2 V}{\partial y^2} = 0. \qquad \text{(L.U.)}$$

12. If $u = x+y$, $v = xy$, and f is a function of x and y, express $\dfrac{\partial f}{\partial x}, \dfrac{\partial f}{\partial y}$ in terms of $\dfrac{\partial f}{\partial u}, \dfrac{\partial f}{\partial v}$ and prove that

$$\frac{\partial^2 f}{\partial x \partial y} = \frac{\partial^2 f}{\partial u^2} + u\frac{\partial^2 f}{\partial u \partial v} + v\frac{\partial^2 f}{\partial v^2} + \frac{\partial f}{\partial v}. \qquad \text{(L.U.)}$$

176

Partial Differentiation [Problems]

13. The independent variables x, y are transformed into new variables X, Y given by the equations

$$X = xy, \quad Y = \frac{1}{y}.$$

If a function $f(x, y)$ is thus transformed into $F(X, Y)$, prove that

(i) $y \dfrac{\partial f}{\partial y}\left\{ x \dfrac{\partial f}{\partial x} - y \dfrac{\partial f}{\partial y} \right\} = Y \dfrac{\partial F}{\partial Y}\left\{ X \dfrac{\partial F}{\partial X} - Y \dfrac{\partial F}{\partial Y} \right\},$

(ii) $y \left\{ x \dfrac{\partial^2 f}{\partial x\, \partial y} - y \dfrac{\partial^2 f}{\partial y^2} - \dfrac{\partial f}{\partial y} \right\} = Y \left\{ X \dfrac{\partial^2 F}{\partial X\, \partial Y} - Y \dfrac{\partial^2 F}{\partial Y^2} - \dfrac{\partial F}{\partial Y} \right\}.$

(L.U.)

14. If f is a function of x and y which can, using the substitutions $x = re^\theta$, $y = re^{-\theta}$, be written as a function of r and θ, prove that

(i) $2x \dfrac{\partial f}{\partial x} = r \dfrac{\partial f}{\partial r} + \dfrac{\partial f}{\partial \theta}, \quad 2y \dfrac{\partial f}{\partial y} = r \dfrac{\partial f}{\partial r} - \dfrac{\partial f}{\partial \theta},$

(ii) $2x^2 \dfrac{\partial^2 f}{\partial x^2} + 2y^2 \dfrac{\partial^2 f}{\partial y^2} = r^2 \dfrac{\partial^2 f}{\partial r^2} + \dfrac{\partial^2 f}{\partial \theta^2} - r \dfrac{\partial f}{\partial r}.$

(L.U.)

15. If $\phi(r, \theta) = f(r) \sin \theta$, where f is an arbitrary function and $x = r \cos \theta$, $y = r \sin \theta$, prove that

$$(x^2 + y^2)\left\{ \frac{\partial^2 \phi}{\partial x^2} + \frac{\partial^2 \phi}{\partial y^2} \right\} = \left(r^2 \frac{d^2 f}{dr^2} + r \frac{df}{dr} - f \right) \sin \theta.$$

Hence show that if ϕ satisfies the differential equation

$$\frac{\partial^2 \phi}{\partial x^2} + \frac{\partial^2 \phi}{\partial y^2} = 0,$$

then $f(r) = Ar + \dfrac{B}{r}$, where A and B are constants. (L.U.)

177

16. (i) If $z = f(y \log_e x)$, prove that

$$x \frac{\partial z}{\partial x} + y \frac{\partial z}{\partial y} = xy \frac{\partial^2 z}{\partial x \partial y} - x^2 \log_e x \frac{\partial^2 z}{\partial x^2}.$$

(ii) If $x = u^2 - v^2$, $y = u^2 + v^2$, show that

$$\frac{\partial x}{\partial u} \cdot \frac{\partial y}{\partial v} - \frac{\partial x}{\partial v} \cdot \frac{\partial y}{\partial u} = 8uv = \left\{ \frac{\partial u}{\partial x} \cdot \frac{\partial v}{\partial y} - \frac{\partial u}{\partial y} \cdot \frac{\partial v}{\partial x} \right\}^{-1}. \qquad \text{(L.U.)}$$

17. Using Taylor's series for a function of two variables expand $f(x, y) = e^{xy}$ to three terms about the point $x = 2$, $y = 3$.

18. Find the stationary values of the function

$$f(x, y) = e^{x+y}(x^2 - xy + y^2).$$

Show that it has no maxima and one minimum value. (L.U.)

19. (i) Find the stationary points of the surface

$$z = x^3 + xy + y^2.$$

Determine their nature.

(ii) Find the stationary points of the function

$$V = x^2 + y^2 + z,$$

subject to the condition $x^2 - z^2 = 1$. (L.U.)

20. Show that $f(x, y) = x^3 + y^3 - 2(x^2 + y^2) + 3xy$ has stationary values at $(0, 0)$ and $(\frac{1}{3}, \frac{1}{3})$, and investigate their nature. (L.U.)

21. If $u = \sin x \sin y \sin(x+y)$, show that

$$\frac{\partial u}{\partial x} = \sin y \sin(2x+y),$$

$$\frac{\partial u}{\partial y} = \sin x \sin(2y+x).$$

Show that u has a maximum value $\dfrac{3\sqrt{3}}{8}$ at a point within the square whose sides are the lines

$$x = 0, \quad x = \pi/2, \quad y = 0, \quad y = \pi/2. \qquad \text{(L.U.)}$$

22. Given that the functions $u(x, y)$, $v(x, y)$ satisfy the equations (the Cauchy-Riemann equations)

$$\frac{\partial u}{\partial x} = \frac{\partial v}{\partial y}, \; \frac{\partial u}{\partial y} = -\frac{\partial v}{\partial x},$$

show that both u and v satisfy Laplace's equation

$$\frac{\partial^2 \phi}{\partial x^2} + \frac{\partial^2 \phi}{\partial y^2} = 0.$$

Show also that the systems of curves $u(x, y) = $ constant, $v(x, y) = $ constant, are mutually orthogonal.

Hence show that the orthogonal trajectories of the system of curves $x^2 - y^2 = $ constant are the curves $xy = $ constant.

23. Find the stationary points of $f(x, y) = x^2 + y^2$, subject to the constraint $3x + 2y = 6$.

CHAPTER 10

Differentiation and Integration
of Integrals

10.1 Differentiation of Indefinite Integrals

We now consider integrals whose integrands are functions of both x and a variable parameter α. Suppose $f(x, \alpha)$ is an integrable function of x; then if

$$\int f(x, \alpha)\, dx = F(x, \alpha) \tag{1}$$

we have by definition (see Chapter 4, 4.1)

$$\frac{\partial F(x, \alpha)}{\partial x} = f(x, \alpha). \tag{2}$$

Consequently, assuming that $f(x, \alpha)$ is such that

$$\frac{\partial^2 F(x, \alpha)}{\partial x\, \partial \alpha} = \frac{\partial^2 F(x, \alpha)}{\partial \alpha\, \partial x}, \tag{3}$$

it follows that

$$\frac{\partial}{\partial x}\left(\frac{\partial F(x, \alpha)}{\partial \alpha}\right) = \frac{\partial}{\partial \alpha}\left(\frac{\partial F(x, \alpha)}{\partial x}\right) = \frac{\partial f(x, \alpha)}{\partial \alpha}. \tag{4}$$

Hence, integrating (4), we find

$$\int \frac{\partial f(x, \alpha)}{\partial \alpha}\, dx = \frac{\partial F(x, \alpha)}{\partial \alpha}, \tag{5}$$

which is valid so long as $\dfrac{\partial f(x, \alpha)}{\partial \alpha}$ is continuous in both x and α. This formula is useful in evaluating indefinite integrals from known integrals as shown by the following examples.

Example 1. Since

$$\int e^{ax} dx = \frac{1}{\alpha} e^{ax}, \tag{6}$$

we have, using (5)

$$\int x e^{ax} dx = \frac{\partial}{\partial \alpha}\left(\frac{1}{\alpha} e^{ax}\right) = \left(\frac{x}{\alpha} - \frac{1}{\alpha^2}\right) e^{ax}. \tag{7}$$

Likewise, differentiating (7) under the integral sign with respect to α

$$\int x^2 e^{ax} dx = \frac{\partial}{\partial \alpha}\left\{\left(\frac{x}{\alpha} - \frac{1}{\alpha^2}\right) e^{ax}\right\} = \left(\frac{x^2}{\alpha} - \frac{2x}{\alpha^2} + \frac{2}{\alpha^3}\right) e^{ax}. \tag{8}$$

These results may be verified directly by integration by parts.

Example 2. From the integral

$$\int \frac{dx}{a^2 - x^2} = \frac{1}{a} \tanh^{-1}\left(\frac{x}{a}\right), \quad (|x| < a) \tag{9}$$

we have, differentiating with respect to a,

$$-\int \frac{2a}{(a^2 - x^2)^2} dx = -\frac{1}{a^2} \tanh^{-1}\left(\frac{x}{a}\right) - \frac{x}{a(a^2 - x^2)}, \tag{10}$$

which gives

$$\int \frac{dx}{(a^2 - x^2)^2} = \frac{1}{2a^3} \tanh^{-1}\left(\frac{x}{a}\right) + \frac{x}{2a^2(a^2 - x^2)}. \tag{11}$$

10.2 Differentiation of Definite Integrals

A similar procedure to that adopted in the last section may be applied to definite integrals. Suppose

$$I(\alpha) = \int_a^b f(x, \alpha) \, dx, \tag{12}$$

where $f(x, \alpha)$ is an integrable function of x in the range $a \leqq x \leqq b$,

181

a and b in general being continuous and at least once differentiable functions of α. Then using the previous notation

$$I(\alpha) = \int_a^b f(x, \alpha)\, dx = F(b, \alpha) - F(a, \alpha) \tag{13}$$

and, from (5),

$$\int_a^b \frac{\partial f(x, \alpha)}{\partial \alpha}\, dx = \frac{\partial F(b, \alpha)}{\partial \alpha} - \frac{\partial F(a, \alpha)}{\partial \alpha}. \tag{14}$$

Also, differentiating (13) totally with respect to α, we have

$$\frac{dI(\alpha)}{d\alpha} = \frac{\partial F(b, \alpha)}{\partial b}\frac{db}{d\alpha} + \frac{\partial F(b, \alpha)}{\partial \alpha} - \frac{\partial F(a, \alpha)}{\partial a}\frac{da}{d\alpha} - \frac{\partial F(a, \alpha)}{\partial \alpha} \tag{15}$$

which becomes, using (2) and (14),

$$\frac{dI(\alpha)}{d\alpha} = f(b, \alpha)\frac{db}{d\alpha} - f(a, \alpha)\frac{da}{d\alpha} + \int_a^b \frac{\partial f(x, \alpha)}{\partial \alpha}\, dx. \tag{16}$$

If the limits of integration do not depend on α, then

$$\frac{d}{d\alpha}\left\{ \int_a^b f(x, \alpha)\, dx \right\} = \int_a^b \frac{\partial f(x, \alpha)}{\partial \alpha}\, dx. \tag{17}$$

These results naturally assume that $\dfrac{\partial f(x, \alpha)}{\partial \alpha}$ exists. In fact, no difficulty occurs provided all the functions are continuous and have continuous derivatives up to the required order. Difficulties (connected with the convergence of the integrals) may arise, however, if either a or b (or both) are infinite, but we shall not discuss these cases here.

Nevertheless where an infinite range of integration is implied in the following examples the reader may assume that (16) and (17) apply.

Example 3. If

$$I(\alpha) = \int_1^{e^\alpha} \frac{\sin \alpha x}{x}\, dx \tag{18}$$

then by (16)

$$\frac{dI(\alpha)}{d\alpha} = e^{\alpha}\left(\frac{\sin{(\alpha e^{\alpha})}}{e^{\alpha}}\right) + \int_{1}^{e^{\alpha}} \cos \alpha x \, dx \tag{19}$$

$$= \sin{(\alpha e^{\alpha})} + \left(\frac{\sin \alpha x}{\alpha}\right)_{1}^{e^{\alpha}} \tag{20}$$

$$= \left(1 + \frac{1}{\alpha}\right)\sin{(\alpha e^{\alpha})} - \frac{\sin \alpha}{\alpha}. \tag{21}$$

Example 4. Since

$$\int_{0}^{\infty} e^{-\alpha x} \, dx = \frac{1}{\alpha}, \quad (\alpha > 0), \tag{22}$$

we have, differentiating n times with respect to α,

$$\int_{0}^{\infty} x^{n} e^{-\alpha x} \, dx = (-1)^{n} \frac{d^{n}}{d\alpha^{n}}\left(\frac{1}{\alpha}\right) = \frac{n!}{\alpha^{n+1}}, \quad (n \geqq 0). \tag{23}$$

Example 5. An important result which will be established in Chapter 12 using double integrals is that

$$\int_{0}^{\infty} e^{-x^{2}} \, dx = \frac{1}{2}\sqrt{\pi}. \tag{24}$$

Accepting this integral for the present we easily find that

$$\int_{0}^{\infty} e^{-\alpha x^{2}} \, dx = \frac{1}{2}\left(\frac{\pi}{\alpha}\right)^{\frac{1}{2}}, \tag{25}$$

where α is an arbitrary non-zero positive parameter. Consequently differentiating (25) n times with respect to α by (17) we obtain the integrals

$$\int_{0}^{\infty} x^{2n} e^{-\alpha x^{2} \, dx} = \frac{1 \cdot 3 \cdot 5 \dots (2n-1)\sqrt{\pi}}{2^{n+1} \alpha^{n+\frac{1}{2}}}, \tag{26}$$

where $n \geqq 1$

183

We note here that integrals of the type

$$\int_0^\infty x^{2n+1} e^{-\alpha x^2}\, dx \tag{27}$$

may always be evaluated by putting $x^2 = u$ and then integrating by parts. For example, in this way

$$\int_0^\infty x^3 e^{-\alpha x^2}\, dx \quad \text{becomes} \quad \frac{1}{2}\int_0^\infty u e^{-\alpha u}\, du, \tag{28}$$

which is directly integrable. There is, however, no such simple way of obtaining (26).

Example 6. If

$$I(\alpha) = \int_0^\infty \frac{e^{-\alpha x} \sin x}{x}\, dx, \quad (\alpha \geqq 0), \tag{29}$$

then

$$\frac{dI(\alpha)}{d\alpha} = -\int_0^\infty e^{-\alpha x} \sin x\, dx. \tag{30}$$

Hence integrating by parts we find

$$\frac{dI(\alpha)}{d\alpha} = -\frac{1}{1+\alpha^2} \tag{31}$$

or

$$I(\alpha) = A - \tan^{-1}\alpha, \tag{32}$$

where A is a constant of integration.

Now as $\alpha \to \infty$, $I \to 0$; hence $A = \pi/2$, and consequently

$$I(\alpha) = \int_0^\infty \frac{e^{-\alpha x} \sin x}{x}\, dx = \frac{\pi}{2} - \tan^{-1}\alpha = \cot^{-1}\alpha. \tag{33}$$

By putting $\alpha = 0$ in (33) we obtain the important result

$$\int_0^\infty \frac{\sin x}{x}\, dx = \frac{\pi}{2}, \tag{34}$$

which has already been mentioned in Chapter 4, Example 18.

10.3 Integration of a Definite Integral

If $f(x, \alpha)$ is a *continuous* function of x and α in the ranges $a \leqq x \leqq b$, $\varepsilon \leqq \alpha \leqq \eta$, where a, b, ε, η are constants, and if

$$I(\alpha) = \int_a^b f(x, \alpha)\, dx, \tag{35}$$

then

$$\int_\varepsilon^\eta I(\alpha)\, d\alpha = \int_\varepsilon^\eta \left\{ \int_a^b f(x, \alpha)\, dx \right\} d\alpha. \tag{36}$$

Furthermore provided that the limits of integration are finite the order of integration may be reversed so that

$$\int_\varepsilon^\eta \left\{ \int_a^b f(x, \alpha)\, dx \right\} d\alpha = \int_a^b \left\{ \int_\varepsilon^\eta f(x, \alpha)\, d\alpha \right\} dx. \tag{37}$$

However, as with the differentiation of definite integrals, if the limits of integration are infinite, difficulties concerning the convergence of the integrals may arise and (37) is not then necessarily true.

Example 7. Consider

$$I(\alpha) = \int_0^\infty e^{-\alpha x} \sin x\, dx = \frac{1}{1+\alpha^2}. \tag{38}$$

Then

$$\int_0^\eta I(\alpha)\, d\alpha = \int_0^\eta \left\{ \int_0^\infty e^{-\alpha x} \sin x\, dx \right\} d\alpha = \int_0^\eta \frac{1}{1+\alpha^2}\, d\alpha = \tan^{-1} \eta. \tag{39}$$

Assuming now that the order of integration may be changed in (39) we find

$$\int_0^\infty \left\{ \int_0^\eta e^{-\alpha x} \sin x\, d\alpha \right\} dx = \tan^{-1} \eta, \tag{40}$$

or, integrating with respect to α,

$$\int_0^\infty \left\{ -\frac{e^{-\eta x} \sin x}{x} + \frac{\sin x}{x} \right\} dx = \tan^{-1} \eta. \tag{41}$$

185

This gives (as in Example 6)

$$\int_0^\infty \frac{e^{-\eta x}\sin x}{x}\, dx = \int_0^\infty \frac{\sin x}{x}\, dx - \tan^{-1}\eta = \frac{\pi}{2} - \tan^{-1}\eta = \cot^{-1}\eta$$

(42)

PROBLEMS 10

1. Evaluate the derivatives with respect to x of the following

   (a) $\displaystyle\int_1^\infty \frac{1}{y}\, e^{-xy^2}\, dy$,

   (b) $\displaystyle\int_x^{2x} \frac{1}{y}\, \sqrt{(1+xy^3)}\, dy$.   (L.U.)

2. Prove, by differentiation with respect to α, that

   $$\int_0^\pi \frac{\log\,(1+\cos\alpha\cos\theta)}{\cos\theta}\, d\theta = \pi\left(\frac{\pi}{2} - \alpha\right), \qquad \text{(C.U.)}$$

   where $0 \leqq \alpha \leqq \pi/2$.

3. If $J_0(x) = \dfrac{1}{\pi}\displaystyle\int_0^\pi \cos\,(x\sin\phi)\, d\phi$, prove that

   $$\frac{d^2 J_0(x)}{dx^2} + \frac{1}{x}\frac{dJ_0(x)}{dx} + J_0(x) = 0.$$

4. By differentiating under the integral sign with respect to α, show that

   $$\int_0^\infty e^{-x^2}\cos\alpha x\, dx = \frac{1}{2}\sqrt{\pi}\, e^{-\alpha^2/4}.$$

5. Show that

   $$\frac{d}{dx}\int_0^x e^{-y}(x-y)^{3/2}\, dy = \frac{3}{2}\int_0^x e^{-y}\sqrt{(x-y)}\, dy$$

   for $x > 0$.   (L.U.)

6. Show that

$$C = \int_0^{x/2\sqrt{(kt)}} e^{-u^2} \, du$$

satisfies the equation

$$k \frac{\partial^2 C}{\partial x^2} = \frac{\partial C}{\partial t}.$$

7. Show, by integrating an integral and assuming that the order of integration may be reversed, that

(a) $\displaystyle\int_0^\infty \frac{e^{-x} \sinh bx}{x} \, dx = \frac{1}{2} \log_e \left(\frac{1+b}{1-b}\right),$ for $|b| < 1,$

(b) $\displaystyle\int_0^\infty \frac{e^{-ax^2} - e^{-bx^2}}{x^2} \, dx = \sqrt{\pi} \, (\sqrt{b} - \sqrt{a}),$

(c) $\displaystyle\int_0^1 \frac{x^b - x^a}{\log_e x} \, dx = \log_e \left(\frac{b+1}{a+1}\right).$

CHAPTER 11

Elliptic Integrals

11.1 Definitions

As we have mentioned in Chapter 4, 4.5 the number of integrals which can be evaluated analytically in terms of elementary functions ($\sin x$, $\cos x$, e^x, and so on) is small, and the chances are that when confronted with a non-standard integral some numerical approximation method such as Simpson's rule (see Chapter 14) must be used. However, by inventing new functions the class of integrals which can be evaluated analytically may be widened. For example, the integral

$$\int_0^1 \frac{dx}{\sqrt{[(1-x^2)(2-x^2)]}} \tag{1}$$

can only be evaluated in terms of elliptic functions. Similarly, any integral whose integrand is a rational function of x and $\sqrt{P(x)}$, where $P(x)$ is a polynomial in x of degree 3 or 4, is said to be of elliptic type and, unlike the case when $P(x)$ is of degree 2 or less, cannot be evaluated in terms of elementary functions.

We now define two standard types of elliptic integrals.

(a) 1st kind.

$$F(k, \phi) = \int_0^\phi \frac{d\theta}{\sqrt{(1-k^2 \sin^2 \theta)}}, \tag{2}$$

where $0 \leq \phi \leq \pi/2$ and $0 < k < 1$.

(b) 2nd kind.

$$E(k, \phi) = \int_0^\phi \sqrt{(1-k^2 \sin^2 \theta)} \, d\theta, \tag{3}$$

where $0 \leq \phi \leq \pi/2$ and $0 < k < 1$.

In both cases the integrals are called complete if $\phi = \pi/2$, and $F(k, \pi/2)$, $E(k, \pi/2)$ are respectively denoted by $K(k)$ and $E(k)$.

Alternative forms of (2) and (3) which are often useful are obtained by putting $\sin \theta = u$. Then

$$*F(k, x) = \int_0^x \frac{du}{\sqrt{[(1-u^2)(1-k^2u^2)]}} \qquad (4)$$

and

$$*E(k, x) = \int_0^x \sqrt{\left(\frac{1-k^2u^2}{1-u^2}\right)} \, du, \qquad (5)$$

where $0 < k < 1$ and $0 \leq x \leq 1$.

The value of the functions $F(k, \phi)$, $E(k, \phi)$ are tabulated for a large number of k and ϕ values (see, for instance, Jahnke and Emde, 'Tables of Functions', Dover Publications, 1945) and consequently if an integral can be put into standard form its value can at once be obtained. Graphs of the F and E functions for a few values of k are shown in Fig. 11.1.

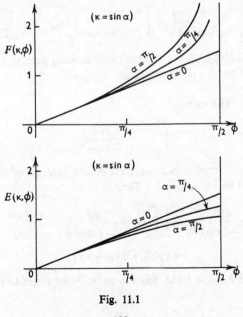

Fig. 11.1

189

We now consider the following examples.

Example 1. To evaluate the integral

$$I = \int_0^\alpha \frac{dx}{\sqrt{(\cos x - \cos \alpha)}}, \quad 0 \leq \alpha \leq \frac{\pi}{2}, \tag{6}$$

we write I as

$$\frac{1}{\sqrt{2}} \int_0^\alpha \frac{dx}{\sqrt{[\sin^2 (\alpha/2) - \sin^2 (x/2)]}} \tag{7}$$

and make the substitution $\sin (x/2) = \sin (\alpha/2) \sin \theta$. Then, since $\cos (x/2)\, dx = 2 \sin (\alpha/2) \cos \theta\, d\theta$, we find

$$I = \sqrt{2} \int_0^{\pi/2} \frac{d\theta}{\sqrt{[1 - \sin^2 (\alpha/2) \sin^2 \theta]}} = \sqrt{2} F\left(\sin \frac{\alpha}{2}, \frac{\pi}{2}\right)$$

$$= \sqrt{2} K\left(\sin \frac{\alpha}{2}\right) \tag{8}$$

By putting $\alpha = \pi/2$ in (6) and (8) we obtain the integral

$$\int_0^{\pi/2} \frac{dx}{\sqrt{(\cos x)}} \left(= \int_0^{\pi/2} \frac{dx}{\sqrt{(\sin x)}}\right) = \sqrt{2} K\left(\frac{1}{\sqrt{2}}\right), \tag{9}$$

which, since $K(1/\sqrt{2}) = 1 \cdot 854$ (to three decimal places), has the value $2 \cdot 62$ correct to two decimal places.

Example 2. The integral

$$\int_0^{\pi/6} \frac{d\theta}{\sqrt{(1 - 4 \sin^2 \theta)}} \tag{10}$$

may be transformed into a complete elliptic integral of the first kind by putting $4 \sin^2 \theta = \sin^2 \psi$. Then

$$\int_0^{\pi/6} \frac{d\theta}{\sqrt{(1 - 4 \sin^2 \theta)}} = \frac{1}{2} \int_0^{\pi/2} \frac{d\psi}{\sqrt{(1 - \frac{1}{4} \sin^2 \psi)}}$$

$$= \tfrac{1}{2} F(\tfrac{1}{2}, \pi/2) = \tfrac{1}{2} K(\tfrac{1}{2}) \tag{11}$$

which, since $K(\tfrac{1}{2}) = 1 \cdot 688$, has the value $0 \cdot 84$ correct to two decimal places.

Example 3. The integral

$$\int_0^{\pi/4} \frac{d\theta}{\sqrt{(1 - \frac{7}{16} \sin^2 \theta)}}, \qquad (12)$$

which was shown in Chapter 4, Example 13 to lie between $\frac{\pi \sqrt{2}}{5}$ and $\pi/4$, is now seen to be an elliptic integral of the first kind, namely $F(\frac{1}{4}\sqrt{7}, \pi/4)$. From tables its value is found to be 0·82 correct to two decimal places.

Example 4. The integral

$$I = \int_0^\infty \sqrt{\left\{ \frac{9 + 8x^2}{(1 + x^2)^3} \right\}} \, dx \qquad (13)$$

becomes, on putting $x = \tan \theta$,

$$\int_0^{\pi/2} \sqrt{\left\{ \frac{9 + 8 \tan^2 \theta}{\sec^6 \theta} \right\}} \sec^2 \theta \, d\theta = \int_0^{\pi/2} \sqrt{(8 + \cos^2 \theta)} \, d\theta$$

$$= \int_0^{\pi/2} \sqrt{(9 - \sin^2 \theta)} \, d\theta = 3 \int_0^{\pi/2} \sqrt{(1 - \tfrac{1}{9} \sin^2 \theta)} \, d\theta$$

$$= 3E\left(\frac{1}{3}, \frac{\pi}{2} \right) = 3 \times 1·525 \ (= 4·57 \text{ to two decimals}). \quad (14)$$

11.2 Jacobian Elliptic Functions

Denoting $F(k, \phi)$ by u, we have from the last section

$$u = \int_0^\phi \frac{d\theta}{\sqrt{(1 - k^2 \sin^2 \theta)}} = \int_0^x \frac{dy}{\sqrt{\{(1 - y^2)(1 - k^2 y^2)\}}}, \qquad (15)$$

where $x = \sin \phi$.

For a given value of k, (15) defines ϕ (or x) as a function of u. It is usual to call ϕ the amplitude of u and to write

$$\phi = \text{am } u, \qquad (16)$$

and to define the first Jacobian elliptic function sn u by

$$x = \sin \phi = \sin (\text{am } u) = \text{sn } u. \qquad (17)$$

Similarly, two other Jacobian elliptic functions cn u and dn u are defined by

$$\text{cn } u = \sqrt{(1-x^2)} = \sqrt{(1-\text{sn}^2 u)}, \tag{18}$$

$$\text{dn } u = \sqrt{(1-k^2x^2)} = \sqrt{(1-k^2 \text{ sn}^2 u)}. \tag{19}$$

It follows from (15), (18) and (19) that

$$\text{sn } 0 = 0, \quad \text{cn } 0 = 1, \quad \text{dn } 0 = 1 \tag{20}$$

and that cn u and dn u are even functions of u. The derivatives of these elliptic functions may readily be obtained from their definitions. For example,

$$\frac{d}{du} \text{ sn } u = \frac{d}{du} \sin \phi = \frac{d}{d\phi} (\sin \phi) \frac{d\phi}{du} = \cos \phi \frac{d\phi}{du} = \text{cn } u \frac{d\phi}{du}. \tag{21}$$

But since by (15)

$$u = \int_0^\phi \frac{d\theta}{\sqrt{(1-k^2 \sin^2 \theta)}}, \tag{22}$$

we have (by differentiating)

$$\frac{du}{d\phi} = \frac{1}{\sqrt{(1-k^2 \sin^2 \phi)}} = \frac{1}{\text{dn} u}. \tag{23}$$

Hence

$$\frac{d}{du} \text{ sn } u = \text{cn } u \text{ dn } u. \tag{24}$$

Similarly,

$$\frac{d}{du} \text{ cn } u = -\text{sn } u \text{ dn } u \tag{25}$$

and

$$\frac{d}{du} \text{ dn } u = -k^2 \text{ sn } u \text{ cn } u. \tag{26}$$

Other properties of the Jacobian elliptic functions are beyond the scope of this book.

PROBLEMS 11

1. Show, using the fact that the integrands of $F(k, \phi)$ and $E(k, \phi)$ are periodic functions with period π, that

$$F(k, \phi + n\pi) = nF(k, \pi) + F(k, \phi),$$
$$E(k, \phi + n\pi) = nE(k, \pi) + E(k, \phi),$$

where $n = 0, 1, 2, \dots$. Hence show that

$$F(k, \phi + n\pi) = 2nK(k) + F(k, \phi),$$
$$E(k, \phi + n\pi) = 2nE(k) + E(k, \phi).$$

2. Establish the following results

(a) $\displaystyle\int_0^{\frac{1}{2}} \frac{dx}{\sqrt{(3 - 4x^2 + x^4)}} = \frac{1}{\sqrt{3}} F\left(\frac{1}{\sqrt{3}}, \frac{\pi}{6}\right),$

(b) $\displaystyle\int_0^{\pi/2} \frac{d\theta}{\sqrt{(1 + k^2 \sin^2 \theta)}} = \frac{1}{\sqrt{(1 + k^2)}} K\left\{\frac{k}{\sqrt{(1 + k^2)}}\right\}.$

3. Verify the following relations

(a) $\operatorname{sn}^2 u + \operatorname{cn}^2 u = 1,$

(b) $\operatorname{dn}^2 u + k^2 \operatorname{sn}^2 u = 1,$

(c) $\dfrac{d}{du}\left\{\log_e\left(\dfrac{\operatorname{dn} u - \operatorname{cn} u}{\operatorname{sn} u}\right)\right\} = \dfrac{1}{\operatorname{sn} u}.$

4. Show that the first two non-vanishing terms in the Maclaurin expansions of the Jacobian elliptic functions are

$$\operatorname{sn} u = u - (1 + k^2)\frac{u^3}{3!} + \dots$$

$$\operatorname{cn} u = 1 - \frac{u^2}{2!} + \dots$$

$$\operatorname{dn} u = 1 - \frac{k^2 u^2}{2!} + \dots .$$

CHAPTER 12

Line Integrals and Multiple Integrals

12.1 Line Integrals in the Plane

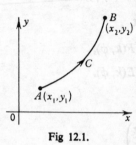

Fig 12.1.

Suppose $y = f(x)$ is a real single-valued monotonic continuous function of x in some interval $x_1 < x < x_2$ as represented by the curve C in Fig. 12.1 (the end points A and B having coordinates (x_1, y_1) and (x_2, y_2), respectively). Then if $P(x, y)$ and $Q(x, y)$ are two real single-valued continuous functions of x and y for all points of C, the integrals

$$\int_C P(x, y)\, dx, \quad \int_C Q(x, y)\, dy \tag{1}$$

and, more frequently, their sum

$$\int_C \left\{ P(x, y)\, dx + Q(x, y)\, dy \right\} \tag{2}$$

are called curvilinear integrals or line integrals, the path of integration C being along the curve $y = f(x)$ from A to B. Since y is expressed in terms of x, each of these integrals is equivalent to an ordinary integral with respect to x.

For example,

$$\int_C P(x, y)\, dx = \int_{x=x_1}^{x=x_2} P(x, f(x))\, dx \tag{3}$$

and

$$\int_C Q(x, y)\, dy = \int_{x=x_1}^{x=x_2} Q(x, f(x)) f'(x)\, dx, \tag{4}$$

(since $dy = f'(x)\, dx$).

194

Alternatively, since $y = f(x)$ is single-valued monotonic and continuous along C, we may express x as a single-valued continuous function of y along C by the relation $x = g(y)$, say. The line integrals (1) may then be expressed as the following ordinary integrals with respect to y:

$$\int_C P(x, y)\, dx = \int_{y=y_1}^{y=y_2} P(g(y), y)g'(y)\, dy \qquad (5)$$

(since $dx = g'(y)\, dy$), and

$$\int_C Q(x, y)\, dy = \int_{y=y_1}^{y=y_2} Q(g(y), y)\, dy. \qquad (6)$$

Similar results apply to (2).

12.2 Some Properties and Examples of Line Integrals

(a) Line integrals may be evaluated in two directions since the path C may be traversed either from A to B or from B to A. However, owing to the equivalence with ordinary integrals we have the result

$$\int_{A \to B} P(x, y)\, dx \equiv \int_{x_1}^{x_2} P(x, f(x))\, dx = -\int_{x_2}^{x_1} P(x, f(x))\, dx$$

$$= -\int_{B \to A} P(x, y)\, dx. \qquad (7)$$

Hence we see that the sign of a line integral of the type given in (1) (and also (2)) is changed when the path of integration is traversed in the opposite direction.

(b) If the path of integration is such that C is a line parallel to the y-axis, say $x = k$, where k is a constant, then

$$\int_C P(x, y)\, dx = 0 \qquad (8)$$

since along C, $dx = 0$.

Similarly, if C is a line parallel to the x-axis, say $y = k$, then

$$\int_C Q(x, y)\, dy = 0 \qquad (9)$$

since along C, $dy = 0$.

(c) If the path C is divided into two parts by the point $E(x_3, y_3)$ (see Fig. 12.2) then in virtue of

$$\int_{x_1}^{x_2} P(x, f(x))\, dx = \int_{x_1}^{x_3} P(x, f(x))\, dx + \int_{x_3}^{x_2} P(x, f(x))\, dx \quad (10)$$

(see Chapter 4, 4.2 (b)), we have

$$\int_C P(x, y)\, dx = \int_{A \to E} P(x, y)\, dx + \int_{E \to B} P(x, y)\, dx. \qquad (11)$$

Similar results apply to $\displaystyle\int_C Q(x, y)\, dy$ and to the linear combination defined in (2).

These results are easily extended to the case when C is divided into any finite number of parts.

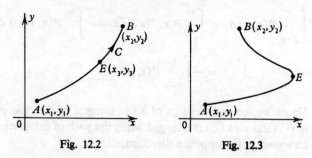

Fig. 12.2   Fig. 12.3

(d) If the path C is such that for some value (or values) of x two distinct values of y are obtained (i.e. y is not single-valued), (see Fig. 12.3), then the line integral from A to B must be written as

$$\int_C P(x, y)\, dx = \int_{A \to E} P(x, y_1)\, dx + \int_{E \to B} P(x, y_2)\, dx, \qquad (12)$$

where $y_1 = f_1(x)$ and $y_2 = f_2(x)$ are single-valued and continuous functions of x representing the parts of the curve AE and EB, respectively.

Similar arguments apply to $\displaystyle\int_C Q(x, y)\, dy$ and to the linear combination (2).

We now illustrate the definitions and properties of line integrals by the following examples.

Example 1. Evaluate $\displaystyle\int_C (x+y)\, dx$ (i) from $A(0, 1)$ to $B(1, 0)$ along the curve $C = C_1$ defined by $y = 1 - x$, (ii) from $O(0, 0)$ to $E(1, 1)$ along the curve $C = C_2$ defined by $y = x^2$; (see Fig. 12.4).

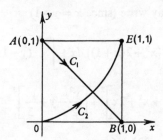

Fig. 12.4

(i) $\displaystyle\int_{C_1} (x+y)\, dx = \int_0^1 (x+1-x)\, dx = \left[x\right]_0^1 = 1.$ \hfill (13)

(ii) $\displaystyle\int_{C_2} (x+y)\, dx = \int_0^1 (x+x^2)\, dx = \left[\frac{x^2}{2}+\frac{x^3}{3}\right]_0^1 = \frac{5}{6}.$ \hfill (14)

Alternatively, expressing x in terms of y (as in (5)), we have

(i) $\displaystyle\int_{C_1} (x+y)\, dx = \int_1^0 (1-y+y)(-dy) = \int_0^1 dy = \left[y\right]_0^1 = 1,$ \hfill (15)

(ii) $\displaystyle\int_{C_2} (x+y)\, dx = \int_0^1 (\sqrt{y}+y)\frac{dy}{2\sqrt{y}} = \left[\tfrac{1}{2}y+\tfrac{1}{3}y^{3/2}\right]_0^1 = \frac{5}{6},$ \hfill (16)

as before.

Example 2. Evaluate the line integral

$$I = \int_C (x^2 + 2y)\,dx + (x + y^2)\,dy \tag{17}$$

from $A(0, 1)$ to $B(2, 3)$ along the curve C defined by $y = x+1$.
Expressing (17) entirely in terms of x we find

$$I = \int_0^2 [\{x^2 + 2(x+1)\}\,dx + \{x + (x+1)^2\}\,dx] \tag{18}$$

$$= \int_0^2 (2x^2 + 5x + 3)\,dx = \frac{64}{3}. \tag{19}$$

Alternatively we may write (since $x = y-1$)

$$I = \int_0^2 \{x^2 + 2(x+1)\}\,dx + \int_1^3 (y - 1 + y^2)\,dy \tag{20}$$

$$= \left[\frac{x^3}{3} + x^2 + 2x\right]_0^2 + \left[\frac{y^2}{2} - y + \frac{y^3}{3}\right]_1^3 \tag{21}$$

$$= \left(\frac{32}{3} + \frac{32}{3}\right) = \frac{64}{3}, \tag{22}$$

as before.

Example 3. Evaluate the line integral

$$I = \int_C \{(x + y)\,dx + xy\,dy\} \tag{23}$$

(i) from $O(0, 0)$ to $B(1, 1)$ along the curve $C = C_1$ defined by $y = x$,
(ii) from $O(0, 0)$ to $A(1, 0)$ along the line $y = 0$, and from $A(1, 0)$ to
$B(1, 1)$ along the line $x = 1$; (see Fig. 12.5).

(i) since $y = x$

$$I_{C_1} = \int_0^1 \{(x + x)\,dx + x^2\,dx\} = \tfrac{4}{3}. \tag{24}$$

(ii) since $y = 0$ from O to A we have $dy = 0$, and hence

$$I_{O \to A} = \int_0^1 x \, dx = \tfrac{1}{2}. \tag{25}$$

From A to B, $x = 1$, $dx = 0$ and hence

$$I_{A \to B} = \int_0^1 y \, dy = \tfrac{1}{2}. \tag{26}$$

Consequently in virtue of the addition property (6)

$$I_{C_2} = I_{O \to A} + I_{A \to B} = \tfrac{1}{2} + \tfrac{1}{2} = 1. \tag{27}$$

We note here that although A and B are the same points for both C_1 and C_2, the value of I taken along C_1 is different from the value taken along C_2. This is a common property of line integrals, although later in this chapter we shall meet line integrals whose values are independent of the path of integration, and which depend only on the coordinates of the initial and final points.

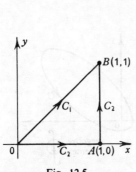

Fig. 12.5

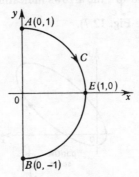

Fig. 12.6

Example 4. Evaluate the line integral

$$I = \int_C (x + y) \, dx \tag{28}$$

from $A(0, 1)$ to $B(0, -1)$ where C is the semicircle $y = \sqrt{(1 - x^2)}$ as shown in Fig. 12.6.

Here since the equation of the path is not a single-valued function of x, we must, by (d), use $y = +\sqrt{(1-x^2)}$ on the first part of the path A to E and $y = -\sqrt{(1-x^2)}$ on the second part E to B. Hence

$$I = \int_0^1 \{x + \sqrt{(1-x^2)}\}\, dx + \int_1^0 \{x - \sqrt{(1-x^2)}\}\, dx \qquad (29)$$

$$= 2\int_0^1 \sqrt{(1-x^2)}\, dx = \frac{\pi}{2}. \qquad (30)$$

12.3 Line Integrals around Closed Plane Curves

To evaluate a line integral along a simple closed plane curve C (i.e. one that does not cut itself) we integrate once around the curve starting and finishing at the same point. In doing this, however, the direction of integration must be specified, the counter clockwise direction usually being called the positive direction and the clockwise direction the negative direction. When integrals are to be evaluated around closed curves it is usual to denote them by the symbols

$\oint_C$ and $\oint_C$, the arrows indicating the direction of integration around C (see Fig. 12.7).

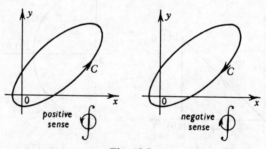

Fig. 12.7

For example, $\oint_C \sin xy\, dx$ requires that $\sin xy$ be integrated around some specified closed curve C in the positive direction, the equation of C defining y in terms of x. Since, however, the equation of a closed curve cannot be such that y is a single-valued function of x we must when integrating with respect to x follow the technique of (d). For example, for the curve (a) of Fig. 12.8 we must use single valued functions $y = f_1(x)$ along AMB and $y = f_2(x)$ along BNA. Similarly for (b) we must divide C into four parts AE, EF, FB and BA, the equation of each part being represented by y as a single-valued function of x.

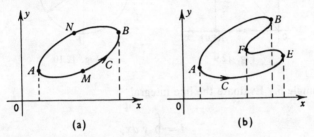

(a)   (b)

Fig. 12.8

Example 5. Evaluate the line integral

$$I = \oint_C (2xy \, dy - x^2 \, dx), \tag{31}$$

where C is made up of the three sides of the triangle with vertices $O(0, 0)$, $A(1, 0)$ and $B(1, 1)$: (see Fig. 12.9). Now from O to A, $y = 0$ and hence (since $dy = 0$)

$$I_{O \to A} = -\int_0^1 x^2 \, dx = -\tfrac{1}{3}. \tag{32}$$

Similarly from A to B, $x = 1$, $dx = 0$ and hence

$$I_{A \to B} = \int_0^1 2y \, dy = 1. \tag{33}$$

Lastly from B to O, $y = x$ and hence (expressing (31) in terms of x)

$$I_{B \to O} = \int_1^0 x^2 \, dx = -\tfrac{1}{3}. \tag{34}$$

Hence finally

$$I = I_{O \to A} + I_{A \to B} + I_{B \to O} = -\tfrac{1}{3} + 1 - \tfrac{1}{3} = \tfrac{1}{3}. \tag{35}$$

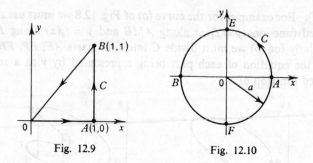

Fig. 12.9   Fig. 12.10

Example 6. Evaluate the line integral

$$I = \oint_C y \, dx, \tag{36}$$

where C is the circle

$$x^2 + y^2 = a^2 \tag{37}$$

(see Fig. 12.10).

Since y is not a single-valued function of x we must divide (36) into two parts with

$$y = +\sqrt{(a^2 - x^2)} \tag{38}$$

for the semicircle AEB which lies above the x-axis, and

$$y = -\sqrt{(a^2 - x^2)} \tag{39}$$

for the semicircle BFA which lies below the x-axis. Hence

$$I = \int_a^{-a} \sqrt{(a^2 - x^2)} \, dx + \int_{-a}^a \{-\sqrt{(a^2 - x^2)}\} \, dx \tag{40}$$

$$= -4 \int_0^a \sqrt{(a^2 - x^2)} \, dx = -\pi a^2. \tag{41}$$

(Integration of (36) in the negative direction just changes the sign of (41).)

One of the important applications of line integrals is to the evalua-

202

tion of the areas of simple closed plane curves. In 12.11 we shall establish the useful result (indicated by the last example)

$$\oint_C x\,dy = -\oint_C y\,dx = \frac{1}{2}\oint_C (x\,dy - y\,dx) = A, \tag{42}$$

where A is the area enclosed by C.

12.4 Line Integrals with respect to Arc Length

Besides the line integrals defined in (1) and (2) we also define the line integral

$$\int_C P(x, y)\,ds, \tag{43}$$

where s is a measure of the arc length of the curve C defined by $y = f(x)$. One of the simplest of such integrals is the elementary formula $\int_A^B ds$ for the length of a curve from the point A to the point B.

As with the other types of line integrals already discussed in this chapter, we may reduce (43) to an ordinary integral in virtue of the elementary results

$$ds = \sqrt{\left\{1 + \left(\frac{dy}{dx}\right)^2\right\}}\,dx, \quad \text{where} \quad y = f(x) \tag{44}$$

and

$$ds = \sqrt{\left\{\left(\frac{dx}{dt}\right)^2 + \left(\frac{dy}{dt}\right)^2\right\}}\,dt, \quad \text{where} \quad \begin{array}{l} x = f(t), \\ y = g(t), \end{array} \tag{45}$$

t being some parameter.

Example 7. Evaluate

$$I = \int_C (3x + 2xy)\,ds, \tag{46}$$

where C is the line $y = x$ from $O(0, 0)$ to $A(1, 1)$. Now in virtue of (44) we have

$$ds = \sqrt{2}\,dx \tag{47}$$

and hence

$$I = \int_0^1 (3x + 2x^2)\sqrt{2}\, dx = \frac{13\sqrt{2}}{6}. \tag{48}$$

Example 8. Evaluate

$$I = \int_C 2xy\, ds, \tag{49}$$

where C is the curve $x = \cos t$, $y = \sin t$ from $t = 0$ to $t = \pi/4$.
Using (45) we have

$$I = \int_{t=0}^{t=\pi/4} 2\cos t \sin t \sqrt{\{(-\sin t)^2 + (\cos t)^2\}}\, dt \tag{50}$$

$$= \int_0^{\pi/4} \sin 2t\, dt = \left[-\frac{\cos 2t}{2} \right]_0^{\pi/4} = \frac{1}{2}. \tag{51}$$

Example 9. Evaluate

$$I = \oint_C (x^2 - y^2)\, ds, \tag{52}$$

where C is the circle $x^2 + y^2 = 4$.

Here $y = +\sqrt{(4-x^2)}$ for the part of C above the x-axis. Consequently

$$\frac{dy}{dx} = -\frac{x}{\sqrt{(4-x^2)}} \tag{53}$$

Similarly $y = -\sqrt{(4-x^2)}$ for the part of C below the x-axis, and hence

$$\frac{dy}{dx} = \frac{x}{\sqrt{(4-x^2)}}. \tag{54}$$

Using (44) we have therefore

$$I = \int_2^{-2} \{x^2 - (4-x^2)\} \sqrt{\left\{1 + \left(\frac{-x}{\sqrt{(4-x^2)}}\right)^2\right\}}(-dx)$$

$$+ \int_{-2}^{2} \{x^2 - (4 - x^2)\} \sqrt{\left\{1 + \left(\frac{x}{\sqrt{(4-x^2)}}\right)^2\right\}} \, dx \tag{55}$$

$$= \int_{2}^{-2} -(2x^2 - 4)\frac{2}{\sqrt{(4-x^2)}} \, dx - \int_{2}^{-2} (2x^2 - 4) \frac{2}{\sqrt{(4-x^2)}} \, dx = 0. \tag{56}$$

by elementary integration.

12.5 Connectivity

In the next section and some that follow it, we shall need the idea of the connectivity of a region or domain. This concept belongs to the subject of topology which is largely concerned with the intrinsic shapes of regions and bodies. A plane region R is said to be simply connected if every simple closed curve C lying in R can be continuously shrunk to a point (see Fig. 12.11 (a)). If, however, R

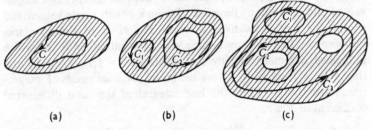

| (a) | (b) | (c) |

Fig. 12.11

contains a hole (Fig. 12.11 (b)) then it is not possible to shrink the curve C_2 to a point without leaving R. Nevertheless the curve C_1 can be shrunk to a point. There are, therefore, two types of curves in this region—those which can, and those which cannot, be continuously shrunk to a point. Such a region is said to be doubly-connected since its boundary consists of two distinct parts. Similarly a region with two holes (Fig. 12.11 (c)) is said to be triply connected since its boundary is in three parts. Here the curves C_2 and C_3 cannot be continuously shrunk to points without leaving R. Generalising this idea a region with $(n - 1)$ holes is therefore said to be n-fold connected (or sometimes just multiply connected). Somewhat similar arguments apply to the surface of a sphere (which is simply connected) and the surface of a torus (which is non-simply connected); (see Fig. 12.12).

 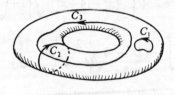

simply connected surface   non-simply connected surface

Fig. 12.12

12.6 Line Integrals Independent of Path

We have already seen in the previous examples that even when the initial and end points of a line integral are kept fixed the value of the integral depends on which path is chosen through these points. However, this is not always the case. Suppose $F(x, y)$ is a single-valued and continuous function of x and y with single-valued and continuous first derivatives in some region R of the xy-plane. Suppose also that C is some curve which lies entirely in R and which is defined parametrically by the equations $x = f(t)$, $y = g(t)$, and that $A(x_1, y_1)$ and $B(x_2, y_2)$ are the initial and final points of integration respectively. Then the line integral of the total differential coefficient of F is

$$\int dF = \int \frac{dF}{dt}\, dt = \int \left(\frac{\partial F}{\partial x}\frac{dx}{dt} + \frac{\partial F}{\partial y}\frac{dy}{dt} \right) dt = \int \frac{\partial F}{\partial x}\, dx + \frac{\partial F}{\partial y}\, dy$$

$$= F(t_2) - F(t_1) = F(x_2, y_2) - F(x_1, y_1). \qquad (57)$$

This result depends only on the value of the function at the initial and final points and not on the equation of the path C. In the special case when C is a simple closed plane curve in R, then A and B are the same point and hence

$$\oint_c \frac{dF}{dt}\, dt = 0, \quad \text{(or more shortly} \oint dF = 0). \qquad (58)$$

We now see that if a function $F(x, y)$ can be found such that

$$P(x, y) = \frac{\partial F(x, y)}{\partial x}, \quad Q(x, y) = \frac{\partial F(x, y)}{\partial y} \qquad (59)$$

206

then the line integral

$$\int_C \{P(x, y)\, dx + Q(x, y)\, dy\} \tag{60}$$

will be independent of the path C.

Furthermore it can be proved that if $P(x, y)$, $Q(x, y)$ have continuous derivatives in R, and R is a simply connected region, then the condition

$$\frac{\partial P(x, y)}{\partial y} = \frac{\partial Q(x, y)}{\partial x} \tag{61}$$

which comes from (59) by eliminating $F(x, y)$, is a necessary and sufficient condition for (60) to be independent of C. However, if R is not simply connected (61) does not guarantee path independence (see Example 11).

We may also prove that if

$$\int \{P(x, y)\, dx + Q(x, y)\, dy\}$$

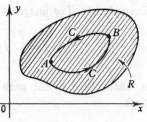

is independent of the path in R then

$$\oint_C \{P(x, y)\, dx + Q(x, y)\, dy\} = 0$$

Fig. 12.13

for all simple closed curves C in R. For (see Fig. 12.13)

$$\oint_C \{P(x, y)\, dx + Q(x, y)\, dy\} = \int_{\substack{A \to B, \\ \text{lower path}}} \{P(x, y)\, dx + Q(x, y)\, dy\}$$

$$+ \int_{\substack{B \to A, \\ \text{upper path}}} \{P(x, y)\, dx + Q(x, y)\, dy\}, \tag{62}$$

$$= \int_{\substack{A \to B, \\ \text{lower path}}} \{P(x, y)\, dx + Q(x, y)\, dy\} - \int_{\substack{A \to B, \\ \text{upper path}}} \{P(x, y)\, dx + Q(x, y)\, dy\}.$$

$$\tag{63}$$

But if the integrals from A to B are independent of which path is taken through these points then the two integrals in (63) are equal. Hence

$$\oint_C \{P(x, y)\, dx + Q(x, y)\, dy\} = 0. \tag{64}$$

Conversely it may be easily proved that if

$$\oint_C \{P(x, y)\, dx + Q(x, y)\, dy\} = 0 \tag{65}$$

for all simple closed paths C in R, then

$$\int \{P(x, y)\, dx + Q(x, y)\, dy\} \tag{66}$$

is independent of path in R.

Example 10. The integral

$$I = \int_A^B (y \cos x\, dx + \sin x\, dy) \tag{67}$$

is independent of the path joining the points $A(0, 0)$, $B(\pi/4, \pi/4)$ since

$$\frac{\partial P(x, y)}{\partial y} = \frac{\partial Q(x, y)}{\partial x} = \cos x. \tag{68}$$

Consequently the integrand of (67) must be an exact differential of some function. It is easily seen that

$$I = \int_{(0,\,0)}^{(\pi/4,\,\pi/4)} d(y \sin x) = \left[y \sin x \right]_{(0,\,0)}^{(\pi/4,\,\pi/4)} = \frac{\pi}{4\sqrt{2}}. \tag{69}$$

Example 11. The integral

$$I = \int_C \left(\frac{-y\, dx + x\, dy}{x^2 + y^2} \right) \tag{70}$$

is independent of the path C in any simply connected region not containing the origin since

$$\frac{\partial P(x, y)}{\partial y} = \frac{\partial Q(x, y)}{\partial x} = \frac{y^2 - x^2}{(x^2 + y^2)^2} \tag{71}$$

except at $(0, 0)$. Consequently in such a region (see (64))

$$I = \oint_C \left(\frac{-y\,dx + x\,dy}{x^2 + y^2} \right) = 0 \tag{72}$$

for all simple closed curves C.

Now in virtue of (71) the integrand of (70) must be an exact differential. It is easily seen, in fact, that

$$I = \int_C d\left(\tan^{-1} \frac{y}{x} \right) = \int_C d\theta = \theta_B - \theta_A, \tag{73}$$

where θ_A and θ_B are the values of the polar coordinates θ at the initial and final points of C (A and B, respectively); (see Fig. 12.14 (a)). For a simple closed curve $\theta_B = \theta_A$, and (73) reproduces (72).

Now consider the doubly connected region shown in Fig. 12.14 (b). If C is any closed curve then along C, θ changes by 2π; hence

$$I = 2\pi. \tag{74}$$

In general, therefore, I is not independent of path in this doubly connected region.

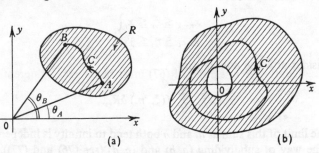

Fig. 12.14

12.7 Line Integrals in Space

The ideas of the last six sections may easily be extended to integrals along lines and curves in a three-dimensional space. We shall not however, give the details here but refer the reader to Problems 6, 7 and 8 at the end of this chapter.

12.8 Double Integrals

The double integral may be defined geometrically in much the same way as the definite Riemann integral was defined in Chapter 4, 4.1 (b). Suppose $f(x, y)$ is a continuous and single-valued function of x and y both inside and on the boundary C of a rectangular region R of the xy-plane. Suppose also that R is bounded by the lines

$$x = a, \quad x = b, \quad y = c, \quad y = d \tag{75}$$

a, b, c, d being constants (see Fig. 12.15 (a)). Then if the interval (a, b) is divided into m parts such that

$$a(= x_0) < x_1 < x_2 < ... < x_{m-1} < b(= x_m) \tag{76}$$

and the interval (c, d) is divided into n parts such that

$$c(= y_0) < y_1 < y_2 < ... < y_{n-1} < d(= y_n), \tag{77}$$

the region R is divided into mn rectangular elements of area δR_{rs} $(r = 1, 2 ... m, s = 1, 2 ... n)$ where

$$\delta R_{rs} = (x_r - x_{r-1})(y_s - y_{s-1}) = \delta x_r \, \delta y_s. \tag{78}$$

We now choose an arbitrary point with coordinates (ξ_r, η_s) lying in the rsth rectangle such that

$$\left. \begin{matrix} x_{r-1} \leqq \xi_r \leqq x_r \\ y_{s-1} \leqq \eta_s \leqq y_s \end{matrix} \right\}. \tag{79}$$

Consider now the double sum

$$\sum_{s=1}^{n} \sum_{r=1}^{m} f(\xi_r, \eta_s) \, \delta R_{rs}. \tag{80}$$

If the limit of this sum as m and n both tend to infinity is independent of the way of subdividing (a, b) and (c, d) (see (76) and (77)), and also the way of choosing the point (ξ_r, η_s) inside δR_{rs}, then, provided $\delta R_{rs} \to 0$,

$$I = \lim_{\substack{m \to \infty \\ n \to \infty}} \sum_{r=1}^{m} \sum_{s=1}^{n} f(\xi_r, \eta_s) \, \delta R_{rs} \tag{81}$$

is called the double integral of the function f over the region R, and is usually written as

$$I = \int_R f(x, y) \, dR = \iint_R f(x, y) \, dx \, dy \qquad (82)$$

(in virtue of (78)). It can be proved that provided the limit (81) exists then it is independent of the order in which m and n tend to infinity. We may therefore proceed in the following two ways:

(i) For any particular value of x lying in the interval $a \leq x \leq b$ we first sum the elemental areas in the vertical strip PQ (see Fig. 12.15 (a)) and then sum over x for all possible vertical strips in R. This corresponds to writing

$$I = \lim_{m \to \infty} \sum_{r=1}^{m} \left\{ \left[\lim_{n \to \infty} \sum_{s=1}^{n} f(\xi_r, \eta_s)(y_s - y_{s-1}) \right] (x_r - x_{r-1}) \right\} \qquad (83)$$

$$= \int_{x=a}^{x=b} \left\{ \int_{y=c}^{y=d} f(x, y) \, dy \right\} dx. \qquad (84)$$

(ii) For any particular value of y lying in the interval $c \leq y \leq d$ we first sum the elemental areas in the horizontal strip ST (see Fig. 12.15 (a)) and then sum over y for all possible horizontal strips in R. This corresponds to writing

$$I = \lim_{n \to \infty} \sum_{s=1}^{n} \left\{ \left[\lim_{m \to \infty} \sum_{r=1}^{m} f(\xi_r, \eta_s)(x_r - x_{r-1}) \right] (y_s - y_{s-1}) \right\}, \qquad (85)$$

$$= \int_{y=c}^{y=d} \left\{ \int_{x=a}^{x=b} f(x, y) \, dx \right\} dy. \qquad (86)$$

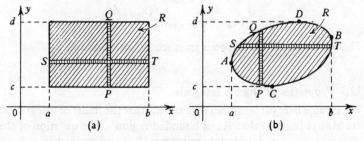

Fig. 12.15

In practice the choice of whether we first integrate $f(x, y)$ with respect to y and then with respect to x, as in (84), or vice-versa, as in (86), is usually governed by which is the simpler operation.

If now the region of integration R is not rectangular but bounded by some closed curve C (see Fig. 12.15 (b)) then we may proceed as follows:

Suppose R is defined by the equations

$$y_1(x) \leqq y \leqq y_2(x) \quad \text{for} \quad a \leqq x \leqq b,$$

$$x_1(y) \leqq x \leqq x_2(y) \quad \text{for} \quad c \leqq y \leqq d,$$

where y_1, y_2, x_1, x_2 are continuous functions of their arguments. Then summing first along the vertical strip PQ and then for all possible vertical strips in R, the analogue of (84) is

$$I = \int_R f(x, y)\, dR = \int_{x=a}^{x=b} \left\{ \int_{y=y_1(x)}^{y=y_2(x)} f(x, y)\, dy \right\} dx. \qquad (87)$$

Likewise summing first along the horizontal strip ST and then for all possible horizontal strips in R, the analogue of (86) is

$$I = \int_R f(x, y)\, dR = \int_{y=c}^{y=d} \left\{ \int_{x=x_1(y)}^{x=x_2(y)} f(x, y)\, dx \right\} dy. \qquad (88)$$

In general, provided $f(x, y)$ is continuous everywhere in R and the boundary of R is a simple shape, the integrals (87) and (88) are the same.

Finally we remark that a notation which avoids the use of brackets in (87) and (88) is to write (87), for example, as

$$\int_{x=a}^{x=b} dx \int_{y=y_1(x)}^{y=y_2(x)} f(x, y)\, dy.$$

This notation will be adopted in much of what follows.

12.9 Properties of Double Integrals

We assume here (unless otherwise stated) that the limits of integration are always finite so that R is a bounded region. The question of the convergence of a double integral when R is unbounded is a much more difficult problem than that of the ordinary definite Riemann integral with an infinite range, and we shall therefore not discuss it here.

If now $f(x, y)$ and $g(x, y)$ are two continuous and single-valued functions over R then the following results may be proved using the basic definition of a double integral given in (81):

(a) $$\iint_R \{f(x, y) + g(x, y)\} \, dx \, dy$$

$$= \iint_R f(x, y) \, dx \, dy + \iint_R g(x, y) \, dx \, dy, \qquad (89)$$

(b) $$\iint_R cf(x, y) \, dx \, dy = c \iint_R f(x, y) \, dx \, dy, \quad (c = \text{constant}), \quad (90)$$

(c) $$\iint_R f(x, y) \, dx \, dy = \iint_{R_1} f(x, y) \, dx \, dy + \iint_{R_2} f(x, y) \, dx \, dy, \tag{91}$$

when R is divided into two parts R_1 and R_2 (see Fig. 12.16).

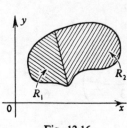
Fig. 12.16

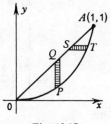
Fig. 12.17

12.10 Some Examples of Double Integrals
Example 12. Evaluate the double integral

$$I = \iint_R (2x^2 + y) \, dx \, dy, \tag{92}$$

where R is the region bounded by the line $y = x$ and the curve $y = x^2$ (see Fig. 12.17).

Method 1. First integrate with respect to y along the strip PQ treating x as a constant, and then integrate with respect to x so that PQ moves from left to right to cover the whole of R. In this way (following (87))

$$I = \int_{x=0}^{x=1} dx \int_{y=x^2}^{y=x} (2x^2+y)\,dy \tag{93}$$

$$I = \int_{x=0}^{x=1} dx \left[2x^2 y + \frac{y^2}{2} \right]_{y=x^2}^{y=x} \tag{94}$$

$$= \int_0^1 \left(2x^3 + \frac{x^2}{2} - \frac{5x^4}{2} \right) dx = \frac{1}{6}. \tag{95}$$

Method 2. Here we first integrate with respect to x along the strip ST treating y as a constant, and then integrate with respect to y so that ST moves vertically to cover the whole of R. In this case (since $x = y$ and $x = \sqrt{y}$ are now the equations of the bounding curves of R with y as the independent variable) we have (following (88))

$$I = \int_{y=0}^{y=1} dy \int_{x=y}^{x=\sqrt{y}} (2x^2+y)\,dx \tag{96}$$

$$= \int_{y=0}^{y=1} dy \left[\frac{2x^3}{3} + xy \right]_{x=y}^{x=\sqrt{y}} \tag{97}$$

$$= \int_0^1 \left(\frac{5y^{3/2}}{3} - \frac{2y^3}{3} - y^2 \right) dy = \frac{1}{6} \tag{98}$$

as before.

Example 13. Evaluate

$$I = \int_0^1 dy \int_0^{\cos^{-1} y} \sec x\,dx, \quad (\cos^{-1} y < \pi/2), \tag{99}$$

by changing the order of integration.

In problems of this type it is always convenient to first sketch the region of integration. Since x ranges from 0 to $\cos^{-1} y$, and y ranges from 0 to 1, the region of integration is clearly bounded by the lines

$x = 0$, $y = 0$ and $y = \cos x$ (see Fig. 12.18). As (99) stands the integration with respect to x is to be performed first along the strip PQ. To change the order of integration, therefore, we first integrate with respect to y along the strip ST and then integrate with respect to x so moving ST from left to right to cover the whole region. In this way (99) becomes

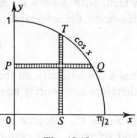

Fig. 12.18

$$I = \int_{x=0}^{x=\pi/2} dx \int_{y=0}^{y=\cos x} \sec x \, dy \qquad (100)$$

$$= \int_{0}^{\pi/2} \left[y \sec x \right]_{y=0}^{y=\cos x} dx = \int_{0}^{\pi/2} dx = \frac{\pi}{2}. \qquad (101)$$

The change of order of integration is often a useful way of avoiding cumbersome integrals as is easily verified here by integrating (99) as it stands.

Example 14. As mentioned earlier the interchange of the order of integration in a double integral is not always permissible and depends on (amongst other things) whether the integrand $f(x, y)$ has any discontinuities in the region of integration R. To illustrate this we give the example of

$$I = \int_{0}^{1} dx \int_{0}^{1} \frac{x - y}{(x + y)^3} \, dy, \qquad (102)$$

where the region of integration R is the rectangular area bounded by the lines $x = 0$, $x = 1$, $y = 0$, $y = 1$. By putting $x + y = u$, it is easily found that $I = \frac{1}{2}$. Now changing the order of integration in (102) we have

$$I' = \int_{0}^{1} dy \int_{0}^{1} \frac{x - y}{(x + y)^3} \, dx, \qquad (103)$$

which, with $x+y = u$ again, gives $I' = -\frac{1}{2}$. Consequently $I \neq I'$. However, the integrand

$$f(x, y) = \frac{x-y}{(x+y)^3} \tag{104}$$

has a discontinuity on the boundary of R at $(0, 0)$ and this result is therefore not entirely unexpected.

Example 15. Double integrals are often useful in evaluating areas and volumes. For example, the integral

$$\iint_R dx\, dy \tag{105}$$

clearly represents the area of R, since $dx\, dy$ is essentially just the area of a typical infinitesimal rectangular element in R. Similarly, if $f(x, y)$ is a continuous and single-valued function of x and y, the volume under the surface $z = f(x, y)$ in a three-dimensional Cartesian coordinate system (see Chapter 9, 9.1) and standing vertically above a region R in the xy-plane is equal to

$$\iint_R z\, dx\, dy = \iint_R f(x, y)\, dx\, dy \tag{106}$$

(volumes above the xy-plane being counted as positive and those below, negative).

To illustrate this use of double integrals consider the following problem: the cylinder $x^2+y^2 = 2ax$, ($a = $ constant) is cut by two planes $z = mx$ and $z = nx$ (m, n constants); find the volume of the wedge lying between these planes and the cylinder (see Fig. 12.19 (a)).

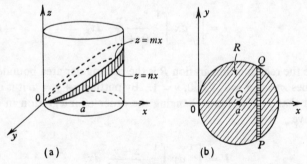

(a)   (b)

Fig. 12.19

216

This is best done in two parts. We first evaluate the volume which lies vertically above the circular region R of the xy-plane defined by $x^2 + y^2 = 2ax$ (centre $(a, 0)$) and below the plane $z = mx$. From this result we then subtract the corresponding result for the volume which lies above R and below $z = nx$. The difference so obtained then represents the volume of the wedge, as required. Proceeding in this way we have

$$I_1 = \iint_R mx \, dx \, dy \tag{107}$$

$$= 2m \int_{x=0}^{x=2a} dx \int_{y=0}^{y=\sqrt{(2ax-x^2)}} x \, dy, \tag{108}$$

(integrating first with respect to y along PQ as shown in Fig. 12.19 (b)).
  Hence

$$I_1 = 2m \int_0^{2a} x\sqrt{(2ax-x^2)} \, dx \tag{109}$$

$$= 2m \int_0^{2a} x\sqrt{\{a^2-(a-x)^2\}} \, dx \tag{110}$$

$$= 2m \int_{-a}^{a} (a-u)\sqrt{(a^2-u^2)} \, du, \tag{111}$$

where $u = a - x$. Putting $u = a \sin \theta$, (111) is easily evaluated to give

$$I_1 = \pi m a^3. \tag{112}$$

Similarly, we find

$$I_2 = \iint_R nx \, dx \, dy = \pi n a^3. \tag{113}$$

Consequently the required volume of the wedge is $\pi(m-n)a^3$ (assuming $m > n$).

12.11 Green's Theorem

Theorem 1. (Green's Theorem). Suppose $P(x, y)$ and $Q(x, y)$ are two functions which are finite and continuous inside and on the boundary C of some region R of the xy-plane. If the first partial derivatives of these functions are continuous inside and on the boundary of R then

$$\iint_R \left\{ \frac{\partial P(x, y)}{\partial y} - \frac{\partial Q(x, y)}{\partial x} \right\} dx\, dy = -\oint_C \{ P(x, y)\, dx + Q(x, y)\, dy \}. \tag{114}$$

The class of regions R to which Green's theorem may be applied is very large and we shall not therefore discuss any instances where it fails. Nevertheless for simplicity we shall only prove (114) for a region R where the boundary C is a simple smooth closed curve (see Fig. 12.15 (b)). Now by (87)

$$\iint_R \frac{\partial P(x, y)}{\partial y} dx\, dy = \int_{x=a}^{x=b} dx \int_{y=y_1(x)}^{y=y_2(x)} \frac{\partial P(x, y)}{\partial y} dy \tag{115}$$

$$= \int_{x=a}^{x=b} \left[P(x, y) \right]_{y=y_1(x)}^{y=y_2(x)} dx \tag{116}$$

$$= \int_{x=a}^{x=b} \{ P(x, y_2(x))\, dx - P(x, y_1(x))\, dx \} \tag{117}$$

$$= -\int_a^b P(x, y_1(x))\, dx - \int_b^a P(x, y_2(x))\, dx \tag{118}$$

$$= -\oint_C P(x, y)\, dx. \tag{119}$$

Similarly by (88)

$$\iint_R \frac{\partial Q(x, y)}{\partial x} dx\, dy = \int_{y=c}^{y=d} dy \int_{x=x_1(y)}^{x=x_2(y)} \frac{\partial Q}{\partial x} dx \tag{120}$$

$$= \int_{y=c}^{y=d} \left[Q(x, y) \right]_{x=x_1(y)}^{x=x_2(y)} dy \tag{121}$$

$$= \int_c^d \{ Q(x_2(y), y)\, dy - Q(x_1(y), y)\, dy \} \tag{122}$$

$$= \int_c^d Q(x_2(y), y) \, dy + \int_d^c Q(x_1(y), y) \, dy \qquad (123)$$

$$= \oint_C Q(x, y) \, dy. \qquad (124)$$

Hence subtracting (124) from (119) we obtain (114) as required.

A special and important case of Green's Theorem is obtained by putting $P(x) = y$, $Q(y) = -x$. In this way (114) gives

$$2 \iint_R dx \, dy = -\oint_C (y \, dx - x \, dy), \qquad (125)$$

or

$$A = \iint_R dx \, dy = \frac{1}{2} \oint_C (x \, dy - y \, dx), \qquad (126)$$

where A is the area enclosed by C.

Similarly (119) and (124) give

$$A = \iint_R dx \, dy = -\oint_C y \, dx \qquad (127)$$

and

$$A = \iint_R dx \, dy = \oint_C x \, dy, \qquad (128)$$

respectively. (These results were given without proof in 12.3 (42).)

The following example illustrates the use of Green's theorem.

Example 16. To evaluate

$$I = \oint_C \{(3x - y) \, dx + (x + 2y) \, dy\} \qquad (129)$$

around the boundary of the ellipse $x^2 + 4y^2 = 9$ we write (since $P = 3x - y$, $Q = x + 2y$)

$$\oint_C \{(3x - y) \, dx + (x + 2y) \, dy\} = -\iint_R (-1 - 1) \, dx \, dy \qquad (130)$$

$$= 2 \iint_R dx \, dy = 2A, \qquad (131)$$

where A is the area of the ellipse. Since the ellipse has semi-axes $a = 3$, $b = \frac{3}{2}$, we have

$$A = \pi ab = \frac{9\pi}{2}, \tag{132}$$

and hence

$$I = 9\pi. \tag{133}$$

12.12 Transformation to Polar Coordinates in Double Integrals

When dealing with double integrals over a region R whose boundary C is defined in terms of polar co-ordinates (r, θ) (where $x = r \cos \theta$, $y = r \sin \theta$) it is usually convenient to subdivide R into elementary non-rectangular areas δR by drawing lines of constant r and θ (see Fig. 12.20). In this way the typical element of area δR which

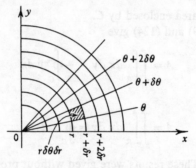

Fig. 12.20

corresponds to the rectangular area $\delta x\, \delta y$ in Cartesian coordinates is given approximately by

$$\delta R \simeq r\, \delta \theta\, \delta r.$$

Consequently if I represents the double integral of some function $f(x, y)$ over a region R_{xy} of the xy-plane then in terms of polar coordinates

$$I = \iint_{R_{xy}} f(x, y)\, dx\, dy = \iint_{R_{r\theta}} f(r \cos \theta, r \sin \theta)\, r\, d\theta\, dr, \tag{134}$$

where $R_{r\theta}$ is the region of the $r\theta$-plane which corresponds to R_{xy}. The correctness of (134), which has been derived here from a purely intuitive point of view, will be shown in the next section. For the time being, therefore, we shall assume its validity and demonstrate its uses by a few examples.

Example 17. Evaluate

$$I = \iint_R \{1 - \sqrt{(x^2 + y^2)}\} \, dx \, dy, \tag{135}$$

where R is the region bounded by the circle $x^2 + y^2 = 1$. Transforming to polar coordinates we have, since $x = r \cos \theta$, $y = r \sin \theta$,

$$I = \iint_{R_{r\theta}} (1 - r)r \, d\theta \, dr \tag{136}$$

$$= \int_{\theta=0}^{\theta=2\pi} d\theta \int_{r=0}^{r=1} (1 - r)r \, dr = (2\pi)\left(\frac{1}{6}\right) = \frac{\pi}{3}. \tag{137}$$

Example 18. Evaluate

$$I = \iint_R x \, dx \, dy, \tag{138}$$

where R is the region bounded by the curve $r = 2a(1 + \cos \theta)$; (see Fig. 12.21).

Expressing I in polar coordinates we have

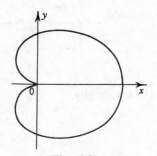

Fig. 12.21

$$I = \iint_R r \cos \theta \, r \, d\theta \, dr = \int_{\theta=0}^{\theta=2\pi} d\theta \int_{r=0}^{r=2a(1+\cos\theta)} r^2 \cos \theta \, dr, \tag{139}$$

$$= \int_{\theta=0}^{\theta=2\pi} \frac{\cos \theta}{3} \left[r^3 \right]_{r=0}^{r=2a(1+\cos\theta)} d\theta, \tag{140}$$

$$= \frac{8a^3}{3} \int_0^{2\pi} (1 + \cos \theta)^3 \cos \theta \, d\theta = 10\pi a^3. \tag{141}$$

Example 19. An important integral which cannot be evaluated using elementary methods is

$$I = \int_0^\infty e^{-x^2}\, dx. \tag{142}$$

One way of proceeding, however, is to write

$$I^2 = \left(\int_0^\infty e^{-x^2}\, dx \right)^2 = \int_0^\infty e^{-x^2}\, dx \int_0^\infty e^{-y^2}\, dy \tag{143}$$

$$= \int_0^\infty dx \int_0^\infty e^{-(x^2+y^2)}\, dy = \iint_{R_{xy}} e^{-(x^2+y^2)}\, dx\, dy, \tag{144}$$

where the region of integration R_{xy} is the whole of the positive quadrant of the xy-plane.

Transforming (144) to polar coordinates we find

$$I^2 = \iint_{R_{r\theta}} e^{-r^2} r\, d\theta\, dr \tag{145}$$

$$= \int_{\theta=0}^{\theta=\pi/2} d\theta \int_{r=0}^{r=\infty} e^{-r^2} r\, dr = \int_0^{\pi/2} \left[-\tfrac{1}{2} e^{-r^2} \right]_0^\infty d\theta \tag{146}$$

$$= \pi/4. \tag{147}$$

Hence, finally,

$$I = \int_0^\infty e^{-x^2}\, dx = \frac{\sqrt{\pi}}{2}. \tag{148}$$

We note here that, unlike the regions considered earlier in this chapter, R_{xy} and $R_{r\theta}$ are infinite in extent (i.e. unbounded). For this reason it is better to proceed more carefully in the following way:

Writing

$$I = \lim_{a \to \infty} \int_0^a e^{-x^2}\, dx \tag{149}$$

we have

$$I^2 = \lim_{a \to \infty} \int_0^a \int_0^a e^{-(x^2+y^2)}\, dx\, dy. \tag{150}$$

Clearly $\int_0^a \int_0^a e^{-(x^2+y^2)} \, dx \, dy$ is just the integral of $e^{-(x^2+y^2)}$ over the square $OABC$ of side a (see Fig. 12.22). Now since the integrand is positive for all points, this integral must have a value which lies somewhere between the value of the integral taken over the sector OAC and the value of the integral taken over the sector OST. Hence

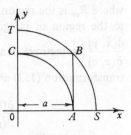

Fig. 12.22

$$\int_{\theta=0}^{\theta=\pi/2} d\theta \int_{r=0}^{r=a} e^{-r^2} r \, dr < \int_0^a \int_0^a e^{-(x^2+y^2)} \, dx \, dy$$

$$< \int_{\theta=0}^{\theta=\pi/2} d\theta \int_{r=0}^{r=a\sqrt{2}} e^{-r^2} r \, dr. \qquad (151)$$

As $a \to \infty$ the integrals over the two sectors both tend to

$$\int_0^{\pi/2} d\theta \int_0^\infty e^{-r^2} r \, dr,$$

and consequently so also does the integral over the square.

12.13 Change of Variables in Double Integrals

The transformation of double integrals from rectangular Cartesian coordinates to polar coordinates given in the last section was based entirely on a geometrical argument. We now give a general analytical method of transforming a double integral from a rectangular Cartesian coordinate system to any other coordinate system. In fact we shall prove that if x and y are related to arbitrary coordinates u and v by continuous differentiable functions so that

$$x = x(u, v), \quad y = y(u, v) \qquad (152)$$

then

$$\iint_{R_{xy}} f(x, y) \, dx \, dy = \iint_{R_{uv}} f[x(u, v), y(u, v)] \left| \frac{\partial(x, y)}{\partial(u, v)} \right| \, du \, dv,$$

$$(153)$$

where R_{uv} is the region of integration in the uv-plane corresponding to the region of integration R_{xy} in the xy-plane. The expression $\dfrac{\partial(x, y)}{\partial(u, v)}\left(\text{sometimes denoted by } J\left(\dfrac{x, y}{u, v}\right)\right)$ is called the Jacobian of the transformation (152) and is defined as

$$\frac{\partial(x, y)}{\partial(u, v)} = \frac{\partial x}{\partial u} \cdot \frac{\partial y}{\partial v} - \frac{\partial x}{\partial v} \cdot \frac{\partial y}{\partial u}, \tag{154}$$

the bars in (153) indicating that the modulus (or numerical value) of the Jacobian is to be taken.

To prove (153) we first put

$$f(x, y) = \frac{\partial F(x, y)}{\partial x}, \tag{155}$$

where $F(x, y)$ is defined inside and on the boundary of R_{xy}. Now by Green's theorem

$$\iint_{R_{xy}} f(x, y) \, dx \, dy = \iint_{R_{xy}} \frac{\partial F(x, y)}{\partial x} \, dx \, dy = \oint_{C_{xy}} F(x, y) \, dy, \tag{156}$$

where C_{xy} is the bounding curve (assumed simple) of R_{xy}. The line integral in (156) may be written as a line integral in the uv-plane by writing

$$\oint_{C_{xy}} F(x, y) \, dy = \pm \oint_{C_{uv}} F[x(u, v), y(u, v)]\left(\frac{\partial y}{\partial u} \, du + \frac{\partial y}{\partial v} \, dv\right), \tag{157}$$

the ambiguity in sign arising in virtue of the fact that the transformation (152) may be such that when C_{xy} is traversed in the positive sense, C_{uv} may be traversed in either the positive or negative sense. We now write (157) in the form

$$\oint_{C_{xy}} F(x, y) \, dy = \pm \oint_{C_{uv}} [P(u, v) \, du + Q(u, v) \, dv], \tag{158}$$

where

$$P(u, v) = F[x(u, v), y(u, v)] \frac{\partial y}{\partial u} \tag{159}$$

and

$$Q(u, v) = F[x(u, v), y(u, v)] \frac{\partial y}{\partial v}. \tag{160}$$

By Green's theorem

$$\oint_{C_{uv}} \{P(u, v)\, du + Q(u, v)\, dv\} = \iint_{R_{uv}} \left\{\frac{\partial Q(u, v)}{\partial u} - \frac{\partial P(u, v)}{\partial v}\right\} du\, dv, \tag{161}$$

which gives on substituting (159) and (160) into the right-hand side of (161)

$$\oint_{C_{uv}} \{P(u, v)\, du + Q(u, v)\, dv\}$$

$$= \iint_{R_{uv}} \left\{\frac{\partial}{\partial x} F[x(u, v), y(u, v)]\right\} \frac{\partial(x, y)}{\partial(u, v)}\, du\, dv, \tag{162}$$

where $\frac{\partial(x, y)}{\partial(u, v)}$ is defined by (154).

Finally using (155) we have

$$\oint_{C_{uv}} \{P(u, v)\, du + Q(u, v)\, dv\} = \iint_{R_{uv}} f[x(u, v), y(u, v)] \frac{\partial(x, y)}{\partial(u, v)}\, du\, dv, \tag{163}$$

and hence from (156) and (158)

$$\iint_{R_{xy}} f(x, y)\, dx\, dy = \pm \iint_{R_{uv}} f[x(u, v), y(u, v)] \frac{\partial(x, y)}{\partial(u, v)}\, du\, dv. \tag{164}$$

Now putting $f(x, y) = 1$, the left-hand side of (164) just gives the area of R_{xy} which is essentially a positive quantity. Hence if the Jacobian is always positive in R_{uv} the plus sign must be taken in (164). Likewise if the Jacobian is always negative in R_{uv} the negative sign must be taken. Both these results are accounted for by writing (164) in the form

$$\iint_{R_{xy}} f(x, y)\, dx\, dy = \iint_{R_{uv}} f[x(u, v), y(u, v)] \left|\frac{\partial(x, y)}{\partial(u, v)}\right|\, du\, dv, \tag{165}$$

which is the result we set out to prove.

It is clear that the infinitesimal element of area $dx\ dy$ in the xy-plane is therefore

$$\left| \frac{\partial(x, y)}{\partial(u, v)} \right| du\ dv. \tag{166}$$

Example 20. Given the transformations

$$x = r\cos\theta, \quad y = r\sin\theta, \tag{167}$$

we have

$$\frac{\partial(x, y)}{\partial(r, \theta)} = \frac{\partial x}{\partial r} \cdot \frac{\partial y}{\partial \theta} - \frac{\partial x}{\partial \theta} \cdot \frac{\partial y}{\partial r} = r\cos^2\theta + r\sin^2\theta = r. \tag{168}$$

Consequently the infinitesimal element of area in the $r\theta$-plane corresponding to $dx\ dy$ in the xy-plane is, by (166),

$$r\ dr\ d\theta \tag{169}$$

as found earlier in 12.12.

Hence, in general,

$$\iint_{R_{xy}} f(x, y)\ dx\ dy = \iint_{R_{r\theta}} f(r\cos\theta, r\sin\theta)\ r\ dr\ d\theta. \tag{170}$$

Example 21. Given the transformations

$$x = u^2 - v^2, \quad y = 4uv \tag{171}$$

we find

$$\frac{\partial(x, y)}{\partial(u, v)} = \frac{\partial x}{\partial u} \cdot \frac{\partial y}{\partial v} - \frac{\partial x}{\partial v} \frac{\partial y}{\partial u} = 2u \cdot 4u + 2v \cdot 4v = 8(u^2 + v^2). \tag{172}$$

Consequently

$$\iint_{R_{xy}} f(x, y)\ dx\ dy = \iint_{R_{uv}} f(u^2 - v^2, 4uv) \cdot 8(u^2 + v^2)\ du\ dv. \tag{173}$$

PROBLEMS 12

1. Evaluate the following line integrals

(a) $\int_C (x^2+2y)\,dx$ from $(0, 1)$ to $(2, 3)$, where C is the line $y = x+1$,

(b) $\int_C x\,dy$ from $(0, 0)$ to $(\pi, 0)$, where C is the curve $y = \sin x$.

2. Evaluate

$$\int_C \{x^2 y\,dx + (x^2 - y^2)\,dy\}$$

from $(0, 0)$ to $(1, 4)$, where (i) C is the curve $y = 4x^2$, (ii) C is the line $y = 4x$.

3. Show that

$$\int_C \{2x \sin y\,dx + x^2 \cos y\,dy\}$$

is independent of the path C, and evaluate it from $(0, 0)$ to $(1, \pi/2)$

4. Evaluate

$$\oint_C (x\,dy - y\,dx)$$

around the curve $x = a \cos^3 t$, $y = a \sin^3 t$, $(a = \text{constant})$.

5. Evaluate

$$\int_C (x^2 - xy)\,ds$$

from $(0, 4)$ to $(4, 0)$, where C is the circle $x^2 + y^2 = 16$.

6. Evaluate

$$\int_C (y^2\,dx + xy\,dy + zx\,dz)$$

227

from $A(0, 0, 0)$ to $B(1, 1, 1)$ where (i) C is the straight line from A to B, (ii) C is a broken line from A to B consisting of parts parallel to the x, y and z axes.

7. Show that, if $P(x, y, z)$, $Q(x, y, z)$ and $R(x, y, z)$ are continuous and single-valued functions with continuous and single-valued partial derivatives then the condition that the line integral

$$\int_C [P(x, y, z)\,dx + Q(x, y, z)\,dy + R(x, y, z)\,dz]$$

shall be independent of path C in some simply connected region R is that

$$\frac{\partial P(x, y, z)}{\partial y} = \frac{\partial Q(x, y, z)}{\partial x}, \quad \frac{\partial Q(x. y, z)}{\partial z} = \frac{\partial R(x, y, z)}{\partial y}$$

$$\frac{\partial R(x, y, z)}{\partial x} = \frac{\partial P(x, y, z)}{\partial z}$$

for every point in R.
Hence show that

$$\int_C \left(\frac{y}{z}\,dx + \frac{x}{z}\,dy - \frac{xy}{z^2}\,dz \right)$$

is independent of path, and evaluate it from $(0, 0, 1)$ to $(1, 1, 1)$.

8. (i) Evaluate the line integral

$$\oint_C \left(\frac{x\,dy - y\,dx}{x^2 + y^2 + 1} \right),$$

where C is the boundary of the minor segment of the circle $x^2 + y^2 = 1$ cut off by the chord $x + y = 1$.

(ii) Show that the line integral

$$\int_{PQ} (yz\,dx + zx\,dy + xy\,dz)$$

is independent of the path of integration, and evaluate it when P and Q are the points $(1, 1, 2)$ and $(3, 2, 1)$ respectively.

(L.U.)

9. Evaluate

   (a) $\iint (x^2 - 4xy - y^2)^2\, dx\, dy$ over the interior of the circle $x^2 + y^2 = a^2$.

   (b) $\iint x\, dx\, dy$ over the first quadrant of the ellipse

$$\frac{x^2}{a^2} + \frac{y^2}{b^2} = 1.$$

10. By suitably transforming the variables x and y to another pair of variables, calculate the integral

$$\int_{-\infty}^{\infty} \int_{-\infty}^{\infty} \frac{x^2\, dx\, dy}{\{1 + \sqrt{(x^2 + y^2)}\}^5} \qquad \text{(C.U.)}$$

11. By putting $x = r \cos \theta$, $y = r \sin \theta$, prove that

$$\int_{0}^{\infty} \int_{0}^{\infty} e^{-(x^2 + 2xy \cos \alpha + y^2)}\, dx\, dy = \frac{\alpha}{2 \sin \alpha}$$

   $(0 \le \alpha < \pi)$. \qquad (C.U.)

12. In each of the following, (a) evaluate the integral, (b) sketch the region of integration, (c) write down the integral with the order of integration reversed, (d) evaluate again and compare with (a):

   (i) $\displaystyle\int_{0}^{a} dx \int_{0}^{a-x} dy$, \qquad (ii) $\displaystyle\int_{0}^{a} dx \int_{0}^{x} (x^2 + y^2)\, dy$,

   (iii) $\displaystyle\int_{0}^{1} dx \int_{x}^{\sqrt{x}} xy^2\, dy$, \qquad (iv) $\displaystyle\int_{-\pi/2}^{\pi/2} d\theta \int_{0}^{2a \cos \theta} r^2 \cos \theta\, dr$.

13. Show, by a diagram, the region over which the following integral extends

$$\int_{0}^{a} dy \int_{y}^{a} \frac{dx}{\sqrt{(a^2 x^2 + b^2 y^2)}}.$$

   Transform to polar coordinates and hence show that the value of the integral is

$$\frac{a}{b} \sinh^{-1}\left(\frac{b}{a}\right). \qquad \text{(L.U.)}$$

229

14. By means of the substitutions

$$x = a \sin \theta \cos \phi,$$
$$y = b \sin \theta \sin \phi,$$

or otherwise, evaluate

$$\iint \frac{x^2 y \, dx \, dy}{\sqrt{\{1-(x^2/a^2)-(y^2/b^2)\}}}$$

over the area inside the ellipse

$$\frac{x^2}{a^2} + \frac{y^2}{b^2} = 1$$

and in the positive quadrant. (C.U.)

15. (i) Transform to polars and evaluate

$$\int_0^{a/\sqrt{2}} dx \int_x^{\sqrt{(a^2-x^2)}} \sqrt{(x^2+y^2)} \, dy.$$

(ii) Reverse the order of integration and hence evaluate

$$\int_0^1 dx \int_x^{2-x} \frac{x}{y} \, dy. \qquad \text{(L.U.)}$$

16. Show by means of a diagram the area over which the double integral

$$\int_0^1 dy \int_y^{2-y} \frac{x+y}{x^2} e^{x+y} \, dx$$

is taken, Apply the transformation of variables $u = x+y$, $v = y/x$ to this integral and hence, or otherwise, evaluate it. (L.U.)

17. Prove that the volume V enclosed by the sphere $x^2+y^2+z^2 = a^2$ and the cylinder $x^2+y^2 = ay$ is given by

$$V = 2 \iint \sqrt{(a^2-x^2-y^2)} \, dx \, dy,$$

the integral being taken over the circle $x^2 + y^2 = ay$. Transform the integral into polar coordinates, and hence show that

$$V = \frac{4}{3} a^3 \left(\frac{\pi}{2} - \frac{2}{3} \right).$$ (L.U.)

18. Show using Green's theorem that if $V(x, y)$ is defined and has continuous first and second partial derivatives inside and on the boundary C of a region R then

$$\oint_C \frac{dV}{dn} ds = \iint_R \left(\frac{\partial^2 V}{\partial x^2} + \frac{\partial^2 V}{\partial y^2} \right) dx \, dy,$$

where $\frac{dV}{dn}$ is the derivative of V along the outward drawn normal at any point of C and s is a measure of the arc length along C from some arbitrary point.

Show also that, if $\frac{\partial^2 V}{\partial x^2} + \frac{\partial^2 V}{\partial y^2} = 0$,

$$\oint_C V \frac{dV}{dn} ds = \iint_R \left\{ \left(\frac{\partial V}{\partial x} \right)^2 + \left(\frac{\partial V}{\partial y} \right)^2 \right\} dx \, dy,$$

and deduce that if $\frac{dV}{dn} = 0$ along C, then V is a constant inside and on the boundary of R.

CHAPTER 13

Gamma-Function and Related Integrals

13.1 The Gamma-Function

As with the elliptic functions discussed in Chapter 11, the gamma-function is usually defined by an integral, although as we shall see in the next section non-integral definitions due to Euler and Weierstrass are often useful. For the moment we define the gamma-function $\Gamma(x)$ by

$$\Gamma(x) = \int_0^\infty t^{x-1} e^{-t} dt, \tag{1}$$

where, for convergence of the integral, $x > 0$ (see Chapter 4, 4.8). From (1) we see that integration by parts gives

$$\Gamma(x+1) = \int_0^\infty t^x e^{-t} dt = \left[-t^x e^{-t} \right]_0^\infty + x \int_0^\infty t^{x-1} e^{-t} dt$$

$$= x\Gamma(x), \quad (x > 0). \tag{2}$$

By using this recurrence relation when $x = n$, n being a positive integer ≥ 1, we have

$$\Gamma(n+1) = n\Gamma(n) = n(n-1)\Gamma(n-1) = \dots = n!\,\Gamma(1), \tag{3}$$

which, since, by (1),

$$\Gamma(1) = \int_0^\infty e^{-t} dt = 1, \tag{4}$$

gives the important result

$$\Gamma(n+1) = n! \tag{5}$$

The Γ-function may therefore be thought of as a generalisation of the factorial function to which it reduces when x is a positive integer.

The value of $\Gamma(\frac{1}{2})$ is often required and may be obtained directly from the definition

$$\Gamma(\tfrac{1}{2}) = \int_0^\infty t^{-\frac{1}{2}} e^{-t}\, dt \tag{6}$$

by putting $t = u^2$ and integrating. Consequently using the result of Chapter 12, Example 19,

$$\Gamma(\tfrac{1}{2}) = 2 \int_0^\infty e^{-u^2}\, du = \sqrt{\pi}. \tag{7}$$

Using this result we may now obtain all other positive half-integral values from the recurrence relation (2).

For example,

$$\Gamma(\tfrac{3}{2}) = \tfrac{1}{2}\Gamma(\tfrac{1}{2}) = \frac{\sqrt{\pi}}{2}, \tag{8}$$

and

$$\Gamma(\tfrac{7}{2}) = \frac{5}{2}\Gamma(\tfrac{5}{2}) = \frac{5}{2}\cdot\frac{3}{2}\Gamma(\tfrac{3}{2}) = \frac{15}{4}\cdot\frac{\sqrt{\pi}}{2}. \tag{9}$$

The recurrence relation (2) is also useful in defining the Γ-function for negative values of x. For re-writing (2) as

$$\Gamma(x) = \frac{\Gamma(x+1)}{x} \tag{10}$$

we have, for example,

$$\Gamma(-\tfrac{3}{2}) = \frac{\Gamma(-\tfrac{1}{2})}{(-\tfrac{3}{2})} = \frac{\Gamma(\tfrac{1}{2})}{(-\tfrac{3}{2})(-\tfrac{1}{2})} = \tfrac{4}{3}\sqrt{\pi}. \tag{11}$$

It also follows directly from (10) that $\Gamma(x)$ becomes infinite at $x = 0$, and consequently also at all negative integral values of x. It is important to emphasise that the values of $\Gamma(x)$ for negative values of x are not given by the integral form (1) but by the recurrence relation (10). The graph of $\Gamma(x)$ for positive and negative values of x is shown in Fig. 13.1. (For tables of the Γ-function see Jahnke and Emde, ' Tables of Functions '. Dover Publications, 1945.)

We now show how certain integrals may be evaluated in terms of the Γ-function.

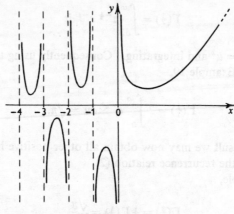

Fig. 13.1

Example 1. To evaluate the integral

$$I = \int_0^\infty x^{5\cdot2} e^{-x^2} \, dx \qquad (12)$$

we put $x^2 = t$ and obtain

$$I = \frac{1}{2}\int_0^\infty t^{2\cdot1} \, e^{-t} \, dt = \frac{1}{2}\Gamma(3\cdot1) = \frac{2\cdot1}{2} \, \Gamma(2\cdot1) = \frac{(2\cdot1)(1\cdot1)}{2}\Gamma(1\cdot1)$$

$$= \frac{1}{20}(2\cdot1)(1\cdot1)\Gamma(0\cdot1). \qquad (13)$$

Since, from tables $\Gamma(0\cdot1) = 9\cdot5135$, we find $I = 1\cdot10$ correct to two decimal places.

Example 2. To evaluate

$$I = \int_0^1 \sqrt{\left\{\log_e\left(\frac{1}{x}\right)\right\}} \, dx \qquad (14)$$

we put $x = e^{-t}$ which gives

$$I = \int_0^\infty t^{\frac{1}{2}}e^{-t} \, dt = \Gamma\left(\frac{3}{2}\right) = \frac{\sqrt{\pi}}{2}, \qquad (15)$$

(by (8)).

234

Example 3. If

$$I = \int_0^{\pi/2} (\tan^3 \theta + \tan^5 \theta) \, e^{-\tan^2 \theta} \, d\theta \qquad (16)$$

then, by putting $\tan^2 \theta = t$, we obtain

$$I = \frac{1}{2} \int_0^\infty t e^{-t} \, dt = \frac{1}{2}\Gamma(2) = \frac{1}{2}. \qquad (17)$$

13.2 Alternative Forms of $\Gamma(x)$

An alternative definition of $\Gamma(x)$ due to Euler is

$$\Gamma(x) = \lim_{n \to \infty} \left\{ \frac{n! \, n^x}{x(x+1) \dots (x+n)} \right\}. \qquad (18)$$

This form is valid for positive and negative x and shows clearly the singularities of $\Gamma(x)$ at $x = 0, -1, -2, \dots$ and so on; it can also be shown to be equivalent to the integral definition (1) for $x > 0$. From (18) we again find

$$\Gamma(x+1) = \lim_{n \to \infty} \left\{ \frac{n! \, n^{x+1}}{(x+1)(x+2) \dots (x+n+1)} \right\}$$

$$= \lim_{n \to \infty} \left\{ \frac{nx}{(x+n+1)} \cdot \frac{n! \, n^x}{x(x+1) \dots (x+n)} \right\}$$

$$= \lim_{n \to \infty} \left\{ \frac{nx}{x+n+1} \right\} \lim_{n \to \infty} \left\{ \frac{n! \, n^x}{x(x+1) \dots (x+n)} \right\}$$

$$= x\Gamma(x). \qquad (19)$$

The Euler definition may also be written in a slightly different form as

$$\Gamma(x) = \frac{1}{x} \lim_{n \to \infty} \left\{ n^x \prod_{m=1}^n \left(1 + \frac{x}{m}\right)^{-1} \right\}, \qquad (20)$$

or

$$\frac{1}{\Gamma(x)} = x \lim_{n \to \infty} \left\{ n^{-x} \prod_{m=1}^n \left(1 + \frac{x}{m}\right) \right\}, \qquad (21)$$

where

$$\prod_{m=1}^n \left(1 + \frac{x}{m}\right) = (1+x)\left(1 + \frac{x}{2}\right)\left(1 + \frac{x}{3}\right) \dots \left(1 + \frac{x}{n}\right). \qquad (22)$$

We now multiply the right-hand side of (21) by unity in the form

$$1 = \left(\lim_{n \to \infty} e^{[1 + (1/2) + (1/3) + \dots + (1/n)]x} \right) \left(\lim_{n \to \infty} \prod_{m=1}^{n} e^{-x/m} \right) \qquad (23)$$

so that

$$\frac{1}{\Gamma(x)} =$$

$$x \left(\lim_{n \to \infty} e^{[1 + (1/2) + (1/3) + \dots + (1/n) - \log_e n]x} \right) \left\{ \lim_{n \to \infty} \prod_{m=1}^{n} \left(1 + \frac{x}{m} \right) e^{-x/m} \right\}. \qquad (24)$$

However, since by Chapter 5, 5.3.

$$\lim_{n \to \infty} \left(1 + \frac{1}{2} + \frac{1}{3} + \dots + \frac{1}{n} - \log_e n \right) = \gamma, \qquad (25)$$

where γ is Euler's constant ($\simeq 0 \cdot 577$), we have

$$\frac{1}{\Gamma(x)} = x e^{\gamma x} \prod_{m=1}^{\infty} \left\{ \left(1 + \frac{x}{m} \right) e^{-x/m} \right\}, \qquad (26)$$

where the infinite product $\prod_{m=1}^{\infty}$ is defined as $\lim_{n \to \infty} \prod_{m=1}^{n}$.

This form of the Γ-function is usually known as Weierstrass's definition.

13.3 The Beta-Function

The beta-function $B(m, n)$ is defined by the integral

$$B(m, n) = \int_0^1 x^{m-1} (1-x)^{n-1} \, dx \qquad (27)$$

which converges if $m > 0$, $n > 0$. It is necessarily symmetric in m and n since by putting $1 - x = u$

$$B(n, m) = \int_0^1 x^{n-1} (1-x)^{m-1} \, dx = - \int_1^0 u^{m-1} (1-u)^{n-1} \, du$$

$$= B(m, n). \qquad (28)$$

236

An alternative form of the beta-function, obtained from (27) by putting $x = \sin^2 \theta$, is

$$B(m, n) = 2\int_0^{\pi/2} \sin^{2m-1} \theta \cos^{2n-1} \theta \, d\theta, \tag{29}$$

from which we may derive a reduction formula relating $B(m, n)$ and $B(m-1, n-1)$. For, by integration by parts, we have

$$\int_0^{\pi/2} \sin^{2m-1} \theta \cos^{2n-1} \theta \, d\theta = \frac{n-1}{m+n-1}\int_0^{\pi/2} \sin^{2m-1} \theta \cos^{2n-3} \theta \, d\theta, \tag{30}$$

and consequently

$$\begin{aligned}
B(m, n) &= 2\frac{n-1}{m+n-1}\int_0^{\pi/2} \sin^{2m-1} \theta \cos^{2n-3} \theta \, d\theta \\
&= 2\frac{n-1}{m+n-1}\int_0^{\pi/2} \cos^{2m-1} \theta \sin^{2n-3} \theta \, d\theta \\
&= 2\frac{(n-1)(m-1)}{(m+n-1)(m+n-2)}\int_0^{\pi/2} \sin^{2n-3} \theta \cos^{2m-3} \theta \, d\theta \quad (31) \\
&= \frac{(m-1)(n-1)}{(m+n-1)(m+n-2)} B(m-1, n-1).
\end{aligned}$$

When m and n are both positive integers repeated application of (31) gives

$$B(m, n) = \frac{(m-1)!(n-1)!}{(m+n-1)!} B(1, 1), \tag{32}$$

where, from (27), $B(1,1) = 1$. (Similarly, from (29), we find $B(\tfrac{1}{2}, \tfrac{1}{2}) = \pi$.)

13.4 Relation between the Gamma- and Beta-Functions
Consider

$$\Gamma(m) = \int_0^\infty t^{m-1}e^{-t} \, dt, \quad \Gamma(n) = \int_0^\infty s^{n-1}e^{-s} \, ds. \tag{33}$$

237

Then

$$\Gamma(m)\Gamma(n) = \int_0^\infty t^{m-1}e^{-t}\, dt \int_0^\infty s^{n-1}e^{-s}\, ds \qquad (34)$$

$$= \iint t^{m-1}s^{n-1}e^{-(s+t)}\, ds\, dt, \qquad (35)$$

where the double integration extends over the first quadrant of the (s, t) plane. Now putting $t = u^2$, $s = v^2$ and using the Jacobian of the transformation (see Chapter 12, 12.13)

$$ds\, dt = \left(\frac{\partial s}{\partial v} \cdot \frac{\partial t}{\partial u} - \frac{\partial s}{\partial u} \cdot \frac{\partial t}{\partial v}\right) du\, dv = 4uv\, du\, dv, \qquad (36)$$

we may transform (35) into

$$\Gamma(m)\Gamma(n) = 4\iint u^{2m-1}v^{2n-1}e^{-(u^2+v^2)}\, du\, dv. \qquad (37)$$

Adopting polar coordinates such that $u = r\cos\theta$, $v = r\sin\theta$, and replacing $du\, dv$ by

$$J\left(\frac{u, v}{r, \theta}\right) dr\, d\theta = r\, dr\, d\theta, \qquad (38)$$

we have

$$\Gamma(m)\Gamma(n) = 4\iint r^{2m+2n-1}\cos^{2m-1}\theta \sin^{2n-1}\theta\, e^{-r^2}\, dr\, d\theta \qquad (39)$$

$$= 4\int_0^\infty r^{2m+2n-1}e^{-r^2}\, dr \int_0^{\pi/2}\cos^{2m-1}\theta \sin^{2n-1}\theta\, d\theta \qquad (40)$$

$$= 2\int_0^\infty y^{m+n-1}e^{-y}\, dy \int_0^{\pi/2}\cos^{2m-1}\theta \sin^{2n-1}\theta\, d\theta \qquad (41)$$

$$= \Gamma(m+n)B(m, n). \qquad (42)$$

Hence finally

$$B(m, n) = \frac{\Gamma(m)\Gamma(n)}{\Gamma(m+n)}. \qquad (43)$$

The following examples illustrate the use of the beta-function in evaluating integrals.

Example 4. To evaluate

$$I = \int_0^1 \frac{dx}{\sqrt{(1-x^4)}} \tag{44}$$

we make the substitution $x^4 = u$; (44) then becomes

$$I = \tfrac{1}{4}\int_0^1 \frac{u^{-3/4}}{\sqrt{(1-u)}}\, du = \tfrac{1}{4}B(\tfrac{1}{4}, \tfrac{1}{2}) = \frac{\Gamma(\tfrac{1}{4})\Gamma(\tfrac{1}{2})}{4\Gamma(\tfrac{3}{4})}. \tag{45}$$

Since, however $\Gamma(\tfrac{1}{2}) = \sqrt{\pi}$, (45) simplifies to give

$$I = \frac{\sqrt{\pi}}{4} \cdot \frac{\Gamma(\tfrac{1}{4})}{\Gamma(\tfrac{3}{4})}. \tag{46}$$

Example 5. To evaluate

$$\int_0^{\pi/2} \sqrt{(\sin\theta)}\, d\theta,$$

we put $\sin^2\theta = u$ to give

$$\int_0^{\pi/2} \sqrt{(\sin\theta)}\, d\theta = \tfrac{1}{2}\int_0^1 \frac{u^{-1/4}}{\sqrt{(1-u)}}\, du = \tfrac{1}{2}B(\tfrac{3}{4}, \tfrac{1}{2}) = \frac{\Gamma(\tfrac{3}{4})\Gamma(\tfrac{1}{2})}{2\Gamma(\tfrac{5}{4})}. \tag{47}$$

Since $\Gamma(\tfrac{5}{4}) = \tfrac{1}{4}\Gamma(\tfrac{1}{4})$ and $\Gamma(\tfrac{1}{2}) = \sqrt{\pi}$, (47) reduces to

$$\int_0^{\pi/2} \sqrt{(\sin\theta)}\, d\theta = 2\sqrt{\pi}\, \frac{\Gamma(\tfrac{3}{4})}{\Gamma(\tfrac{1}{4})} \simeq 1\cdot198. \tag{48}$$

13.5 The Error Function

The error function erf x is defined as

$$\text{erf } x = \frac{2}{\sqrt{\pi}} \int_0^x e^{-u^2}\, du \tag{49}$$

and clearly represents (apart from the factor $2/\sqrt{\pi}$) the area under the curve e^{-u^2} from $u = 0$ to $u = x$; we see immediately that erf $(0) = 0$ and that erf $(\infty) = 1$ $\left(\text{since, by (7), } \int_0^{\infty} e^{-u^2}\, du = \frac{\sqrt{\pi}}{2}\right)$. The graph of the error function is shown in Fig. 13.2.

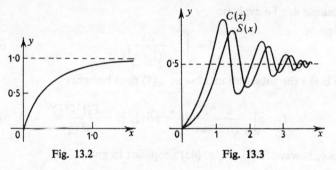

Fig. 13.2   Fig. 13.3

Closely related to this function are the functions $C(x)$ and $S(x)$ (the Fresnel integrals) defined by

$$C(x) = \int_0^x \cos \frac{\pi u^2}{2} \, du, \tag{50}$$

and

$$S(x) = \int_0^x \sin \frac{\pi u^2}{2} \, du. \tag{51}$$

Consequently

$$C(x) - iS(x) = \int_0^x e^{-i(\pi/2)u^2} \, du. \tag{52}$$

Now assuming that

$$\int_0^\infty e^{-au^2} \, du = \frac{1}{2}\sqrt{\frac{\pi}{a}}, \tag{53}$$

holds when a is complex, we may evaluate $C(\infty)$ and $S(\infty)$ by writing

$$\int_0^\infty \cos \frac{\pi}{2} u^2 \, du - i \int_0^\infty \sin \frac{\pi}{2} u^2 \, du = \int_0^\infty e^{-i(\pi/2)u^2} \, du = \frac{1}{\sqrt{(2i)}}$$

$$= \tfrac{1}{2}(1-i). \tag{54}$$

Hence comparing the real and imaginary parts of each side of (54) we have

$$\int_0^\infty \cos \frac{\pi}{2} u^2 \, du = \int_0^\infty \sin \frac{\pi}{2} u^2 \, du = \frac{1}{2}. \tag{55}$$

Graphs of $C(x)$ and $S(x)$ are shown in Fig. 13.3.

13.6 Stirling's Formula

We now obtain an approximation to $\Gamma(x+1)$ for large x. Consider

$$\Gamma(x+1) = \int_0^\infty t^x e^{-t}\, dt \tag{56}$$

and put $t = x + \tau\sqrt{x}$. Then

$$\Gamma(x+1) = \int_{-\sqrt{x}}^\infty (x+\tau\sqrt{x})^x e^{-(x+\tau\sqrt{x})} \cdot \sqrt{x}\, d\tau, \tag{57}$$

or

$$\frac{\Gamma(x+1)}{e^{-x}x^{x+\frac{1}{2}}} = \int_{-\sqrt{x}}^\infty e^{-\tau\sqrt{x}}\left(1 + \frac{\tau}{\sqrt{x}}\right)^x d\tau = \int_{-\sqrt{x}}^\infty e^{-\tau\sqrt{x} + x\log_e[1+(\tau/\sqrt{x})]}\, d\tau. \tag{58}$$

Hence expanding the logarithmic term we have

$$\frac{\Gamma(x+1)}{e^{-x}x^{x+\frac{1}{2}}} = \int_{-\sqrt{x}}^{\sqrt{x}} e^{-\tau\sqrt{x} + x[(\tau/\sqrt{x}) - (\tau^2/2x) + \ldots]}\, d\tau$$

$$+ \int_{\sqrt{x}}^\infty e^{-\tau\sqrt{x} + x\log_e[1+(\tau/\sqrt{x})]}\, d\tau, \tag{59}$$

which for large x may be expressed as

$$\frac{\Gamma(x+1)}{e^{-x}x^{x+\frac{1}{2}}} \simeq \int_{-\infty}^\infty e^{-\frac{1}{2}\tau^2}\, d\tau = \sqrt{(2\pi)}. \tag{60}$$

If now $x = n$, where n is a positive integer, then since $\Gamma(n+1) = n!$ we get

$$n! \simeq \sqrt{(2\pi)} e^{-n} n^{n+\frac{1}{2}}. \tag{61}$$

This is Stirling's formula for $n!$ when n is large. The alternative form

$$n! = \sqrt{(2\pi)} e^{-n} n^{n+\frac{1}{2}}(1 + \delta_n), \tag{62}$$

where $\delta_n \to 0$ as $n \to \infty$, may be obtained by more refined techniques.

PROBLEMS 13

1. If $\Gamma(1\cdot1) = 0\cdot951$, find $\Gamma(4\cdot1)$ and $\Gamma(-3\cdot9)$.

2. Prove that

$$\int_0^\infty \frac{t^{m-1}\,dt}{(1+t)^{m+n}} = B(m, n), \quad (m, n > 0),$$

and hence show that

$$\int_0^{\pi/2} \sqrt{(\tan \theta)}\,d\theta = \tfrac{1}{2}B(\tfrac{3}{4}, \tfrac{1}{4}).$$

3. Show that

(a) $\displaystyle\int_0^\infty \text{sech}^8 x\,dx = \tfrac{16}{35}$,   (b) $\displaystyle\int_0^1 \frac{dx}{\sqrt{(1-x^3)}} = \tfrac{1}{3}B(\tfrac{1}{2}, \tfrac{1}{3})$,

(c) $\displaystyle\int_0^{\pi/2} \sin^n x\,dx = \frac{\sqrt{\pi}}{2} \cdot \frac{\Gamma\left(\dfrac{n+1}{2}\right)}{\Gamma\left(\dfrac{n}{2}+1\right)}.$

4. Show that the value of the double integral

$$\iint x^{2/3} y^{2/5}dx\,dy$$

taken over the circle $x^2 + y^2 = a^2$ is

$$\frac{a^{46/15}\,\Gamma(\tfrac{5}{6})\Gamma(\tfrac{7}{10})}{\Gamma(\tfrac{38}{15})}. \qquad\qquad \text{(L.U)}$$

5. Given that for non-zero and non-integral values of x

$$\frac{\sin \pi x}{\pi x} = \prod_{m=1}^\infty \left(1 - \frac{x^2}{m^2}\right),$$

show, using the Euler's definition of $\Gamma(x)$, that

$$\Gamma(x)\Gamma(1-x) = \frac{\pi}{\sin \pi x}.$$

6. From the Euler definition of $\Gamma(x)$ show that

$$\frac{d}{dx}\{\log_e \Gamma(x+1)\} = \lim_{n \to \infty}\left(\log_e n - \frac{1}{x+1} - \frac{1}{x+2} \cdots - \frac{1}{x+n+1}\right),$$

and hence that

$$\gamma = -\left\{\frac{d}{dx}\Gamma(x+1)\right\}_{x=0} = \int_\infty^0 (\log_e t)e^{-t}\, dt.$$

7. Show, using the beta-function, that

$$\left\{\Gamma\left(\frac{1}{4}\right)\right\}^2 = 4\sqrt{(\pi)}F\left(\frac{1}{\sqrt{2}}\right),$$

where

$$F\left(\frac{1}{\sqrt{2}}\right) = \int_0^{\pi/2} \frac{d\theta}{\sqrt{(1 - \frac{1}{2}\sin^2\theta)}}$$

is a complete elliptic integral of the first kind.

8. If

$$\phi(x) = \int_0^x e^{-t^2}\, dt \quad \text{and} \quad \phi(\infty) = \frac{\sqrt{\pi}}{2},$$

prove the following results

(a) $\phi(x) = x - \dfrac{x^3}{3.1!} + \dfrac{x^5}{5.2!} - \dfrac{x^7}{7.3!} + \cdots,$

(b) $\phi(x) = \dfrac{\sqrt{\pi}}{2} + \displaystyle\int_\infty^x e^{-t^2}\, dt,$

(c) $\displaystyle\int_\infty^x e^{-t^2}\, dt = -\dfrac{1}{2x}e^{-x^2} - \dfrac{1}{2}\int_\infty^x \dfrac{1}{t^2}e^{-t^2}\, dt.$

Hence, or otherwise, deduce that

$$\phi(x) = \frac{\sqrt{\pi}}{2} - e^{-x^2}\left(\frac{1}{2x} - \frac{1}{2^2 x^3} + \frac{3}{2^3 x^5} - \cdots\right). \qquad \text{(L.U.)}$$

243

9. By putting $x = y$ in the relation

$$\frac{\Gamma(x)\Gamma(y)}{\Gamma(x+y)} = B(x, y) = \int_0^1 u^{x-1}(1-u)^{y-1} \, du,$$

show that

$$\frac{[\Gamma(x)]^2}{\Gamma(2x)} = \tfrac{1}{2}\int_0^1 (\tfrac{1}{4} - \tfrac{1}{4}\tau)^{x-1}\tau^{-\frac{1}{2}} \, d\tau = 2^{1-2x}\frac{\Gamma(x)\Gamma(\tfrac{1}{2})}{\Gamma(x+\tfrac{1}{2})}.$$

Hence using the result

$$\Gamma(x)\Gamma(x+\tfrac{1}{2}) = 2^{1-2x}\sqrt{\pi}\,\Gamma(2x)$$

(known as the ' duplication formula '), show that

$$\Gamma(\tfrac{1}{4})\Gamma(\tfrac{3}{4}) = \pi\sqrt{2}.$$

CHAPTER 14

Numerical Integration

14.1 Trapezium Rule

As mentioned in earlier chapters the evaluation of an integral in terms of known functions is often impossible. Furthermore, in some cases the integrand may only be defined by a set of tabulated values. To meet these difficulties some numerical procedure is required which will give a good approximation to the value of the integral. Clearly one of the simplest methods of doing this is to interpret the integral $\int_a^b f(x)\,dx$ graphically (see Chapter 4) as the area between the curve $y = f(x)$, the x-axis, and the lines $x = a$, $x = b$, and to estimate this area as accurately as possible. Consider, for example, the curve $y = f(x)$ as shown in Fig. 14.1. Then to obtain an approximation

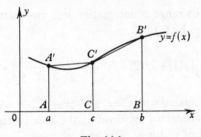

Fig. 14.1

to the required area we may draw in the straight lines $A'C'$ and $C'B'$ and evaluate the sum of the areas of the two trapeziums $ACC'A'$ and $CBB'C'$. If now the point C is chosen to be the mid-point of the range (a, b) such that $AC = CB = h$, then

$$\text{area } ACC'A' = \frac{h}{2}(AA' + CC') = \frac{h}{2}\left\{f(a) + f\left(\frac{a+b}{2}\right)\right\} \quad (1)$$

and

$$\text{area } CBB'C' = \frac{h}{2}(CC' + BB') = \frac{h}{2}\left\{f\left(\frac{a+b}{2}\right) + f(b)\right\}. \quad (2)$$

245

Hence by adding (1) and (2) we have

$$\int_a^b f(x)\,dx \simeq \frac{h}{2}\left\{f(a)+2f\left(\frac{a+b}{2}\right)+f(b)\right\}. \tag{3}$$

This formula, usually known as the trapezium rule, gives a good approximation to the value of the integral when the curve $y = f(x)$ deviates only slightly from the straight lines $A'C'$, $C'B'$. When violent deviations occur, however, the accuracy may usually be improved by dividing the area under the curve into a larger (even) number of trapeziums of smaller width and applying (3) to each pair. As an example of the trapezium rule we now consider the numerical evaluation of a simple integral whose value is known exactly.

Example 1. If

$$I = \int_a^b f(x)\,dx = \int_1^3 \frac{dx}{x^2}, \tag{4}$$

then dividing the range of integration into two parts each of width $h\,(=1)$, we have by (3)

$$I \simeq \frac{1}{2}\{f(1)+2f(2)+f(3)\} = \frac{1}{2}\left(1+\frac{2}{4}+\frac{1}{9}\right) = 0\!\cdot\!81. \tag{5}$$

This is to be compared with the exact value of $2/3$. A better approximation may be obtained by dividing the area under the curve into four parts each of width $h\,(=\tfrac{1}{2})$ and applying (3) to each pair. In this way we find

$$I = \tfrac{1}{2}\cdot\tfrac{1}{2}\{f(1)+2f(1\!\cdot\!5)+f(2)\} + \tfrac{1}{2}\cdot\tfrac{1}{2}\{f(2)+2f(2\!\cdot\!5)+f(3)\} \tag{6}$$

$$= \frac{1}{4}\left\{1+\frac{2}{2\!\cdot\!25}+\frac{1}{4}\right\} + \frac{1}{4}\left\{\frac{1}{4}+\frac{2}{6\!\cdot\!25}+\frac{1}{9}\right\}, \tag{7}$$

$$= 0\!\cdot\!70.$$

14.2 Simpson's Rule

A better approximation to the area indicated in Fig. 14.1 may be obtained in the following way. Suppose $x = c$ is the coordinate of

246

the point C such that $a = c - h$, $b = c + h$. Then writing $x = c + y$, expanding by Taylor's series, and integrating term-by-term we obtain

$$\int_a^b f(x)\, dx = \int_{c-h}^{c+h} f(x)\, dx = \int_{-h}^h f(c+y)\, dy \tag{8}$$

$$= \int_{-h}^h \left\{ f(c) + y f^{(1)}(c) + \frac{y^2}{2!} f^{(2)}(c) + \ldots + \frac{y^r}{r!} f^{(r)}(c) + \ldots \right\} dy \tag{9}$$

$$= 2h \left\{ f(c) + \frac{h^2 f^{(2)}(c)}{3!} + \frac{h^4 f^{(4)}(c)}{5!} + \ldots + \frac{h^{2r} f^{(2r)}(c)}{(2r+1)!} + \ldots \right\}, \tag{10}$$

where, in general, $f^{(r)}(c)$ is the value of $\dfrac{d^r f}{dx^r}$ at $x = c$.

Now, since by Taylor's series

$$f(c+h) = f(c) + h f^{(1)}(c) + \frac{h^2}{2!} f^{(2)}(c) + \ldots + \frac{h^r}{r!} f^{(r)}(c) + \ldots \tag{11}$$

$$f(c-h) = f(c) - h f^{(1)}(c) + \frac{h^2}{2!} f^{(2)}(c) + \ldots + \frac{(-1)^r h^r}{r!} f^{(r)}(c) + \ldots, \tag{12}$$

we also have

$$f(c+h) + f(c-h) = 2 \left\{ f(c) + \frac{h^2}{2!} f^{(2)}(c) + \frac{h^4}{4!} f^{(4)}(c) + \ldots \right.$$

$$\left. + \frac{h^{2r}}{(2r)!} f^{(2r)}(c) + \ldots \right\}. \tag{13}$$

Hence neglecting terms involving h^4 and higher powers of h in (10) and (13), and eliminating $f^{(2)}(c)$, we finally obtain Simpson's formula

$$\int_a^b f(x)\, dx \simeq 2h \left\{ f(c) + \frac{h^2}{3!} f^{(2)}(c) \right\} \tag{14}$$

$$\simeq 2h \left\{ f(c) + \frac{h^2}{3!} \left(\frac{f(c+h) + f(c-h) - 2f(c)}{h^2} \right) \right\}, \tag{15}$$

$$= \frac{h}{3} \{ f(c-h) + 4f(c) + f(c+h) \}, \tag{16}$$

$$= \frac{h}{3}\left\{ f(a) + 4f\left(\frac{a+b}{2}\right) + f(b) \right\}. \qquad (17)$$

The error involved here by approximating to the integral in this way is such that if

$$\int_a^b f(x)\, dx = \frac{h}{3}\left\{ f(a) + 4f\left(\frac{a+b}{2}\right) + f(b) \right\} + E, \qquad (18)$$

then

$$E \simeq -\frac{h^4}{180}[f^{(3)}(b) - f^{(3)}(a)], \qquad (19)$$

where, as before, $f^{(3)}(b)$ and $f^{(3)}(a)$ mean the values of $\dfrac{d^3 f}{dx^3}$ at $x = b$ and $x = a$ respectively.

As with the trapezium rule it is usually possible to obtain a more accurate result by first dividing the area under the curve between $x = a$ and $x = b$ into a larger (even) number of strips and then applying (17) to each successive pair. In this way, if $f_0, f_1, f_2, f_3 \ldots f_{n-1}, f_n$ are the values of $f(x)$ at $x = a,\ a+h,\ a+2h,\ \ldots\ a+(n-1)h,\ a+nh(=b)$, where n is an even integer, then

$$\int_a^b f(x)\, dx \simeq \frac{h}{3}\{f_0 + f_n + 4(f_1 + f_3 + \ldots + f_{n-1})$$
$$+ 2(f_2 + f_4 + \ldots + f_{n-2})\}. \qquad (20)$$

For example, Simpson's rule with five ordinates (i.e. four strips) is

$$\int_a^b f(x)\, dx \simeq \frac{h}{3}\{f_0 + f_4 + 4(f_1 + f_3) + 2f_2\} \qquad (21)$$

(see Fig. 14.2).

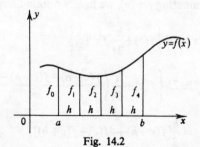

Fig. 14.2

248

14.3 Applications of Simpson's Rule

Example 2. We now consider the numerical evaluation of the integral

$$I = \int_0^{\pi/2} \frac{d\theta}{\sqrt{(1 - \frac{1}{2} \sin^2 \theta)}} \tag{22}$$

using Simpson's rule with five ordinates. As defined in Chapter 11, 11.1, this integral is the complete elliptic integral of the first kind $K(1/\sqrt{2})$ whose tabulated value is 1·854. To apply Simpson's rule we now divide the range of integration $(0, \pi/2)$ into four parts such that $h = \pi/8$ and evaluate the integrand $f(\theta) = (1 - \frac{1}{2} \sin^2 \theta)^{-\frac{1}{2}}$ at the five points $\theta = 0, \pi/8, \pi/4, 3\pi/8$ and $\pi/2$.

The values are given below:

θ	0	$\pi/8$	$\pi/4$	$3\pi/8$	$\pi/2$
$f(\theta)$	1·000	1·0387	1·1547	1·3206	1·4142

Hence using (21) we have

$$\int_0^{\pi/2} \frac{d\theta}{\sqrt{(1 - \frac{1}{2} \sin^2 \theta)}}$$

$$\simeq \frac{1}{3} \cdot \frac{\pi}{8} \{1 + 1\cdot4142 + 4(1\cdot0387 + 1\cdot3206) + 2(1\cdot1547)\} \tag{23}$$

$$= \frac{\pi}{24} \times 14\cdot1608,$$

$$= 1\cdot854. \tag{24}$$

Example 3. Given the following nine pairs of (x, y) values

x	1	2	3	4	5	6	7	8	9.
y	2·061	2·312	2·819	3·106	3·670	4·721	6·103	7·950	9·942

we may easily estimate $\int_1^9 y\,dx$ using (20). Since $h = 1$ we have

$$\int_1^9 y\,dx \simeq \frac{1}{3}\{2\!\cdot\!061 + 9\!\cdot\!942 + 4(2\!\cdot\!312 + 3\!\cdot\!106 + 4\!\cdot\!721 + 7\!\cdot\!950)$$
$$+ 2(2\!\cdot\!819 + 3\!\cdot\!670 + 6\!\cdot\!103)\} = 36\!\cdot\!514. \qquad (25)$$

14.4 Series Expansion Method

When a function $f(x)$ can be expanded as a power series in x, term-by-term integration is permissible (see Chapter 5, 5.10) and the evaluation of $\int f(x)\,dx$ is reduced to the summation of a series. This is illustrated by the following examples.

Example 4. To evaluate

$$I = \int_0^1 \frac{\sin x}{x}\,dx \qquad (26)$$

we use the Maclaurin expansion

$$\sin x = x - \frac{x^3}{3!} + \frac{x^5}{5!} - \frac{x^7}{7!} + \ldots \qquad (27)$$

and write

$$I = \int_0^1 \left(1 - \frac{x^2}{3!} + \frac{x^4}{5!} - \frac{x^6}{7!} + \ldots\right) dx \qquad (28)$$

$$= \left[x - \frac{x^3}{18} + \frac{x^5}{600} - \frac{x^7}{35280} + \ldots\right]_0^1 \qquad (29)$$

$$= \left[1 - \frac{1}{18} + \frac{1}{600} \ldots\right] \simeq 0\!\cdot\!946. \qquad (30)$$

Greater accuracy may be obtained by summing more terms of the series.

Example 5. To evaluate

$$I = \int_0^2 \sqrt{(8 + x^3)}\,dx \qquad (31)$$

we expand the integrand by the binomial theorem to give

$$I = 2\sqrt{2}\int_0^2 \sqrt{\left\{1+\left(\frac{x}{2}\right)^3\right\}}\, dx \tag{32}$$

$$= 2\sqrt{2}\int_0^2 \left\{1+\frac{1}{2}\left(\frac{x}{2}\right)^3+\frac{1}{2}\left(-\frac{1}{2}\right)\frac{1}{2!}\left(\frac{x}{2}\right)^5+\ldots\right\} dx. \tag{33}$$

Hence integrating term-by-term we have

$$I = 2\sqrt{2}\left[x+\frac{x^4}{64}-\frac{x^7}{3584}+\ldots\right]_0^2 \tag{34}$$

$$= 2\sqrt{2}\left[2+\frac{1}{4}-\frac{1}{28}+\ldots\right] \simeq 6\cdot25. \tag{35}$$

Example 6. The elliptic integral discussed in Example 2 may also be evaluated by the series expansion method. To do this we expand the integrand by the binomial theorem to give

$$I = \int_0^{\pi/2} \frac{d\theta}{\sqrt{(1-\frac{1}{2}\sin^2\theta)}} \tag{36}$$

$$= \int_0^{\pi/2} \left\{1+\left(-\frac{1}{2}\right)\left(-\frac{\sin^2\theta}{2}\right)+\frac{(-\frac{1}{2})(-\frac{3}{2})}{2!}\left(-\frac{\sin^2\theta}{2}\right)^2\right.$$

$$\left.+\frac{(-\frac{1}{2})(-\frac{3}{2})(-\frac{5}{2})}{3!}\left(-\frac{\sin^2\theta}{2}\right)^3+\ldots\right\} d\theta. \tag{37}$$

Hence, integrating term-by-term, and using Wallis's formula

$$\int_0^{\pi/2} \sin^n x\, dx = \frac{(n-1)(n-3)(n-5)\ldots 3\cdot1}{n(n-2)(n-4)\ldots 4\cdot2}\cdot\frac{\pi}{2}, \tag{38}$$

where n is an even integer, we have

$$I = \frac{\pi}{2}\left\{1+\frac{1}{8}+\frac{9}{256}+\frac{25}{2048}+\ldots\right\}. \tag{39}$$

Taking the first four terms only we find $I\simeq 1\cdot843$, which is in close agreement with the exact value of $1\cdot845$ (to three places of decimals).

In this example, the term-by-term integration should really be justified since the series in (37) is not a power series in θ. However, as this requires the concept of uniform convergence (see Chapter 5, 5.11) we shall accept the validity of it here without proof.

PROBLEMS 14

1. Evaluate the following integrals to three decimal places using Simpson's rule with five ordinates

(a) $\displaystyle\int_0^{\pi/2} \sqrt{(\sin \theta)}\, d\theta,$

(b) $\displaystyle\frac{1}{\pi}\int_0^{\pi/2} \sqrt{(4+\sin^2 \theta)}\, d\theta,$

(c) $\displaystyle\int_0^{0\cdot8} e^{-x^2}\, dx,$

(d) $\displaystyle\int_2^3 \frac{dx}{1+x^4},$

(e) $\displaystyle\int_0^{\pi/2} \cos(\cos \theta)\, d\theta,$

(f) $\displaystyle\int_0^{\pi} \sqrt{(3+\cos \theta)}\, d\theta.$

Also evaluate (b), (c), (e) and (f) to the same accuracy by the series expansion method.

2. Express

$$\int_0^{\theta} (\cos x)^{3/2}\, dx$$

as a power series in θ up to and including terms in θ^5. Hence evaluate the integral to three decimal places when $\theta = 2/3$ radian.

(C.U.)

3. Using Simpson's rule, estimate

$$\int_0^2 y\, dx$$

from the following pairs of (x, y) values

x	0	0·25	0·50	0·75	1·00	1·25	1·50	1·75	2·00
y	1·31	2·41	3·04	2·97	2·16	1·80	0·75	0·13	0·04

4. The Euler-Maclaurin formula gives (using the notation of Simpson's formula (20))

$$\int_a^b f(x)\, dx = h\left\{\frac{f_0}{2} + f_1 + f_2 + \cdots + f_{n-1} + \frac{f_n}{2}\right\}$$
$$- \sum_{\substack{\text{odd} \\ r}}\left\{\frac{h^{r+1}}{(r+1)!}\, B_{r+1}\,(f_n^{(r)} - f_0^{(r)})\right\},$$

where B_r is the rth Bernoulli number defined in Chapter 5, Problem 12, and where $f_n^{(r)}, f_0^{(r)}$ are the values of $d^r f/dx^r$ evaluated at $x = b$ and $x = a$ respectively.

By taking $h = 1$ and neglecting terms involving Bernoulli numbers beyond B_4, evaluate

$$\int_4^8 \frac{1}{x^2}\, dx.$$

Compare this result with the exact value of the integral.

CHAPTER 15

Fourier Series

15.1 Introduction

In Chapter 6 we have seen how certain functions may be represented as power series by means of the Taylor and Maclaurin expansions. These functions must be continuous and infinitely differentiable within the interval of convergence of the power series. We now show how functions which may be neither differentiable nor continuous at certain points in a given interval can be represented by a trigonometric series of the type

$$\frac{a_0}{2} + a_1 \cos x + a_2 \cos 2x + \dots + b_1 \sin x + b_2 \sin 2x + \dots$$

$$= \frac{a_0}{2} + \sum_{r=1}^{\infty} (a_r \cos rx + b_r \sin rx), \quad (1)$$

where r takes integral values and a_0, a_r, b_r are constants. Since (1) is unchanged by replacing x by $x + 2k\pi$, where k is an integer, it necessarily represents a periodic function in x of period 2π (see Chapter 1, 1.3 (i)). Consequently in discussing series of this type it is sufficient to consider any interval of width 2π; we choose here the interval

$$-\pi < x \leqq \pi. \quad (2)$$

Suppose now $f(x)$ is an arbitrarily defined function in the interval (2). Then if the coefficients a_0, a_r, b_r in (1) are defined by

$$a_0 = \frac{1}{\pi} \int_{-\pi}^{\pi} f(x)\, dx, \quad (3)$$

$$a_r = \frac{1}{\pi} \int_{-\pi}^{\pi} f(x) \cos rx\, dx, \quad (r = 1, 2, 3 \dots), \quad (4)$$

$$b_r = \frac{1}{\pi} \int_{-\pi}^{\pi} f(x) \sin rx\, dx, \quad (r = 1, 2, 3 \dots), \quad (5)$$

the resulting series is called the Fourier series of $f(x)$ and the coefficients so defined are the Fourier coefficients. The sum of a Fourier series, however, is not necessarily equal to the function from which it is derived and the conditions under which the Fourier series of $f(x)$ converges to $f(x)$ in the sense that

$$f(x) = \frac{a_0}{2} + \sum_{r=1}^{\infty} (a_r \cos rx + b_r \sin rx) \tag{6}$$

depend very much on the form of the particular function chosen. We now state in the form of a theorem a set of sufficient conditions (Dirichlet's conditions) which $f(x)$ must satisfy for (6) to be valid.

Theorem 1. Suppose $f(x)$ is defined arbitrarily in the interval $-\pi < x \leq \pi$ and extended to other values of x by the periodicity relation $f(x+2k\pi) = f(x)$, where k is an integer. Then, if in $-\pi < x \leq \pi, f(x)$ is single-valued and continuous except for a finite number of points of finite discontinuities, and has only a finite number of maxima and minima, its Fourier series converges to $f(x)$ at all points in this interval where $f(x)$ is continuous. At a point of finite discontinuity, say $x = x_0$, the Fourier series converges to the value

$$\tfrac{1}{2} \lim_{\delta \to 0} \{f(x_0+\delta)+f(x_0-\delta)\}, \tag{7}$$

which is just the mean of the two limiting values of $f(x)$ as x approaches x_0 from the right and left-hand sides. To illustrate the meaning of (7) more clearly we consider a function $f(x)$ defined as $f(x) = 0$ for $-\pi < x < 0$, and $f(x) = 1$ for

Fig. 15.1

$0 < x < \pi$. Then $x_0 = 0$ is a point of finite discontinuity (see Fig. 15.1) and hence

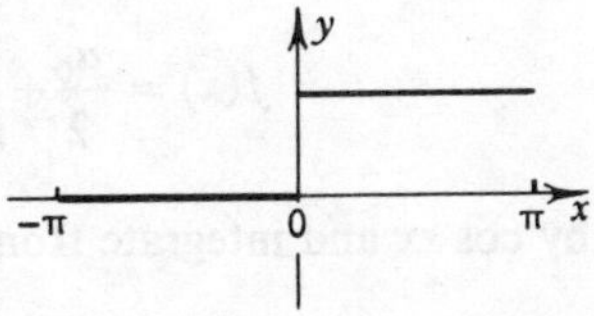

$$\tfrac{1}{2} \lim_{\delta \to 0} \{f(x_0+\delta)+f(x_0-\delta)\} = \tfrac{1}{2}(1+0) = \tfrac{1}{2}. \tag{8}$$

Functions satisfying Dirichlet's conditions (as stated in Theorem 1) are called piecewise regular functions.

15.2 Fourier Coefficients

We shall now justify the forms of the Fourier coefficients a_0, a_r, b_r given by (3), (4) and (5). To do this we make use of the following results which may be obtained by simple integration.

If r and s are positive integers or zero, then

$$\int_{-\pi}^{\pi} \cos rx \cos sx \, dx = \begin{cases} 0 & \text{for } r \neq s, \\ 2\pi & \text{for } r = s = 0, \\ \pi & \text{for } r = s > 0, \end{cases} \tag{9}$$

$$\int_{-\pi}^{\pi} \sin rx \sin sx \, dx = \begin{cases} 0 & \text{for } r \neq s, \\ 0 & \text{for } r = s = 0, \\ \pi & \text{for } r = s > 0, \end{cases} \tag{10}$$

$$\int_{-\pi}^{\pi} \sin rx \cos sx \, dx = 0 \quad \text{for all } r \text{ and } s, \tag{11}$$

$$\int_{-\pi}^{\pi} \cos rx \, dx = \begin{cases} 0 & \text{for } r > 0, \\ 2\pi & \text{for } r = 0, \end{cases} \tag{12}$$

$$\int_{-\pi}^{\pi} \sin rx \, dx = 0 \quad \text{for all } r. \tag{13}$$

We now multiply the series (6)

$$f(x) = \frac{a_0}{2} + \sum_{r=1}^{\infty} (a_r \cos rx + b_r \sin rx) \tag{14}$$

by $\cos sx$ and integrate from $x = -\pi$ to $x = \pi$. Then

$$\int_{-\pi}^{\pi} f(x) \cos sx \, dx = \frac{a_0}{2} \int_{-\pi}^{\pi} \cos sx \, dx + \int_{-\pi}^{\pi} \cos sx \left\{ \sum_{r=1}^{\infty} a_r \cos rx \right\} dx$$

$$+ \int_{-\pi}^{\pi} \cos sx \left\{ \sum_{r=1}^{\infty} b_r \sin rx \right\} dx. \tag{15}$$

On the assumption that we may interchange the order of the integral and summation signs and then integrate the series term-by-term (see the remarks on uniform convergence in Chapter 5, 5.11), we have

$$\int_{-\pi}^{\pi} f(x) \cos sx \, dx = \frac{a_0}{2} \int_{-\pi}^{\pi} \cos sx \, dx + \sum_{r=1}^{\infty} \left\{ a_r \int_{-\pi}^{\pi} \cos sx \cos rx \, dx \right\}$$
$$+ \sum_{r=1}^{\infty} \left\{ b_r \int_{-\pi}^{\pi} \sin rx \cos sx \, dx \right\}. \qquad (16)$$

Using the properties of the integrals given in (9) to (13) we see that when $s = 0$, the only non-vanishing integral on the right-hand side of (16) is the first. Hence (16) gives

$$\int_{-\pi}^{\pi} f(x) \, dx = \frac{a_0}{2} \cdot 2\pi \quad \text{or} \quad a_0 = \frac{1}{\pi} \int_{-\pi}^{\pi} f(x) \, dx \qquad (17)$$

as in (3). When s is non-zero the only non-vanishing integral on the right-hand side of (16) occurs in the first summation term when $s = r$. Hence, using (9), (16) gives

$$\int_{-\pi}^{\pi} f(x) \cos rx \, dx = a_r \int_{-\pi}^{\pi} \cos rx \cos rx \, dx = \pi a_r, \qquad (18)$$

which is (4).

To obtain the Fourier coefficients b_r we multiply (14) by $\sin sx$ and integrate from $x = -\pi$ to $x = \pi$. Proceeding as for a_r we easily find

$$\int_{-\pi}^{\pi} f(x) \sin rx \, dx = b_r \int_{-\pi}^{\pi} \sin rx \sin rx \, dx = \pi b_r, \qquad (19)$$

which gives (5).

We note that the Fourier coefficients a_0 and a_r as given by (3) and (4) may be expressed more compactly by the single formula

$$a_r = \frac{1}{\pi} \int_{-\pi}^{\pi} f(x) \cos rx \, dx, \qquad (20)$$

where now $r = 0, 1, 2, \dots$.

In deriving the Fourier coefficients in this way we have implicitly assumed that $f(x)$ is continuous in $-\pi < x < \pi$. If $f(x)$ is discontinuous at $x = x_0$ then the Fourier coefficients must be evaluated by summing the integrals over the continuous parts of $f(x)$. For example, a_r would be given by

$$a_r = \frac{1}{\pi} \int_{-\pi}^{x_0} f(x) \cos rx \, dx + \frac{1}{\pi} \int_{x_0}^{\pi} f(x) \cos rx \, dx \qquad (21)$$

with similar expressions for b_r and a_0.

Finally when $f(x)$ is either an even or an odd function of x in $-\pi < x \leqq \pi$ then the Fourier coefficients simplify in virtue of the properties of definite integrals given in Chapter 4, 4.3. We have

(a) $f(x)$ even; $f(x) = f(-x)$

$$a_0 = \frac{1}{\pi} \int_{-\pi}^{\pi} f(x)\, dx = \frac{2}{\pi} \int_0^{\pi} f(x)\, dx, \tag{22}$$

$$a_r = \frac{1}{\pi} \int_{-\pi}^{\pi} f(x) \cos rx\, dx = \frac{2}{\pi} \int_0^{\pi} f(x) \cos rx\, dx, \tag{23}$$

$$(r = 1, 2, 3 \dots)$$

$$b_r = \frac{1}{\pi} \int_{-\pi}^{\pi} f(x) \sin rx\, dx = 0, \quad \text{for all } r. \tag{24}$$

Hence if $f(x)$ is an even function its Fourier series reduces to a cosine series.

(b) $f(x)$ odd; $f(x) = -f(-x)$

$$a_0 = \frac{1}{\pi} \int_{-\pi}^{\pi} f(x)\, dx = 0, \tag{25}$$

$$a_r = \frac{1}{\pi} \int_{-\pi}^{\pi} f(x) \cos rx\, dx = 0 \quad \text{for all } r, \tag{26}$$

$$b_r = \frac{1}{\pi} \int_{-\pi}^{\pi} f(x) \sin rx\, dx = \frac{2}{\pi} \int_0^{\pi} f(x) \sin rx\, dx. \tag{27}$$

$$(r = 1, 2, 3 \dots).$$

Hence if $f(x)$ is an odd function its Fourier series reduces to a sine series.

15.3 Fourier Expansions

Example 1. Find the Fourier series which represents the non-periodic function $f(x) = x^2$ in the interval $-\pi \leqq x \leqq \pi$. First we extend $f(x)$ periodically (as required by the Dirichlet conditions) by letting $f(x+2k\pi) = f(x)$ (see Fig. 15.2). Then since $f(x) = x^2$ is an even function of x in $-\pi < x < \pi$ we have, by (22), (23) and (24),

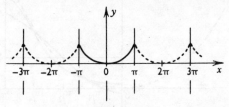

Fig. 15.2

$$a_0 = \frac{2}{\pi} \int_0^\pi x^2 \, dx = \frac{2\pi^2}{3}, \tag{28}$$

$$a_r = \frac{2}{\pi} \int_0^\pi x^2 \cos rx \, dx \quad (r = 1, 2, 3 \ldots) \tag{29}$$

$$= \frac{2}{\pi} \left\{ \left[\frac{x^2 \sin rx}{r} \right]_0^\pi - \frac{2}{r} \int_0^\pi x \sin rx \, dx \right\} \tag{30}$$

$$= \frac{4 \cos r\pi}{r^2} = \frac{4}{r^2} (-1)^r, \tag{31}$$

and

$$b_r = 0 \quad \text{for all } r. \tag{32}$$

Hence the required Fourier series is, by (6),

$$x^2 = \frac{\pi^2}{3} - 4 \left(\frac{\cos x}{1^2} - \frac{\cos 2x}{2^2} + \frac{\cos 3x}{3^2} \ldots \right) \tag{33}$$

$$= \frac{\pi^2}{3} + 4 \sum_{r=1}^\infty (-1)^r \frac{\cos rx}{r^2}. \tag{34}$$

The convergence of (34) to x^2 is best seen graphically by plotting its first few partial sums, say

$$S_1 = \frac{\pi^2}{3}, \quad S_2 = \frac{\pi^2}{3} - 4 \cos x, \quad S_3 = \frac{\pi^2}{3} - 4 \cos x + \cos 2x. \tag{35}$$

These are shown in Fig. 15.3.

259

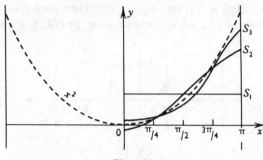

Fig. 15.3

Since $f(x)$ is continuous at all points in $-\pi \leqq x \leqq \pi$, (34) is valid everywhere in this interval. Outside this interval, however (34) (which is periodic in x as shown by Fig. 15.2) clearly does not represent x^2.

The sums of series can often be obtained from Fourier expansions by considering particular values of x within the interval of definition. For example, putting $x = \pm\pi$ in (34) we have

$$\pi^2 = \frac{\pi^2}{3} + 4\left(\frac{1}{1^2} + \frac{1}{2^2} + \frac{1}{3^2} + \dots + \frac{1}{r^2} + \dots\right) \tag{36}$$

or

$$\frac{\pi^2}{6} = 1 + \frac{1}{2^2} + \frac{1}{3^2} + \dots = \sum_{r=1}^{\infty} \frac{1}{r^2}. \tag{37}$$

Similarly, by putting $x = 0$, we find

$$\frac{\pi^2}{12} = 1 - \frac{1}{2^2} + \frac{1}{3^2} - \frac{1}{4^2} + \dots = \sum_{r=1}^{\infty} \frac{(-1)^{r+1}}{r^2}. \tag{38}$$

Example 2. Find the Fourier series which represents the function $f(x)$ defined by

$$f(x) = \begin{cases} -\cos x & \text{for} \quad -\pi \leqq x < 0 \\ \cos x & \text{for} \quad 0 < x \leqq \pi. \end{cases}$$

This function is periodic in x (see Fig. 15.4) and therefore already satisfies the periodicity relation $f(x+2k\pi) = f(x)$ as required by

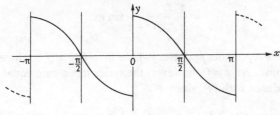

Fig. 15.4

Theorem 1. In addition it is an odd function of x in $-\pi \leqq x \leqq \pi$, and in this interval has a finite discontinuity at $x = 0$. Hence

$$a_0 = \frac{1}{\pi} \int_{-\pi}^{0} (-\cos x) \, dx + \frac{1}{\pi} \int_{0}^{\pi} \cos x \, dx = 0, \tag{39}$$

$$a_r = \frac{1}{\pi} \int_{-\pi}^{0} (-\cos x) \cos rx \, dx + \frac{1}{\pi} \int_{0}^{\pi} \cos x \cos rx \, dx = 0$$
$$\text{for all } r, \tag{40}$$

and

$$b_r = \frac{1}{\pi} \int_{-\pi}^{0} (-\cos x) \sin rx \, dx + \frac{1}{\pi} \int_{0}^{\pi} \cos x \sin rx \, dx \tag{41}$$

$$= \frac{1}{2\pi} \left[\frac{\cos (r+1)x}{r+1} + \frac{\cos (r-1)x}{r-1} \right]_{-\pi}^{0}$$

$$\qquad\qquad - \frac{1}{2\pi} \left[\frac{\cos (r+1)x}{r+1} + \frac{\cos (r-1)x}{r-1} \right]_{0}^{\pi} \tag{42}$$

$$= \frac{1}{\pi} \left[\frac{1-(-1)^{r+1}}{r+1} + \frac{1-(-1)^{r-1}}{r-1} \right] \tag{43}$$

For odd values of r, therefore, (43) gives $b_r = 0$, whilst for even values of r, (43) gives

$$b_r = \frac{4}{\pi} \left(\frac{r}{r^2 - 1} \right), \quad (r = 2, 4, 6 \ldots). \tag{44}$$

Hence the required Fourier series is

$$f(x) = \frac{8}{\pi} \left(\frac{1}{1.3} \sin 2x + \frac{2}{3.5} \sin 4x + \frac{3}{5.7} \sin 6x + \ldots \right) \tag{45}$$

$$= \frac{4}{\pi} \sum_{\substack{\text{even} \\ r}}^{\infty} \frac{r}{r^2 - 1} \sin rx. \qquad (46)$$

This series represents $f(x)$ everywhere in $-\pi \leqq x \leqq \pi$ where $f(x)$ is continuous. At $x = 0$, however, the series has a sum equal to zero in accordance with (7), since

$$\tfrac{1}{2} \lim_{\delta \to 0} \{f(x_0 + \delta) + f(x_0 - \delta)\} = \tfrac{1}{2} \lim_{\delta \to 0} \{\cos \delta - \cos(-\delta)\} \qquad (47)$$

$$= \tfrac{1}{2}(1 - 1) = 0.$$

As in the last example, we may use (46) to obtain the sum of a series. For by putting $x = \pi/4$ we have

$$f\left(\frac{\pi}{4}\right) = \cos\frac{\pi}{4} = \frac{1}{\sqrt{2}} = \frac{8}{\pi}\left(\frac{1}{1.3} - \frac{3}{5.7} + \frac{5}{9.11} - \dots\right) \qquad (48)$$

or

$$\frac{\pi\sqrt{2}}{16} = \frac{1}{1.3 \cdot 5.7} + \frac{5}{9.11} - \dots . \qquad (49)$$

Example 3. Find the Fourier series representing the function $f(x) = e^x$ in $-\pi < x < \pi$. As in Example 1 we first extend the function periodically (see Fig. 15.5) as required by Theorem 1.

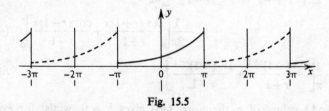

Fig. 15.5

Finite discontinuities then occur at $x = \pm\pi$. Evaluating the Fourier coefficients we have

$$a_0 = \frac{1}{\pi}\int_{-\pi}^{\pi} e^x \, dx = \frac{e^\pi - e^{-\pi}}{\pi}; \qquad (50)$$

$$a_r = \frac{1}{\pi}\int_{-\pi}^{\pi} e^x \cos rx \, dx = \frac{1}{\pi(1 + r^2)}\left[e^x(\cos rx + r \sin rx)\right]_{-\pi}^{\pi} \qquad (51)$$

$$= \frac{1}{\pi} \cdot \frac{(-1)^r}{1+r^2} (e^\pi - e^{-\pi}), \tag{52}$$

$$b_r = \frac{1}{\pi} \int_{-\pi}^{\pi} e^x \sin rx \, dx = \frac{1}{\pi(1+r^2)} \left[e^x (\sin rx - r \cos rx) \right]_{-\pi}^{\pi} \tag{53}$$

$$= \frac{1}{\pi} \cdot \frac{(-1)^{r+1} r}{1+r^2} (e^\pi - e^{-\pi}). \tag{54}$$

Hence in $-\pi < x < \pi$

$$e^x = \frac{\sinh \pi}{\pi} + \frac{2 \sinh \pi}{\pi} \sum_{r=1}^{\infty} \frac{(-1)^r \cos rx}{1+r^2}$$

$$+ \frac{2 \sinh \pi}{\pi} \sum_{r=1}^{\infty} \frac{(-1)^{r+1} r \sin rx}{1+r^2} \tag{55}$$

$$= \frac{2 \sinh \pi}{\pi} \left\{ \frac{1}{2} + \sum_{r=1}^{\infty} \frac{(-1)^r \cos rx}{1+r^2} + \sum_{r=1}^{\infty} \frac{(-1)^{r+1} r \sin rx}{1+r^2} \right\}. \tag{56}$$

At the points of finite discontinuity, $x = \pm \pi$, we know that the Fourier series must converge to

$$\tfrac{1}{2} \lim_{\delta \to 0} \{ e^{\pi+\delta} + e^{-\pi-\delta} \} = \cosh \pi, \tag{57}$$

(using the periodicity relation).

Hence

$$\cosh \pi = \frac{2 \sinh \pi}{\pi} \left\{ \frac{1}{2} + \sum_{r=1}^{\infty} \frac{(-1)^{2r}}{1+r^2} \right\} \tag{58}$$

or

$$\frac{\pi}{2} \coth \pi = \frac{1}{2} + \sum_{r=1}^{\infty} \frac{1}{1+r^2}. \tag{59}$$

15.4 Cosine and Sine Series

It is sometimes required to expand a function $f(x)$ in a Fourier series in the interval $0 \leq x \leq \pi$. This may be done using either a series of sine terms or a series of cosine terms, whichever may be required.

(a) Cosine Series

Suppose $f(x)$ is defined arbitrarily in the interval $0 \leq x \leq \pi$. We now define a function $F(x)$ such that

$$
\left.
\begin{aligned}
F(x) &\equiv f(x), \quad 0 \leq x \leq \pi, \\
F(x) &\equiv f(-x), \quad -\pi \leq x \leq 0,
\end{aligned}
\right\}
\tag{60}
$$

and

$$
F(x+2\pi k) = F(x).
$$

Then $F(x)$ is clearly an even periodic function of x in $-\pi \leq x \leq \pi$ and is identical with $f(x)$ in $0 \leq x \leq \pi$. Hence by (22)–(24) the Fourier series of $F(x)$ in $-\pi \leq x \leq \pi$ is the cosine series

$$
F(x) = \frac{a_0}{2} + a_1 \cos x + a_2 \cos 2x + \dots + a_r \cos rx + \dots, \tag{61}
$$

where

$$
a_0 = \frac{2}{\pi} \int_0^\pi F(x)\, dx \tag{62}
$$

$$
a_r = \frac{2}{\pi} \int_0^\pi F(x) \cos rx\, dx \quad (r = 1, 2, 3 \dots). \tag{63}
$$

But in $0 \leq x \leq \pi$, $F(x)$ is identical (by definition) with $f(x)$ and hence in this interval

$$
f(x) = \frac{a_0}{2} + a_1 \cos x + a_2 \cos 2x + \dots + a_r \cos rx + \dots, \tag{64}
$$

where

$$
a_0 = \frac{2}{\pi} \int_0^\pi f(x)\, dx, \tag{65}
$$

$$
a_r = \frac{2}{\pi} \int_0^\pi f(x) \cos rx\, dx \quad (r = 1, 2, 3 \dots). \tag{66}
$$

264

(*b*) *Sine Series*

Suppose now we define $F(x)$ such that

$$\left.\begin{array}{l} F(x) \equiv f(x) \quad \text{in} \quad 0 \leq x \leq \pi, \\[2mm] F(x) \equiv -f(-x) \quad \text{in} \quad -\pi < x < 0, \\[4mm] F(x+2\pi k) = F(x). \end{array}\right\} \qquad (67)$$

and

Then $F(x)$ is an odd periodic function of x in $-\pi < x \leq \pi$ and is identical with $f(x)$ in $0 \leq x \leq \pi$. Hence by (25)–(27) the Fourier series of $F(x)$ in $-\pi < x \leq \pi$ is now

$$F(x) = b_1 \sin x + b_2 \sin 2x + \ldots + b_r \sin rx + \ldots, \qquad (68)$$

where

$$b_r = \frac{2}{\pi} \int_0^\pi F(x) \sin rx \, dx \quad (r = 1, 2, 3 \ldots). \qquad (69)$$

But since $F(x)$ is identical with $f(x)$ in $0 \leq x \leq \pi$ we have the sine series

$$f(x) = b_1 \sin x + b_2 \sin 2x + \ldots + b_r \sin rx + \ldots, \qquad (70)$$

where

$$b_r = \frac{2}{\pi} \int_0^\pi f(x) \sin rx \, dx \quad (r = 1, 2, 3 \ldots). \qquad (71)$$

Example 4. Find a Fourier sine series to represent the function $f(x) = x^2$ in the interval $0 \leq x < \pi$.

Here, according to (*b*) above, $f(x)$ must be extended periodically by the odd function $F(x)$ where

$$\left.\begin{array}{l} F(x) = x^2, \quad 0 \leq x < \pi, \\[2mm] F(x) = -x^2, \quad -\pi < x \leq 0, \end{array}\right\} \qquad (72)$$

and $F(x+2\pi k) = F(x)$, (see Fig. 15.6).

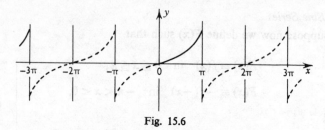

Fig. 15.6

Hence, by (70) and (71), the required series is

$$x^2 = b_1 \sin x + b_2 \sin 2x + \ldots + b_r \sin rx + \ldots, \qquad (73)$$

where

$$b_r = \frac{2}{\pi} \int_0^\pi x^2 \sin rx \, dx, \quad (r = 1, 2, 3 \ldots), \qquad (74)$$

$$= \frac{2\pi(-1)^{r+1}}{r} + \frac{4}{\pi r^3} \{(-1)^r - 1\}. \qquad (75)$$

Consequently the series becomes

$$x^2 = \frac{2}{\pi}\left\{\left(\frac{\pi^2}{1} - \frac{4}{1^3}\right)\sin x - \frac{\pi^2}{2}\sin 2x + \left(\frac{\pi^2}{3} - \frac{4}{3^3}\right)\sin 3x \right.$$

$$\left. - \frac{\pi^2}{4}\sin 4x + \left(\frac{\pi^2}{5} - \frac{4}{5^3}\right)\sin 5x \ldots\right\}. \qquad (76)$$

15.5 Change of Interval

Instead of expanding a function in the interval $-\pi < x \leqq \pi$ (or, for that matter, any interval of width 2π) it is sometimes desirable to obtain an expansion valid in the interval $-l < x \leqq l$, where l is a given number. Suppose $f(x)$ is a given piecewise regular function in $-l < x \leqq l$ which is defined outside this interval by the periodicity relation $f(x+2lk) = f(x)$, where k is an integer.

Then putting $z = \pi x/l$ we have

$$f(x) = f\left(\frac{lz}{\pi}\right) = F(z), \qquad (77)$$

where $F(z)$ is now a periodic function of z of period 2π. Hence in $-\pi < z \leqq \pi$

$$F(z) = \frac{a_0}{2} + \sum_{r=1}^{\infty} (a_r \cos rz + b_r \sin rz), \qquad (78)$$

where

$$a_r = \frac{1}{\pi} \int_{-\pi}^{\pi} F(z) \cos rz \, dz, \quad (r = 0, 1, 2 \dots) \qquad (79)$$

$$b_r = \frac{1}{\pi} \int_{-\pi}^{\pi} F(z) \sin rz \, dz, \quad (r = 1, 2, 3 \dots). \qquad (80)$$

Consequently putting $z = \pi x/l$ in these results we have

$$f(x) = \frac{a_0}{2} + \sum_{r=1}^{\infty} \left(a_r \cos\frac{\pi x r}{l} + b_r \sin\frac{\pi x r}{l} \right) \qquad (81)$$

where

$$a_r = \frac{1}{l} \int_{-l}^{l} f(x) \cos\frac{\pi x r}{l} \, dx, \quad (r = 0, 1, 2 \dots) \qquad (82)$$

and

$$b_r = \frac{1}{l} \int_{-l}^{l} f(x) \sin\frac{\pi x r}{l} \, dx, \quad (r = 1, 2, 3 \dots). \qquad (83)$$

A similar approach may be applied to the results of 15.4 to obtain the cosine and sine series for a function in the interval $0 \leqq x \leqq l$.

Example 5. Find the Fourier series representing the function $f(x) = x$ in the interval $-1 < x < 1$.

From (82) and (83)

$$a_r = \int_{-1}^{1} x \cos \pi rx \, dx = 0 \qquad (84)$$

for all $r(= 0, 1, 2 \dots)$, and

$$b_r = \int_{-1}^{1} x \sin \pi rx \, dx \qquad (85)$$

$$= -\frac{2}{\pi r}(-1)^r. \qquad (86)$$

Hence in $-1 < x < 1$

$$x = \frac{2}{\pi}\left(\sin \pi x - \frac{1}{2}\sin 2\pi x + \frac{1}{3}\sin 3\pi x - \dots\right), \qquad (87)$$

$$= -\frac{2}{\pi}\sum_{r=1}^{\infty}(-1)^r\frac{\sin \pi r x}{r}. \qquad (88)$$

At $x = \pm 1$, finite discontinuities occur (in virtue of the periodic extension of $f(x)$ by the relation $f(x+2) = f(x)$ (see Fig. 15.7).

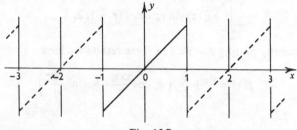

Fig. 15.7

Hence at these points the series does not represent x but converges to the value

$$\tfrac{1}{2}\lim_{\delta\to 0}\{f(x_0+\delta)+f(x_0-\delta)\} = \tfrac{1}{2}\{1+(-1)\} = 0. \qquad (89)$$

15.6 Integration and Differentiation of a Fourier Series

In Chapter 5, 5.11 we briefly mentioned some of the difficulties connected with term-by-term differentiation and integration of an infinite series of functions. We now state without proof an important theorem on the term-by-term integration of a Fourier series.

Theorem 2. The Fourier series of a function $f(x)$ may always be integrated term-by-term to give a new series which converges to the integral of $f(x)$.

In other words if we are considering an interval of definition $-\pi < x \leqq \pi$, say, then for $-\pi < x_1 < x_2 \leqq \pi$

$$\int_{x_1}^{x_2} f(x)\, dx = \int_{x_1}^{x_2}\frac{a_0}{2}\, dx + \int_{x_1}^{x_2}\sum_{r=1}^{\infty}(a_r\cos rx + b_r\sin rx)\, dx$$

$$= \frac{1}{2}a_0(x_2 - x_1) + \sum_{r=1}^{\infty}\int_{x_1}^{x_2}(a_r\cos rx + b_r\sin rx)\, dx, \qquad (90)$$

where a_0, a_r and b_r are defined by (3), (4) and (5). For example, since the Fourier expansion of $f(x) = x$ in the interval $-\pi < x < \pi$ is

$$x = 2(\sin x - \tfrac{1}{2} \sin 2x + \tfrac{1}{3} \sin 3x - \ldots)$$

$$= -2 \sum_{r=1}^{\infty} (-1)^r \frac{\sin rx}{r}, \tag{91}$$

we have, integrating term-by-term from 0 to x,

$$\frac{x^2}{2} = -2 \left[\cos x - \frac{1}{2^2} \cos 2x + \frac{1}{3^2} \cos 3x - \ldots \right]_0^x \tag{92}$$

or

$$\frac{x^2}{2} = 2 \sum_{r=1}^{\infty} (-1)^r \frac{\cos rx}{r^2} + 2 \left(1 - \frac{1}{2^2} + \frac{1}{3^2} - \frac{1}{4^2} + \ldots \right). \tag{93}$$

Now by (38)

$$1 - \frac{1}{2^2} + \frac{1}{3^2} - \frac{1}{4^2} + \ldots = \frac{\pi^2}{12}. \tag{94}$$

Hence (93) becomes

$$x^2 = \frac{\pi^2}{3} + 4 \sum_{r=1}^{\infty} (-1)^r \frac{\cos rx}{r^2}. \tag{95}$$

This result is easily seen to be correct since it is just the Fourier series of x^2 (see Example 1). In general, however, the term-by-term integration of a Fourier series does not produce another Fourier series owing to the presence of the $\tfrac{1}{2}a_0(x_2 - x_1)$ term in (90).

We now see that unlike term-by-term integration, term-by-term differentiation is not always permissible in the sense that the differentiated series does not necessarily converge to the differentiated function. For, although in $-\pi < x < \pi$

$$x = 2(\sin x - \tfrac{1}{2} \sin 2x + \tfrac{1}{3} \sin 3x \ldots), \tag{96}$$

it is not true that the differentiated series

$$2(\cos x - \cos 2x + \cos 3x - \ldots) \tag{97}$$

is equal to unity. In fact (97) diverges for all x since $\lim_{r \to \infty} \cos rx \neq 0$.

Nevertheless term-by-term differentiation of a Fourier series is permissible provided the conditions stated in the following theorem are satisfied.

Theorem 3. If $f(x)$ is a *continuous* function of x for all x, and is periodic (of period 2π) outside the interval $-\pi < x < \pi$, then term-by-term differentiation of the Fourier series of $f(x)$ leads to the Fourier series of $f'(x)$, provided $f'(x)$ satisfies Dirichlet's conditions.

To prove this theorem we suppose as usual that in $-\pi < x \leqq \pi$

$$f(x) = \frac{1}{2} a_0 + \sum_{r=1}^{\infty} (a_r \cos rx + b_r \sin rx), \tag{98}$$

a_0, a_r, b_r being defined by (3), (4) and (5). Now suppose the Fourier series of $f'(x)$ is

$$f'(x) = \frac{1}{2} \alpha_0 + \sum_{r=1}^{\infty} (\alpha_r \cos rx + \beta_r \sin rx). \tag{99}$$

Then

$$\alpha_0 = \frac{1}{\pi} \int_{-\pi}^{\pi} f'(x) \, dx = \frac{1}{\pi} \{ f(\pi) - f(-\pi) \} = 0 \tag{100}$$

(since $f(\pi) = f(-\pi)$),

$$\alpha_r = \frac{1}{\pi} \int_{-\pi}^{\pi} f'(x) \cos rx \, dx = \frac{r}{\pi} \int_{-\pi}^{\pi} f(x) \sin rx \, dx = rb_r, \tag{101}$$

$$(r = 1, 2, 3 \ldots)$$

and

$$\beta_r = \frac{1}{\pi} \int_{-\pi}^{\pi} f'(x) \sin rx \, dx = -\frac{r}{\pi} \int_{-\pi}^{\pi} f(x) \cos rx \, dx = -ra_r, \tag{102}$$

$$(r = 1, 2, 3 \ldots).$$

Consequently (99) becomes

$$f'(x) = \sum_{r=1}^{\infty} (-ra_r \sin rx + rb_r \cos rx), \tag{103}$$

which is just the series which would be obtained by differentiating (98) term-by-term.

According to this theorem we now see that since $f(x) = x^2$ (when extended periodically) is a continuous function of x for all x (see Fig. 15.2) the Fourier series of x^2 may be differentiated term-by-term

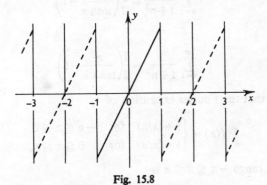

Fig. 15.8

to give the Fourier series of $2x$. However, since $f(x) = x$ (when extended periodically) is not a continuous function of x for all x (see Fig. 15.8) its Fourier series (as we have noticed already in (96) and (97)) cannot be differentiated term-by-term to give the Fourier series of its derivative.

PROBLEMS 15

1. Show that
$$|\sin x| = \frac{2}{\pi} - \frac{4}{\pi} \sum_{\substack{r \text{ even} \\ r}}^{\infty} \frac{\cos rx}{r^2 - 1}.$$

2. Given that
$$f(x) = \begin{cases} -1 & \text{for } -\pi < x < 0, \\ 0 & \text{for } x = 0, \\ +1 & \text{for } 0 < x < \pi, \end{cases}$$

show that in $-\pi < x < \pi$
$$f(x) = \frac{4}{\pi}\left(\sin x + \frac{1}{3}\sin 3x + \frac{1}{5}\sin 5x + \ldots\right)$$

271

3. Find the Fourier series of period 2π which gives a function equal to 0 for $-\pi < x < 0$ and equal to $\cosh x$ for $0 < x < \pi$. By considering the values of the series at the points of discontinuity, deduce that

$$\sum_{n=1}^{\infty} \frac{(-1)^n}{1+n^2} = \frac{1}{2}\left(\frac{\pi}{\sinh \pi} - 1\right),$$

and

$$\sum_{n=1}^{\infty} \frac{1}{1+n^2} = \frac{1}{2}\left(\frac{\pi}{\tanh \pi} - 1\right). \qquad \text{(C.U.)}$$

4. Show that the Fourier expansion of

$$f(x) = \begin{cases} 1+(x/\pi) & \text{for} \quad -\pi \leqq x \leqq 0, \\ 1-(x/\pi) & \text{for} \quad 0 \leqq x \leqq \pi, \end{cases}$$

in the range $-\pi \leqq x \leqq \pi$ is

$$f(x) = \frac{1}{2} + \frac{4}{\pi^2}\left(\cos x + \frac{1}{3^2}\cos 3x + \frac{1}{5^2}\cos 5x + \ldots\right).$$

Hence deduce that

$$\frac{\pi^2}{8} = 1 + \frac{1}{3^2} + \frac{1}{5^2} + \ldots = \sum_{r=1}^{\infty} \frac{1}{(2r-1)^2}.$$

5. Show that in the range $-\pi < x < \pi$

$$\sinh kx = \frac{2}{\pi}\sinh k\pi \sum_{r=1}^{\infty} (-1)^{r-1}\left(\frac{r}{r^2+k^2}\right)\sin rx,$$

where k is a constant.

6. Show that, if $-\pi \leqq x \leqq \pi$, and k is non-integral,

$$\cos kx = \frac{2k}{\pi}\sin k\pi \left\{\frac{1}{2k^2} + \sum_{r=1}^{\infty} (-1)^{r-1}\frac{\cos rx}{(r^2-k^2)}\right\}.$$

By putting $x = \pi$, and $k\pi = \theta$, show that

$$\cot \theta = \frac{1}{\theta} - \sum_{r=1}^{\infty} \frac{2\theta}{r^2\pi^2-\theta^2}.$$

7. Express the function $f(x) = 1 + x$ as a sine series valid in the range $0 < x \le \pi$.

8. If, in $0 < x < \pi$, $f(x)$ is defined by

$$f(x) = \begin{cases} 1, & 0 < x < \pi/2, \\ 0, & x = \pi/2, \\ -1, & \pi/2 < x < \pi, \end{cases}$$

show that

$$f(x) = \frac{4}{\pi}\left(\cos x - \frac{1}{3}\cos 3x + \frac{1}{5}\cos 5x - \dots\right).$$

9. Expand the function

$$f(x) = \begin{cases} x, & 0 \le x \le \pi/2, \\ \pi/2, & \pi/2 \le x < \pi, \end{cases}$$

in (i) a sine series, and (ii) a cosine series valid in $0 \le x < \pi$. Sketch the graphs of the functions represented by these series in the range $-2\pi < x < 2\pi$. (L.U.)

10. Show that for $0 \le \dot{x} \le \pi$

$$\sin x = \frac{4}{\pi}\left\{\frac{1}{2} - \frac{1}{1.3}\cos 2x - \frac{1}{3.5}\cos 4x - \frac{1}{5.7}\cos 6x - \dots\right\}.$$

Sketch the curve which the series represents for values of x from $-\pi/2$ to $3\pi/2$.
Show that

$$\frac{1}{1.3} - \frac{1}{3.5} + \frac{1}{5.7} - \dots = \frac{\pi}{4} - \frac{1}{2}. \qquad \text{(L.U.)}$$

11. If $f(x)$ is an odd function of x, of period 2π, and

$$f(x) = \begin{cases} x, & 0 \le x \le \pi/3; \\ \pi/3, & \pi/3 \le x \le 2\pi/3, \\ \pi - x, & 2\pi/3 \le x \le \pi, \end{cases}$$

show that the Fourier series of $f(x)$ may be expressed in the form

$$\frac{4}{\pi} \sum_{n=1}^{\infty} \frac{1}{n^2} \sin \frac{n\pi}{3} \sin nx,$$

where n is an odd positive integer, and determine the first three non-zero terms. (L.U.)

12. Find a Fourier sine series for $f(x)$ over the range $0 \leq x \leq \pi$ when $f(x) = x$, $0 \leq x \leq \pi/2$ and $f(x) = \pi - x$, $\pi/2 \leq x \leq \pi$. Draw a sketch to indicate the form of the function represented by the series in the range $-2\pi \leq x \leq 2\pi$. (L.U.)

13. If $f(x)$ is defined arbitrarily in $0 < x < \pi$ and extended into $-\pi < x < \pi$ by the function $F(x)$, where

$$F(x) = \begin{cases} f(x), & 0 < x < \pi, \\ -f(x+\pi), & -\pi < x < 0, \end{cases}$$

show that the Fourier series of $f(x)$ in $0 < x < \pi$ contains (apart from a constant term) only cosines and sines of odd multiples of x.

14. The function $y(x)$ is defined in the range $-1 \leq x \leq 1$ by

$$y = \begin{cases} 1/2\varepsilon & \text{in} \quad -\varepsilon < x < \varepsilon, \\ 0 & \text{in} \quad -1 \leq x < -\varepsilon, \text{ and } \varepsilon < x \leq 1. \end{cases}$$

Show that the Fourier expansion of y in the range $-1 \leq x \leq 1$ is given by

$$y = \frac{1}{2} + \sum_{n=1}^{\infty} \frac{\sin n\pi\varepsilon}{n\pi\varepsilon} \cos n\pi x. \qquad \text{(C.U.)}$$

15. Show that the general term in the Fourier half-range sine series for $1 + (x/l)$, $(0 < x < l)$ is

$$\frac{2}{n\pi}(1 - 2\cos n\pi) \sin \frac{n\pi x}{l}.$$

By considering the function of x represented by this series in the range $l < x < 2l$, and the sum of the series at $x = 3l/2$, prove that

$$\frac{\pi}{4} = 1 - \frac{1}{3} + \frac{1}{5} - \frac{1}{7} + \dots .$$

(L.U.)

16. Any set of functions $\phi_1(x)$, $\phi_2(x)$, ..., $\phi_n(x)$, ... having the property

$$\int_a^b \phi_r(x)\phi_s(x)\, dx = \begin{cases} A, & \text{for } r = s, \\ 0, & \text{for } r \neq s, \end{cases}$$

where $r, s = 1, 2, \dots$, and A is a constant, is called an orthogonal set of functions in the range $a \leq x \leq b$. If, in addition, the functions satisfy the condition

$$\int_a^b [\phi_r(x)]^2\, dx = 1, \quad \text{for all } r.$$

then they are said to be normalised to unity, and the set is termed an orthonormal set. Verify that the set of functions

$$\frac{1}{\sqrt{2\pi}}, \frac{1}{\sqrt{\pi}} \cos x, \frac{1}{\sqrt{\pi}} \sin x, \dots, \frac{1}{\sqrt{\pi}} \cos nx, \frac{1}{\sqrt{\pi}} \sin nx, \dots$$

is orthonormal in the range $0 \leq x \leq 2\pi$.

CHAPTER 16

Determinants

16.1 Definitions

Determinants arise naturally in the solution of a set of linear equations. Suppose, for example, we have the equations

$$a_1x + b_1y = c_1,$$
$$a_2x + b_2y = c_2,$$
(1)

where x and y are to be found, and a_1, b_1, c_1, a_2, b_2, c_2 are given constants. Solving for x and y in the usual way we find

$$x = \frac{c_1b_2 - c_2b_1}{a_1b_2 - a_2b_1},$$
(2)

$$y = \frac{a_1c_2 - a_2c_1}{a_1b_2 - a_2b_1},$$
(3)

provided $a_1b_2 - a_2b_1 \neq 0$. If we now define the symbol

$$D_2 = \begin{vmatrix} a_1 & b_1 \\ a_2 & b_2 \end{vmatrix} \equiv a_1b_2 - a_2b_1,$$
(4)

then these solutions may be written as

$$\frac{x}{\begin{vmatrix} c_1 & b_1 \\ c_2 & b_2 \end{vmatrix}} = \frac{y}{\begin{vmatrix} a_1 & c_1 \\ a_2 & c_2 \end{vmatrix}} = \frac{1}{\begin{vmatrix} a_1 & b_1 \\ a_2 & b_2 \end{vmatrix}}.$$
(5)

It is usual to call the symbol defined by (4) a determinant and to speak of it as being of the second order since it has two rows and two columns. The diagonal containing the numbers a_1 and b_2 is called the leading diagonal of the determinant. As an example of a second order determinant we give

$$D_2 = \begin{vmatrix} 6 & 2 \\ 3 & 4 \end{vmatrix} = 6.4 - 3.2 = 18.$$
(6)

We now consider the set of three linear equations

$$\left.\begin{aligned} a_1x + b_1y + c_1z &= d_1, \\ a_2x + b_2y + c_2z &= d_2, \\ a_3x + b_3y + c_3z &= d_3, \end{aligned}\right\} \tag{7}$$

where x, y and z are unknown and $a_1, b_1 \ldots c_3, d_3$ are given constants Then

$$x = \frac{d_1b_2c_3 - d_1b_3c_2 + d_2b_3c_1 - d_2b_1c_3 + d_3b_1c_2 - d_3b_2c_1}{a_1b_2c_3 - a_1b_3c_2 + a_2b_3c_1 - a_2b_1c_3 + a_3b_1c_2 - a_3b_2c_1} \tag{8}$$

with similar expressions for y and z.

Here again if we define

$$D_3 = \begin{vmatrix} a_1 & b_1 & c_1 \\ a_2 & b_2 & c_2 \\ a_3 & b_3 & c_3 \end{vmatrix} = \begin{aligned} & a_1b_2c_3 - a_1b_3c_2 + a_2b_3c_1 - a_2b_1c_3 \\ & \quad + a_3b_1c_2 - a_3b_2c_1, \end{aligned} \tag{9}$$

then the solutions of (7) may be written as

$$\frac{x}{\begin{vmatrix} d_1 & b_1 & c_1 \\ d_2 & b_2 & c_2 \\ d_3 & b_3 & c_3 \end{vmatrix}} = \frac{y}{\begin{vmatrix} a_1 & d_1 & c_1 \\ a_2 & d_2 & c_2 \\ a_3 & d_3 & c_3 \end{vmatrix}} = \frac{z}{\begin{vmatrix} a_1 & b_1 & d_1 \\ a_2 & b_2 & d_2 \\ a_3 & b_3 & d_3 \end{vmatrix}} = \frac{1}{\begin{vmatrix} a_1 & b_1 & c_1 \\ a_2 & b_2 & c_2 \\ a_3 & b_3 & c_3 \end{vmatrix}}$$

$$\tag{10}$$

The symbol defined by (9) is called a determinant of the third order in that it has three rows and three columns. From this definition it is easily seen that

$$D_3 = a_1 \begin{vmatrix} b_2 & c_2 \\ b_3 & c_3 \end{vmatrix} - a_2 \begin{vmatrix} b_1 & c_1 \\ b_3 & c_3 \end{vmatrix} + a_3 \begin{vmatrix} b_1 & c_1 \\ b_2 & c_2 \end{vmatrix} \tag{11}$$

$$= a_1\alpha_1 - a_2\alpha_2 + a_3\alpha_3, \tag{12}$$

where α_1, α_2 and α_3 are the second order determinants left in (9) when the row and column containing a_1, a_2, a_3, respectively, are deleted. These determinants are called the minors of a_1, a_2 and a_3. Similarly we have

$$D_3 = -b_1\beta_1 + b_2\beta_2 - b_3\beta_3, \tag{13}$$

and

$$D_3 = c_1\gamma_1 - c_2\gamma_2 + c_3\gamma_3, \tag{14}$$

where β_1, β_2, β_3 are respectively the minors of b_1, b_2 and b_3, and γ_1, γ_2, γ_3 are respectively the minors of c_1, c_2 and c_3.

Likewise comparing with (9) we also have

$$D_3 = a_1\alpha_1 - b_1\beta_1 + c_1\gamma_1, \tag{15}$$

$$D_3 = -a_2\alpha_2 + b_2\beta_2 - c_2\gamma_2, \tag{16}$$

and

$$D_3 = a_3\alpha_3 - b_3\beta_3 + c_3\gamma_3. \tag{17}$$

We see from equations (12)–(17) therefore that D_3 may be expanded down any of its columns or along any of its rows by suitably combining products of elements and their minors. The sign to be attached to each product may be easily found from the rule of alternating signs

$$\begin{vmatrix} + & - & + \\ - & + & - \\ + & - & + \end{vmatrix} \tag{18}$$

For example, expanding

$$D_3 = \begin{vmatrix} 1 & 4 & 7 \\ 2 & 5 & 8 \\ 3 & 6 & 9 \end{vmatrix} \tag{19}$$

along the second row we have

$$D_3 = -2 \begin{vmatrix} 4 & 7 \\ 6 & 9 \end{vmatrix} + 5 \begin{vmatrix} 1 & 7 \\ 3 & 9 \end{vmatrix} - 8 \begin{vmatrix} 1 & 4 \\ 3 & 6 \end{vmatrix} \tag{20}$$

$$= 12 + (-60) - (-48) = 0. \tag{21}$$

Similarly, expanding down the third column

$$D_3 = 7 \begin{vmatrix} 2 & 5 \\ 3 & 6 \end{vmatrix} - 8 \begin{vmatrix} 1 & 4 \\ 3 & 6 \end{vmatrix} + 9 \begin{vmatrix} 1 & 4 \\ 2 & 5 \end{vmatrix} \tag{22}$$

$$= -21 - (-48) + (-27) = 0. \tag{23}$$

Now, since by (12) we have

$$D_3 = \sum_{r=1}^{3} (-1)^{r+1} a_r \alpha_r,\tag{24}$$

it is possible to conceive of an nth order determinant associated with the solution of a set of n linear equations in n unknowns and denoted by the square array

$$D_n = \begin{vmatrix} a_1 b_1 & \ldots & w_1 \\ a_2 b_2 & \ldots & w_2 \\ \cdot & & \cdot \\ \cdot & & \cdot \\ \cdot & & \cdot \\ a_n b_n & \ldots & w_n \end{vmatrix}\tag{25}$$

such that when expanded down the first column

$$D_n = \sum_{r=1}^{n} (-1)^{r+1} a_r \alpha_r.\tag{26}$$

This expression is an inductive definition of an nth order determinant and reproduces (with $n = 2$ and 3, respectively) the expansions down the first columns of the second and third order determinants already given in (4) and (12). However, (26) is by no means the only way of defining the general nth order determinant, but it is the most convenient one for our purposes here. Determinants of order $n > 3$ will be discussed again in a later section of this chapter.

16.2 Properties of Determinants

It is important to note that the following statements apply to all finite determinants whatever their order. However, we shall verify them only for determinants of order three.

(a) The value of a determinant is unaltered by interchanging the elements of all corresponding rows and columns.

For if

$$D = \begin{vmatrix} a_1 & b_1 & c_1 \\ a_2 & b_2 & c_2 \\ a_3 & b_3 & c_3 \end{vmatrix}\tag{27}$$

then the determinant $\bar{D}$ formed from D by interchanging the roles of rows and columns is given by

$$\bar{D} = \begin{vmatrix} a_1 & a_2 & a_3 \\ b_1 & b_2 & b_3 \\ c_1 & c_2 & c_3 \end{vmatrix} \tag{28}$$

$$= a_1(b_2c_3 - b_3c_2) - b_1(a_2c_3 - a_3c_2) + c_1(a_2b_3 - a_3b_2) \tag{29}$$

$$= a_1(b_2c_3 - b_3c_2) - a_2(b_1c_3 - b_3c_1) + a_3(b_1c_2 - b_2c_1) \tag{30}$$

$$= \begin{vmatrix} a_1 & b_1 & c_1 \\ a_2 & b_2 & c_2 \\ a_3 & b_3 & c_3 \end{vmatrix} = D. \tag{31}$$

This is illustrated by the following example.

Example 1.

$$\begin{vmatrix} 1 & 2 & 3 \\ 1 & 2 & 1 \\ 1 & 3 & 4 \end{vmatrix} = \begin{vmatrix} 1 & 1 & 1 \\ 2 & 2 & 3 \\ 3 & 1 & 4 \end{vmatrix} = 2. \tag{32}$$

Since the rows and columns of a determinant are interchangeable, any statement which is true for the rows is also true for the columns, and vice-versa.

(b) The sign of a determinant is reversed by interchanging any two of its rows (or columns).

For if

$$D = \begin{vmatrix} a_1 & b_1 & c_1 \\ a_2 & b_2 & c_2 \\ a_3 & b_3 & c_3 \end{vmatrix} \tag{33}$$

then the determinant D' formed by interchanging the first and second rows is

$$D' = \begin{vmatrix} a_2 & b_2 & c_2 \\ a_1 & b_1 & c_1 \\ a_3 & b_3 & c_3 \end{vmatrix} \tag{34}$$

$$= a_2(b_1c_3 - b_3c_1) - a_1(b_2c_3 - b_3c_2) + a_3(b_2c_1 - b_1c_2) \tag{35}$$

$$= -a_1(b_2c_3 - b_3c_2) + a_2(b_1c_3 - b_3c_1) - a_3(b_1c_2 - b_2c_1) \quad (36)$$

$$= - \begin{vmatrix} a_1 & b_1 & c_1 \\ a_2 & b_2 & c_2 \\ a_3 & b_3 & c_3 \end{vmatrix} = -D. \quad (37)$$

Similar arguments apply to the interchange of other rows (or columns).

Example 2.

$$\begin{vmatrix} \sin x & x & 1 \\ 0 & 2 & 3 \\ 1 & 2 & 4 \end{vmatrix} = - \begin{vmatrix} 0 & 2 & 3 \\ \sin x & x & 1 \\ 1 & 2 & 4 \end{vmatrix} \quad (38)$$

(c) The value of a determinant is zero if any two of its rows (or columns) are identical.

This is easily proved using the previous result. If two identical rows (or columns) are interchanged then according to (b) the determinant changes sign. But clearly the interchange of two identical rows (or columns) can make no difference to the determinant. Hence

$$D = -D, \quad \text{i.e. } D = 0. \quad (39)$$

Example 3.

$$\begin{vmatrix} 1 & 2 & 4 \\ 1 & 2 & 4 \\ x & y & z \end{vmatrix} = 0 \quad (40)$$

for all x, y, z. This is easily verified by expansion.

(d) If all the elements of any one row (or column) are multiplied by a common factor, the value of the determinant is multiplied by this factor.

Suppose

$$D = \begin{vmatrix} a_1 & b_1 & c_1 \\ a_2 & b_2 & c_2 \\ a_3 & b_3 & c_3 \end{vmatrix} \quad (41)$$

and

$$E = \begin{vmatrix} \lambda a_1 & b_1 & c_1 \\ \lambda a_2 & b_2 & c_2 \\ \lambda a_3 & b_3 & c_3 \end{vmatrix}. \tag{42}$$

Then by expansion

$$E = \lambda a_1(b_2 c_3 - b_3 c_2) - \lambda a_2(b_1 c_3 - b_3 c_1) + \lambda a_3(b_1 c_2 - b_2 c_1) \tag{43}$$

$$= \lambda \{ a_1(b_2 c_3 - b_3 c_2) - a_2(b_1 c_3 - b_3 c_1) + a_3(b_1 c_2 - b_2 c_1) \} \tag{44}$$

$$= \lambda D. \tag{45}$$

Example 4.

$$\begin{vmatrix} 1 & 2 & 1 \\ 2\lambda & 3\lambda & 4\lambda \\ 3 & 3 & 2 \end{vmatrix} = \lambda \begin{vmatrix} 1 & 2 & 1 \\ 2 & 3 & 4 \\ 3 & 3 & 2 \end{vmatrix}. \tag{46}$$

(e) If any two rows (or columns) of a determinant have proportional elements, the value of the determinant is zero.

This result follows directly from (c) and (d) for if corresponding elements of the first and second are in the ratio $\lambda : 1$ then

$$\begin{vmatrix} \lambda a_1 & a_1 & c_1 \\ \lambda a_2 & a_2 & c_2 \\ \lambda a_3 & a_3 & c_3 \end{vmatrix} = \lambda \begin{vmatrix} a_1 & a_1 & c_1 \\ a_2 & a_2 & c_2 \\ a_3 & a_3 & c_3 \end{vmatrix} = 0. \tag{47}$$

Example 5.

$$\begin{vmatrix} 1 & -1 & 3 \\ 2 & -2 & 4 \\ 3 & -3 & 5 \end{vmatrix} = 0, \tag{48}$$

since corresponding elements of the first two columns are proportional. Similarly,

$$\begin{vmatrix} \cos y & \sin x \cos y & 2 \cos y \\ 1 & \sin x & 2 \\ 3 & x & y \end{vmatrix} = 0, \tag{49}$$

since corresponding elements of the first two rows are proportional.

(*f*) If the elements of any row (or column) are the sums or differences of two or more terms, the determinant may be written as the sum or difference of two or more determinants.

This statement is easily verified for a third order determinant by direct expansion. We find, for example, that

$$\begin{vmatrix} a_1+x & b_1 & c_1 \\ a_2+y & b_2 & c_2 \\ a_3+z & b_3 & c_3 \end{vmatrix} = \begin{vmatrix} a_1 & b_1 & c_1 \\ a_2 & b_2 & c_2 \\ a_3 & b_3 & c_3 \end{vmatrix} + \begin{vmatrix} x & b_1 & c_1 \\ y & b_2 & c_2 \\ z & b_3 & c_3 \end{vmatrix} \qquad (50)$$

(*g*) The value of a determinant is unchanged if equal multiples of the elements of any row (or column) are added to the corresponding elements of any other row (or column).

Suppose to the elements of the first column of the determinant

$$D = \begin{vmatrix} a_1 & b_1 & c_1 \\ a_2 & b_2 & c_2 \\ a_3 & b_3 & c_3 \end{vmatrix} \qquad (51)$$

we add a constant multiple (λ) of the elements of the second column. If we denote the resulting determinant by E, we have, by (*f*),

$$E = \begin{vmatrix} a_1+\lambda b_1 & b_1 & c_1 \\ a_2+\lambda b_2 & b_2 & c_2 \\ a_3+\lambda b_3 & b_3 & c_3 \end{vmatrix} = \begin{vmatrix} a_1 & b_1 & c_1 \\ a_2 & b_2 & c_2 \\ a_3 & b_3 & c_3 \end{vmatrix} + \begin{vmatrix} \lambda b_1 & b_1 & c_1 \\ \lambda b_2 & b_2 & c_2 \\ \lambda b_3 & b_3 & c_3 \end{vmatrix} \qquad (52)$$

But by (*e*) the last determinant in (52) is zero since two of its columns have corresponding elements proportional. Hence $E = D$.

Example 6. If to the second row of the determinant

$$D = \begin{vmatrix} x & y & z \\ 1-2x & 2-2y & 3-2z \\ 2 & 3 & 4 \end{vmatrix} \qquad (53)$$

we add twice the first row, then

$$D = \begin{vmatrix} x & y & z \\ 1 & 2 & 3 \\ 2 & 3 & 4 \end{vmatrix} \qquad (54)$$

16.3 Evaluation of Determinants

The results of the last section may be used in the evaluation of determinants as shown by the following examples.

Example 7. To evaluate

$$D = \begin{vmatrix} 3 & 4 & 1 \\ 2 & 5 & 2 \\ 4 & 1 & 3 \end{vmatrix} \tag{55}$$

we first subtract the third column from the first column to give

$$D = \begin{vmatrix} 2 & 4 & 1 \\ 0 & 5 & 2 \\ 1 & 1 & 3 \end{vmatrix} \tag{56}$$

and then subtract twice the third row from the first row in (56) to give

$$D = \begin{vmatrix} 0 & 2 & -5 \\ 0 & 5 & 2 \\ 1 & 1 & 3 \end{vmatrix} = 1 \begin{vmatrix} 2 & -5 \\ 5 & 2 \end{vmatrix} = 29. \tag{57}$$

Example 8. To factorise

$$D = \begin{vmatrix} 1 & 1 & 1 \\ \lambda & \mu & v \\ \lambda^2 & \mu^2 & v^2 \end{vmatrix} \tag{58}$$

we first subtract the third column from the second column, and then subtract the first column from the third column. In this way we find

$$D = \begin{vmatrix} 1 & 0 & 0 \\ \lambda & \mu - v & v - \lambda \\ \lambda^2 & \mu^2 - v^2 & v^2 - \lambda^2 \end{vmatrix} \tag{59}$$

$$= (\mu - v)(v^2 - \lambda^2) - (v - \lambda)(\mu^2 - v^2) \tag{60}$$

$$= (\mu - v)(v - \lambda)(\lambda - \mu). \tag{61}$$

Alternatively we notice that by putting $\lambda = \mu$ two columns of D become identical. Hence $\lambda - \mu$ must be a factor. Similarly $\mu - v$ and $v - \lambda$ can be seen to be factors. Consequently, since D is a second degree expression in λ, the product of these three factors (which is also of second degree in λ) must represent D uniquely (apart from a possible numerical factor which is easily seen to be unity). Hence we again obtain (61).

Example 9. To show that

$$D = \begin{vmatrix} bc & a & a^2 \\ ca & b & b^2 \\ ab & c & c^2 \end{vmatrix} = \begin{vmatrix} 1 & a^2 & a^3 \\ 1 & b^2 & b^3 \\ 1 & c^2 & c^3 \end{vmatrix} \tag{62}$$

we multiply the first row of the first determinant by a, the second row by b, and the third row by c. Hence

$$D = \frac{1}{abc} \begin{vmatrix} abc & a^2 & a^3 \\ bca & b^2 & b^3 \\ cab & c^2 & c^3 \end{vmatrix} = \begin{vmatrix} 1 & a^2 & a^3 \\ 1 & b^2 & b^3 \\ 1 & c^2 & c^3 \end{vmatrix} \tag{63}$$

as required.

16.4 Alternative Notation

When dealing with determinants of order $n \geq 4$ it is usually convenient to denote the elements by a common letter with two suffices attached such that a_{ik} (say) denotes the element in the ith row and kth column. The general nth order determinant then takes the form

$$D_n = \begin{vmatrix} a_{11} & a_{12} & a_{13} \dots a_{1n} \\ a_{21} & a_{22} & \dots\dots a_{2n} \\ \cdot & \cdot & \cdot \\ \cdot & \cdot & \cdot \\ \cdot & \cdot & \cdot \\ a_{n1} & a_{n2} & \dots\dots a_{nn} \end{vmatrix} \cdot \tag{64}$$

If we now denote the minor of a typical element a_{ik} by α_{ik} then expanding down the first column of (64) we have (using the rule of alternating signs)

$$D_n = a_{11}\alpha_{11} - a_{21}\alpha_{21} + a_{31}\alpha_{31} \dots + (-1)^{n+1}a_{n1}\alpha_{n1} \tag{65}$$

$$= \sum_{r=1}^{n} (-1)^{r+1}a_{r1}\alpha_{r1}, \tag{66}$$

which is just an alternative form of (26).

In general

$$D_n = \sum_{r=1}^{n} (-1)^{r+s} a_{rs} \alpha_{rs}, \tag{67}$$

where s represents the column down which the expansion is carried out.

Similarly, expanding along the rth row we have

$$D_n = \sum_{s=1}^{n} (-1)^{r+s} a_{rs} \alpha_{rs}. \tag{68}$$

It is usually convenient to define the cofactor A_{ik} of the element a_{ik} as its minor with the correct sign attached such that

$$A_{ik} = (-1)^{i+k} \alpha_{ik}, \tag{69}$$

and to write (67) and (68) respectively as

$$D_n = \sum_{r=1}^{n} a_{rs} A_{rs} \quad (s \text{ being fixed}) \tag{70}$$

and

$$D_n = \sum_{s=1}^{n} a_{rs} A_{rs} \quad (r \text{ being fixed}). \tag{71}$$

An important result that can now be easily deduced is

$$\sum_{r=1}^{n} a_{rs} A_{rp} = 0 \quad (s \neq p). \tag{72}$$

Each term of this summation represents the product of an element in the sth column and the corresponding cofactor in the pth column. But this is just the expansion of a determinant whose sth and pth columns are identical. By 16.2 (c) the value of such a determinant is zero, and hence (72) is proved. Similarly we may show that

$$\sum_{s=1}^{n} a_{rs} A_{ps} = 0 \quad (r \neq p). \tag{73}$$

Example 10. Consider the third order determinant of Example 7.

$$D = \begin{vmatrix} 3 & 4 & 1 \\ 2 & 5 & 2 \\ 4 & 1 & 3 \end{vmatrix} = 29. \tag{74}$$

Then, for example,

$$A_{11} = (-1)^{1+1} (5.3 - 1.2) = 13, \tag{75}$$

$$A_{21} = (-1)^{2+1} (4.3 - 1.1) = -11, \tag{76}$$

$$A_{31} = (-1)^{3+1} (4.2 - 5.1) = 3, \tag{77}$$

$$A_{22} = (-1)^{2+2} (3.3 - 4.1) = 5, \tag{78}$$

$$A_{23} = (-1)^{2+3} (3.1 - 4.4) = 13, \tag{79}$$

and so on.

Hence by (70) (with $s = 1$)

$$D_3 = \sum_{r=1}^{3} a_{r1} A_{r1} = 3.13 + 2(-11) + 4.3 = 29 \tag{80}$$

as before.

To verify (73) we write

$$\sum_{s=1}^{3} a_{rs} A_{ps} = a_{r1} A_{p1} + a_{r2} A_{p2} + a_{r3} A_{p3} \tag{81}$$

and choose some value of r (1, 2 or 3) not equal to p, say $r = 1$, $p = 2$.

Then (81) becomes

$$a_{11} A_{21} + a_{12} A_{22} + a_{13} A_{23} = 3(-11) + 4.5 + 1.13 = 0 \tag{82}$$

as required.

16.5 Symmetric and Skew-symmetric Determinants

Suppose D_3 is an arbitrary third order determinant given by

$$D_3 = \begin{vmatrix} a_{11} & a_{12} & a_{13} \\ a_{21} & a_{22} & a_{23} \\ a_{31} & a_{32} & a_{33} \end{vmatrix}. \tag{83}$$

If the elements are such that

$$a_{ik} = a_{kl} \quad \text{for all } i, k \tag{84}$$

the determinant is said to be symmetric. Since (84) implies $a_{12} = a_{21}$, $a_{31} = a_{13}$, and so on, the elements above the leading diagonal are just the mirror images of those below. For example, the determinant

$$D_3 = \begin{vmatrix} 1 & x & y \\ x & 2 & z \\ y & z & 3 \end{vmatrix} \tag{85}$$

is symmetric.

If, however, the elements of (83) are such that

$$a_{ik} = -a_{ki} \quad \text{for all } i, k \tag{86}$$

the determinant is said to be skew-symmetric. The leading diagonals of such determinants always have zero elements since according to (86)

$$a_{11} = -a_{11}, \quad \text{or} \quad a_{11} = 0, \quad \text{etc.} \tag{87}$$

For example, the determinant

$$D_3 = \begin{vmatrix} 0 & -a & -b \\ a & 0 & -c \\ b & c & 0 \end{vmatrix} \tag{88}$$

is skew-symmetric.

It is easily seen that skew-symmetric determinants of an odd order are identically zero, for by interchanging the rows and columns in (88) we have

$$D_3 = \begin{vmatrix} 0 & -a & -b \\ a & 0 & -c \\ b & c & 0 \end{vmatrix} = \begin{vmatrix} 0 & a & b \\ -a & 0 & c \\ -b & -c & 0 \end{vmatrix} = \bar{D}_3. \tag{89}$$

But $\bar{D}_3$ can also be obtained by multiplying each row of D_3 by -1. Hence

$$D_3 = \bar{D}_3 = (-1)^3 D_3 = -D_3, \tag{90}$$

and consequently $D_3 = 0$.

16.6 Product of Two Determinants

If A and B are two nth order determinants with elements a_{ik} and b_{ik} respectively, their product AB is the nth order determinant C with elements c_{ik} given by

$$c_{ik} = \sum_{r=1}^{n} a_{ir} b_{rk}. \tag{91}$$

For example, if

$$A = \begin{vmatrix} a_{11} & a_{12} \\ a_{21} & a_{22} \end{vmatrix}, \qquad B = \begin{vmatrix} b_{11} & b_{12} \\ b_{21} & b_{22} \end{vmatrix} \qquad (92)$$

then

$$AB = \begin{vmatrix} a_{11}b_{11} + a_{12}b_{21} & a_{11}b_{12} + a_{12}b_{22} \\ a_{21}b_{11} + a_{22}b_{21} & a_{21}b_{12} + a_{22}b_{22} \end{vmatrix}. \qquad (93)$$

16.7 Self-consistent Equations

Suppose we have a set of three linear homogeneous equations

$$\left. \begin{array}{l} a_1 x + b_1 y + c_1 z = 0, \\ a_2 x + b_2 y + c_2 z = 0, \\ a_3 x + b_3 y + c_3 z = 0, \end{array} \right\} \qquad (94)$$

where x, y, z are unknowns and a_1, b_1, ..., b_3, c_3 are given constants. Besides the trivial solution $x = y = z = 0$, it may be possible to find a non-trivial solution giving the ratio $x : y : z$. Applying the results of 16.1 (equation (10)) to (94) we have

$$\begin{vmatrix} a_1 & b_1 & c_1 \\ a_2 & b_2 & c_2 \\ a_3 & b_3 & c_3 \end{vmatrix} x = \begin{vmatrix} 0 & b_1 & c_1 \\ 0 & b_2 & c_2 \\ 0 & b_3 & c_3 \end{vmatrix} = 0, \qquad (95)$$

with similar results for y and z. Hence if x is to be non-zero, (95) can only be satisfied if the determinant of the coefficients

$$\begin{vmatrix} a_1 & b_1 & c_1 \\ a_2 & b_2 & c_2 \\ a_3 & b_3 & c_3 \end{vmatrix} = 0. \qquad (96)$$

This is the condition for the existence of a non-trivial solution of (94).

Example 11. The value of λ which makes the equations

$$\begin{array}{l} x + 5y + 3z = 0, \\ 5x + y - \lambda z = 0, \\ x + 2y + \lambda z = 0, \end{array} \qquad (97)$$

consistent is given by the solution of the equation

$$\begin{vmatrix} 1 & 5 & 3 \\ 5 & 1 & -\lambda \\ 1 & 2 & \lambda \end{vmatrix} = 0. \qquad (98)$$

Evaluating (98) we find directly that $\lambda = 1$. Hence substituting this value of λ back into (97) we find

$$\frac{x}{-1} = \frac{y}{2} = \frac{z}{-3}. \tag{99}$$

When $\lambda = 1$, therefore, (97) possesses an infinity of non-trivial solutions as given by (99); on the other hand when $\lambda \neq 1$, the only solution of (97) is the trivial one $x = y = z = 0$.

16.8 Jacobians and Wronskians

Suppose $f(x, y)$ and $g(x, y)$ are two continuous functions of x and y with continuous first partial derivatives. If there exists a relation of the type

$$F\{f(x, y), g(x, y)\} = 0, \tag{100}$$

which is true for all x and y in the region of definition of $f(x, y)$ and $g(x, y)$, then $f(x, y)$ and $g(x, y)$ are said to be functionally dependent. If there is no relation of this type $f(x, y)$ and $g(x, y)$ are said to be functionally independent.

For example, the functions

$$f(x, y) = e^x \cos y, \quad g(x, y) = x + \log_e \cos y \tag{101}$$

are functionally dependent since

$$\log_e f(x, y) = g(x, y) \tag{102}$$

for all x and y for which $f(x, y)$ and $g(x, y)$ are defined.

We now assume that the arbitrary functions $f(x, y)$, $g(x, y)$ are functionally dependent. Then differentiating (100) partially we have

$$\frac{\partial F}{\partial f} \frac{\partial f}{\partial x} + \frac{\partial F}{\partial g} \frac{\partial g}{\partial x} = 0, \tag{103}$$

$$\frac{\partial F}{\partial f} \frac{\partial f}{\partial y} + \frac{\partial F}{\partial g} \frac{\partial g}{\partial y} = 0. \tag{104}$$

By the results of 16.7 these two equations have non-trivial solutions $\partial F/\partial f \neq 0$, $\partial F/\partial g \neq 0$ only if the determinant of the coefficients is zero. Hence the condition for the functional dependence of $f(x, y)$ and $g(x, y)$ is

$$\begin{vmatrix} \dfrac{\partial f}{\partial x} & \dfrac{\partial g}{\partial x} \\[2ex] \dfrac{\partial f}{\partial y} & \dfrac{\partial g}{\partial y} \end{vmatrix} = 0. \tag{105}$$

The left-hand side of (105) is nothing more than the Jacobian (see also Chapter 12, 12.13)

$$\frac{\partial(f, g)}{\partial(x, y)}\left(= J\left(\frac{f, g}{x, y}\right)\right). \tag{106}$$

Two functions are functionally dependent therefore if their Jacobian (or functional determinant as it is often called) vanishes. Conversely if their Jacobian is non-zero the functions are functionally independent. For example, the functions defined in (101) have the Jacobian

$$\frac{\partial(f, g)}{\partial(x, y)} = \begin{vmatrix} \dfrac{\partial f}{\partial x} & \dfrac{\partial g}{\partial x} \\[2ex] \dfrac{\partial f}{\partial y} & \dfrac{\partial g}{\partial y} \end{vmatrix} = \begin{vmatrix} e^x \cos y & 1 \\[1ex] -e^x \sin y & -\tan y \end{vmatrix} = 0, \tag{107}$$

which demonstrates their functional dependence.

Similar arguments apply to functions of three or more variables.

We now consider the meaning of linear independence of a set of functions $y_1(x)$, $y_2(x)$, ..., $y_n(x)$. These functions are said to be linearly independent if the identity

$$c_1 y_1(x) + c_2 y_2(x) + \ldots + c_n y_n(x) = 0 \tag{108}$$

can be satisfied only by taking $c_1 = c_2 = \ldots = c_n = 0$. Now since (108) is to be satisfied identically we also have (differentiating (108))

$$c_1 y_1'(x) \quad + c_2 y_2'(x) \quad + \ldots + c_n y_n'(x) \quad = 0,$$
$$c_1 y_1''(x) \quad + c_2 y_2''(x) \quad + \ldots + c_n y_n''(x) \quad = 0,$$
$$\vdots \qquad\qquad \vdots \qquad\qquad\qquad \vdots$$
$$c_1 y_1^{(n-1)}(x) + c_2 y_2^{(n-1)}(x) + \ldots + c_n y_n^{(n-1)}(x) = 0.$$
$$(109)$$

These n equations ((108) and (109)) have the trivial solution $c_1 = c_2 = \ldots = c_n$ only when the determinant

$$\begin{vmatrix} y_1(x) & y_2(x) & \ldots\ldots y_n(x) \\ y_1'(x) & y_2'(x) & \ldots\ldots y_n'(x) \\ \cdot & \cdot & \cdot \\ \cdot & \cdot & \cdot \\ \cdot & \cdot & \cdot \\ y_1^{(n-1)}(x) & y_2^{(n-1)}(x) & \ldots y_n^{(n-1)}(x) \end{vmatrix} \neq 0. \qquad (110)$$

This determinant is called the Wronskian of the set of functions $y_1(x), y_2(x), \ldots, y_n(x)$. The condition for the linear independence of these functions therefore is that their Wronskian be non-zero.

Example 12. The three functions e^x, e^{2x}, e^{3x} are linearly independent since their Wronskian

$$\begin{vmatrix} e^x & e^{2x} & e^{3x} \\ e^x & 2e^{2x} & 3e^{3x} \\ e^x & 4e^{2x} & 9e^{3x} \end{vmatrix} = e^{6x} \begin{vmatrix} 1 & 1 & 1 \\ 1 & 2 & 3 \\ 1 & 4 & 9 \end{vmatrix} = 2e^{6x} \qquad (111)$$

is not identically zero.

Example 13. The three functions $e^x, e^{-x}, \cosh x$ are linearly dependent since their Wronskian

$$\begin{vmatrix} e^x & e^{-x} & \cosh x \\ e^x & -e^{-x} & \sinh x \\ e^x & e^{-x} & \cosh x \end{vmatrix} = 0 \qquad (112)$$

in virtue of two rows of the determinant being identical. This linear dependence shows itself by way of the obvious identity

$$\tfrac{1}{2}e^x + \tfrac{1}{2}e^{-x} - \cosh x = 0. \qquad (113)$$

292

PROBLEMS 16

1. Evaluate the determinants

$$\begin{vmatrix} 1 & 1 & 1 \\ \lambda & \mu & \nu \\ \lambda^3 & \mu^3 & \nu^3 \end{vmatrix}, \qquad \begin{vmatrix} 7 & 11 & 4 \\ 13 & 15 & 10 \\ 3 & 9 & 6 \end{vmatrix}.$$

2. Show that

(a) $\begin{vmatrix} 0 & x & y & z \\ -x & 0 & u & v \\ -y & -u & 0 & w \\ -z & -v & -w & 0 \end{vmatrix} = (xw - yv + uz)^2,$

(b) $\begin{vmatrix} (b+c)^2 & a^2 & a^2 \\ b^2 & (c+a)^2 & b^2 \\ c^2 & c^2 & (a+b)^2 \end{vmatrix} = 2abc(a+b+c)^3,$

(c) $\begin{vmatrix} 1 & a & a^2 & a^3+bcd \\ 1 & b & b^2 & b^3+cda \\ 1 & c & c^2 & c^3+abd \\ 1 & d & d^2 & d^3+abc \end{vmatrix} = 0,$

(d) $\begin{vmatrix} 1 & \cos \alpha & \sin \alpha \\ 1 & \cos \beta & \sin \beta \\ 1 & \cos \gamma & \sin \gamma \end{vmatrix}$

$$= 4 \sin\left(\frac{\beta-\gamma}{2}\right) \sin\left(\frac{\gamma-\alpha}{2}\right) \sin\left(\frac{\alpha-\beta}{2}\right).$$

3. Show that the equation

$$\begin{vmatrix} 10-\lambda & -6 & 2 \\ -6 & 9-\lambda & -4 \\ 2 & -4 & 5-\lambda \end{vmatrix} = 0$$

has $\lambda = 2$ as a root, and find the others.

4. Find the values of λ which satisfy the equation

$$\begin{vmatrix} -\lambda & 1 & 0 \\ 1 & -\lambda & 1 \\ 0 & 1 & -\lambda \end{vmatrix} = 0.$$

293

5. If $a+b+c=0$, prove that the roots of the equation

$$\begin{vmatrix} a-x & b & c \\ b & c-x & a \\ c & a & b-x \end{vmatrix} = 0$$

are $x = 0$ and $x = \pm\sqrt{\{-3(ab+bc+ca)\}}$.

6. Show that

$$D = \begin{vmatrix} 1 & z & z^2 & 0 \\ 0 & 1 & z & z^2 \\ z^2 & 0 & 1 & z \\ z & z^2 & 0 & 1 \end{vmatrix} = 1+z^4+z^8,$$

and plot the solutions of $D = 0$ on the Argand diagram.

(L.U.)

7. Show that the determinant

$$\begin{vmatrix} x & x+a+b+c & x+a+b+c & x+a+b+c \\ y & y-a & y-b & y-c \\ z & z-b & z-c & z-a \\ t & t-c & t-a & t-b \end{vmatrix}$$

has the value $(x+y+z+t)(a^3+b^3+c^3-3abc)$.   (C.U.)

8. (i) Prove that

$$\begin{vmatrix} a^3 & b^3 & c^3 \\ a^2 & b^2 & c^2 \\ b+c & c+a & a+b \end{vmatrix} = -(a^2-b^2)(b^2-c^2)(c^2-a^2).$$

(ii) Solve the equation

$$\begin{vmatrix} x+1 & w & w^2 \\ w & x+w^2 & 1 \\ w^2 & 1 & x+w \end{vmatrix} = 0,$$

where w is a cube root of unity, considering the two cases
(a) w real, (b) w complex.

(L.U.)

9. The variables x, y can be expressed as functions of ξ, η, and ξ, η can be expressed as functions of λ, μ. Show that the Jacobian

$$\frac{\partial(x, y)}{\partial(\lambda, \mu)} = \begin{vmatrix} \dfrac{\partial x}{\partial \lambda} & \dfrac{\partial y}{\partial \lambda} \\ \dfrac{\partial x}{\partial \mu} & \dfrac{\partial y}{\partial \mu} \end{vmatrix}$$

satisfies the equation

$$\frac{\partial(x, y)}{\partial(\lambda, \mu)} = \frac{\partial(x, y)}{\partial(\xi, \eta)} \cdot \frac{\partial(\xi, \eta)}{\partial(\lambda, \mu)},$$

and deduce that

$$\frac{\partial(x, y)}{\partial(\xi, \eta)} = 1 \bigg/ \left(\frac{\partial(\xi, \eta)}{\partial(x, y)}\right).$$

Find $\dfrac{\partial(\xi, \eta)}{\partial(x, y)}$ in the case where

$$x = c \cosh \xi \cos \eta, \quad y = c \sinh \xi \sin \eta. \qquad \text{(C.U.)}$$

10. Solve, using determinants, the equations

$$x + y + z = 1,$$
$$2x - 3y + z = 2,$$
$$2x - 2y + z = 3.$$

11. Show that the three equations

$$-2x + y + z = a,$$
$$x - 2y + z = b,$$
$$x + y - 2z = c,$$

have no solutions unless $a + b + c = 0$, in which case they have infinitely many solutions.

12. Show that there are two values of k for which the equations

$$kx + 3y + 2z = 1,$$
$$x + (k-1)y = 4,$$
$$10y + 3z = -2,$$
$$2x - ky - z = 5,$$

are consistent. Find their common solution for that value of k which is an integer. (L.U.)

13. If the elements a_{ik} of an nth order determinant D_n are functions of a parameter t, then

$$\frac{dD_n}{dt} = \sum_{i=1}^{n} \sum_{k=1}^{n} A_{ik} \frac{da_{ik}}{dt}.$$

Verify this formula for the second-order determinant

$$D_2 = \begin{vmatrix} t^2 & 1 \\ 2t & \sin t \end{vmatrix}$$

by first expanding D_2 and then differentiating the result.

CHAPTER 17

Matrices

17.1 Definition of a Matrix

A matrix is a set of mn quantities arranged in a rectangular array of m rows and n columns. In writing down matrices it is usual to enclose this array by large brackets and to denote the matrix by a single letter, say A, such that

$$A = \begin{pmatrix} a_{11} & a_{12} \dots a_{1n} \\ a_{21} & a_{22} \dots a_{2n} \\ \cdot & \cdot & \cdot \\ \cdot & \cdot & \cdot \\ \cdot & \cdot & \cdot \\ a_{m1} & a_{m2} \dots a_{mn} \end{pmatrix} \tag{1}$$

The individual quantities a_{ik} are called the elements of the matrix. A matrix of m rows and n columns is said to be of order $(m \times n)$ and to be square if $m = n$. Unlike determinants, matrices have no algebraic value — in fact they must be regarded as operators in the same sense that d/dx is an operator. Before showing the use of matrices we must first set up an algebra defining the various operations of addition, subtraction, multiplication, and so on.

17.2 Algebra of Matrices

(a) Addition and Subtraction

In order that these operations may be carried out, the matrices concerned must be of the same order. If A and B are two matrices of the same order with elements a_{ik} and b_{ik} respectively then their sum $A + B$ is defined as a matrix C whose elements $c_{ik} = a_{ik} + b_{ik}$. Clearly C has the same order as A and B. For example, if

$$A = \begin{pmatrix} 1 & 2 & 3 \\ 0 & -1 & 2 \end{pmatrix} \quad \text{and} \quad B = \begin{pmatrix} 3 & 4 & 5 \\ 1 & 2 & 3 \end{pmatrix} \tag{2}$$

then

$$C = A + B = \begin{pmatrix} 4 & 6 & 8 \\ 1 & 1 & 5 \end{pmatrix}. \tag{3}$$

Subtraction of matrices is defined in a similar way in that the difference of two matrices A and B of the same order is a matrix D whose elements $d_{ik} = a_{ik} - b_{ik}$. For example, using the matrices A and B given in (2) we have

$$D = A - B = \begin{pmatrix} -2 & -2 & -2 \\ -1 & -3 & -1 \end{pmatrix}. \tag{4}$$

We see that provided the elements of matrices are real or complex numbers the laws of addition and subtraction of elementary algebra also apply to matrices since their sum and difference is defined directly in terms of the addition and subtraction of their elements. In other words, addition and subtraction of matrices are both associative and commutative in that

$$A + B = B + A \tag{5}$$

and

$$(A + B) + C = (A + B) + C. \tag{6}$$

(b) Equality of Matrices

Two matrices A and B with elements a_{ik} and b_{ik} respectively are equal only if they are of the same order, and if all their corresponding elements are equal (i.e. $a_{ik} = b_{ik}$).

(c) Multiplication by a Number

The result of multiplying a matrix A (with elements a_{ik}) by a number k (real or complex) is defined as a matrix B whose elements b_{ik} are k times the elements of A. For example, if

$$A = \begin{pmatrix} 1 & 3 \\ 2 & 4 \end{pmatrix} \quad \text{then} \quad kA = \begin{pmatrix} k & 3k \\ 2k & 4k \end{pmatrix}. \tag{7}$$

From this definition it follows that the distributive law of elementary algebra holds also for matrices in that

$$k(A \pm B) = kA \pm kB. \tag{8}$$

Furthermore if we define

$$kA = Ak, \tag{9}$$

then multiplication by a number is commutative. It is important to emphasise that k must be a number and not, for instance, another matrix. Multiplication of matrices by matrices will be defined in the next sub-section.

(d) Matrix Multiplication

The definition of matrix multiplication is such that two matrices A and B can only be multiplied together to form their product AB when the number of columns of A is equal to the number of rows of B. Such matrices are called conformable matrices. Suppose A is a matrix of order $(m \times p)$ with elements a_{ik} and B is a matrix of order $(p \times n)$ with elements b_{ik}. Then their product AB is a matrix C of order $(m \times n)$ with elements c_{ik} defined by

$$c_{ik} = \sum_{s=1}^{p} a_{is} b_{sk}. \tag{10}$$

For example, if A and B are (3×2) and (2×2) matrices, respectively, given by

$$A = \begin{pmatrix} a_{11} & a_{12} \\ a_{21} & a_{22} \\ a_{31} & a_{32} \end{pmatrix}, \qquad B = \begin{pmatrix} b_{11} & b_{12} \\ b_{21} & b_{22} \end{pmatrix} \tag{11}$$

then the product $C = AB$ is a (3×2) matrix defined as

$$C = \begin{pmatrix} a_{11}b_{11} + a_{12}b_{21} & a_{11}b_{12} + a_{12}b_{22} \\ a_{21}b_{11} + a_{22}b_{21} & a_{21}b_{12} + a_{22}b_{22} \\ a_{31}b_{11} + a_{32}b_{21} & a_{31}b_{12} + a_{32}b_{22} \end{pmatrix}. \tag{12}$$

As another example, we take

$$A(2 \times 3) = \begin{pmatrix} 3 & 1 & 2 \\ 2 & 1 & 3 \end{pmatrix}, \qquad B(3 \times 2) = \begin{pmatrix} 1 & 2 \\ 3 & 1 \\ 2 & 3 \end{pmatrix}. \tag{13}$$

Then, from (10),

$$C(2 \times 2) = A(2 \times 3)B(3 \times 2) = \begin{pmatrix} 10 & 13 \\ 11 & 14 \end{pmatrix}. \tag{14}$$

On the other hand A and B are also conformable for the product $B(3 \times 2)A(2 \times 3)$. This defines a (3×3) matrix D given by

$$D = \begin{pmatrix} 7 & 3 & 8 \\ 11 & 4 & 9 \\ 12 & 5 & 13 \end{pmatrix}. \tag{15}$$

Clearly $AB \neq BA$, since the orders of the matrices representing the two products are different. This non-commutative property of matrix multiplication appears even when A and B are such that their two products AB and BA are matrices of the same order. For example, if

$$A = \begin{pmatrix} 0 & 1 \\ 1 & 1 \end{pmatrix}, \qquad B = \begin{pmatrix} 0 & -1 \\ 1 & 0 \end{pmatrix} \tag{16}$$

then

$$AB = \begin{pmatrix} 1 & 0 \\ 1 & -1 \end{pmatrix} \quad \text{and} \quad BA = \begin{pmatrix} -1 & -1 \\ 0 & 1 \end{pmatrix}. \tag{17}$$

Again $AB \neq BA$. However apart from this non-commutative law of multiplication, matrices satisfy the associative and distributive laws of multiplication in that

$$(AB)C = A(BC), \tag{18}$$

and

$$(A+B)C = AC+BC, \tag{19}$$

provided that the products are defined.

17.3 Special Types of Matrices

(a) Row Vector

A set of n quantities arranged in a row is a matrix of order $(1 \times n)$. Such a matrix is usually called a row matrix or row vector and is denoted by

$$[A] = (a_1, a_2, a_3, ..., a_n). \tag{20}$$

(b) Column Vector

A set of m quantities arranged in a column is a matrix of order $(m \times 1)$. Such a matrix is called a column matrix or column vector and is denoted by

$$\{A\} = \begin{pmatrix} a_1 \\ a_2 \\ a_3 \\ \cdot \\ \cdot \\ \cdot \\ a_m \end{pmatrix} \tag{21}$$

(c) Null-Matrix

Any matrix of order $(m \times n)$ with all its elements equal to zero is called a null matrix of order $(m \times n)$.

(d) Square Matrix

A square matrix has the same number of rows as columns; for example

$$A = \begin{pmatrix} a_{11} & a_{12} \dots a_{1n} \\ a_{21} & a_{22} \dots a_{2n} \\ \cdot & \cdot & \cdot \\ \cdot & \cdot & \cdot \\ \cdot & \cdot & \cdot \\ a_{n1} & a_{n2} \dots a_{nn} \end{pmatrix} \tag{22}$$

is a square matrix of order $(n \times n)$. The diagonal containing the elements $a_{11}, a_{22}, a_{33}, \dots, a_{nn}$ is called the leading diagonal and the sum of these elements is called the trace (or spur) of the matrix. This sum is usually denoted by Tr (or Sp) for short. For example, if

$$A = \begin{pmatrix} 1 & 2 & 3 \\ 4 & 5 & 6 \\ 7 & 8 & 9 \end{pmatrix} \tag{23}$$

then

$$\text{Tr } A \text{ (or Sp } A) = 1 + 5 + 9 = 15. \tag{24}$$

(e) Diagonal Matrix

A square matrix with zero elements everywhere except in the leading diagonal is called a diagonal matrix. In other words, if a_{ik} are the elements of a diagonal matrix we must have $a_{ik} = 0$ for $i \neq k$. A typical (4×4) diagonal matrix is

$$A = \begin{pmatrix} 1 & 0 & 0 & 0 \\ 0 & 2 & 0 & 0 \\ 0 & 0 & 3 & 0 \\ 0 & 0 & 0 & 4 \end{pmatrix}. \tag{25}$$

Clearly all diagonal matrices of the same order commute under multiplication.

(f) Unit Matrix

The unit matrix is a diagonal matrix with all its diagonal elements equal to unity. It is usual to denote such matrices by the letter I (or sometimes E); for example, the (3×3) unit matrix is

$$I = \begin{pmatrix} 1 & 0 & 0 \\ 0 & 1 & 0 \\ 0 & 0 & 1 \end{pmatrix}. \tag{26}$$

In general, if A is a square matrix of order $(m \times m)$ and I is the unit matrix of the same order then

$$IA = AI = A. \tag{27}$$

Likewise

$$I = I^2 = I^3 = \ldots = I^k \ldots, \tag{28}$$

where k is any positive integer. In fact multiplying any matrix by a unit matrix leaves the matrix unchanged provided the product is defined. For example,

$$\begin{pmatrix} 1 & 0 & 0 \\ 0 & 1 & 0 \\ 0 & 0 & 1 \end{pmatrix} \begin{pmatrix} 1 \\ 2 \\ 3 \end{pmatrix} = \begin{pmatrix} 1 \\ 2 \\ 3 \end{pmatrix}. \tag{29}$$

(g) Determinant of a Matrix

The determinant of a matrix A is only defined when A is square and is then just the determinant of the matrix elements. It is usually denoted by $|A|$ or det A. For example, if

$$A = \begin{pmatrix} 1 & 0 & 1 \\ 2 & 3 & 1 \\ 3 & 1 & 2 \end{pmatrix} \tag{30}$$

then

$$|A| = \begin{vmatrix} 1 & 0 & 1 \\ 2 & 3 & 1 \\ 3 & 1 & 2 \end{vmatrix} = -2. \qquad (31)$$

When a square matrix is such that its determinant is zero it is called a singular matrix (otherwise it is non-singular). For example, the matrix

$$\begin{pmatrix} 3 & 0 & 3 \\ 4 & 1 & 4 \\ 5 & 2 & 5 \end{pmatrix} \qquad (32)$$

is singular since its determinant vanishes in virtue of two columns being identical.

(h) The Transposed Matrix

In the case of determinants we have seen that interchanging the roles of rows and columns leaves the value of the determinant unaltered. However, if the rows and columns of a matrix are interchanged a new matrix called the transposed matrix is obtained. For example, if A is a (3×2) matrix given by

$$A = \begin{pmatrix} a_{11} & a_{12} \\ a_{21} & a_{22} \\ a_{31} & a_{32} \end{pmatrix} \qquad (33)$$

then its transpose (denoted by A') is the (2×3) matrix

$$A' = \begin{pmatrix} a_{11} & a_{21} & a_{31} \\ a_{12} & a_{22} & a_{32} \end{pmatrix}. \qquad (34)$$

(We remark here that in some books the transpose of a matrix A is denoted by $\tilde{A}$ or A^T.)

Clearly the transpose of a column vector, say $\{A\} = \begin{pmatrix} a_1 \\ a_2 \\ a_3 \end{pmatrix}$ is a

row vector since

$$\{A\}' = (a_1 \ a_2 \ a_3) = [A]. \qquad (35)$$

303

Similarly the transpose of a row vector $[A]$ is the column vector $\{A\}$. It follows that

$$\{A\}'\{A\} = [A][A]' = a_1^2 + a_2^2 + a_3^2. \tag{36}$$

In general we see that if A is of order $(m \times n)$ then A' is of order $(n \times m)$, and hence A and A' are conformable to both products AA' and $A'A$ (i.e. both products exist but are of different orders unless A is square).

We now show that if A and B are two matrices conformable to the product $AB = C$, then

$$C' = (AB)' = B'A'. \tag{37}$$

Suppose A is of order $(m \times p)$ with elements a_{ik} and B is of order $(p \times n)$ with elements b_{ik}. Then C is a matrix of order $(m \times n)$ with elements c_{ik} given by

$$c_{ik} = \sum_{s=1}^{p} a_{is}b_{sk}. \tag{38}$$

Consequently

$$c'_{ik} = c_{ki} = \sum_{s=1}^{p} a_{ks}b_{si} = \sum_{s=1}^{p} b'_{is}a'_{sk} \tag{39}$$

and therefore

$$C' = B'A'. \tag{40}$$

Similarly we may show that

$$(ABC)' = C'B'A', \tag{41}$$

and so on for a finite number of matrices. In other words, in taking the transpose of matrix products the order of the matrices forming the product must be reversed.

For example, if

$$A = \begin{pmatrix} 1 & 2 \\ 3 & 4 \end{pmatrix}, \qquad B = \begin{pmatrix} 2 \\ 1 \end{pmatrix} \tag{42}$$

then

$$A' = \begin{pmatrix} 1 & 3 \\ 2 & 4 \end{pmatrix}, \qquad B' = (2 \quad 1), \tag{43}$$

and

$$AB = \begin{pmatrix} 4 \\ 10 \end{pmatrix}, \qquad B'A' = (4 \quad 10). \tag{44}$$

Clearly $(AB)' = B'A'$ as required.

(i) Symmetric and Skew-symmetric Matrices

A symmetric matrix A is square and has elements a_{ik} such that $a_{ik} = a_{ki}$. In other words, the elements above and below the leading diagonal are mirror images of each other. For example, the matrix

$$A = \begin{pmatrix} 1 & x & y \\ x & 3 & z \\ y & z & 4 \end{pmatrix} \tag{45}$$

is symmetric. Clearly every symmetric matrix is its own transpose in that $A = A'$. Conversely any matrix which is its own transpose is necessarily symmetric.

A skew-symmetric matrix is also square but has elements a_{ik} such that $a_{ik} = -a_{ki}$ from which it follows that the leading diagonal has zero elements (since $a_{11} = -a_{11}$, etc.). A typical skew-symmetric matrix is

$$A = \begin{pmatrix} 0 & 2 & 1 \\ -2 & 0 & 3 \\ -1 & -3 & 0 \end{pmatrix}. \tag{46}$$

Such matrices are the negatives of their own transposes (i.e. $A = -A'$).

Any square matrix A can be resolved into the sum of a symmetric matrix $\underline{A}$ and a skew-symmetric matrix $\underset{v}{A}$ since if the elements of A are a_{ik} then

$$a_{ik} = \tfrac{1}{2}(a_{ik} + a_{ki}) + \tfrac{1}{2}(a_{ik} - a_{ki}), \tag{47}$$

or

$$A = \underline{A} + \underset{v}{A}. \tag{48}$$

305

For example, if

$$A = \begin{pmatrix} 3 & 1 & 2 \\ 2 & 1 & 3 \\ 1 & 2 & 1 \end{pmatrix} \tag{49}$$

then

$$\underline{A} = \begin{pmatrix} 3 & \frac{3}{2} & \frac{3}{2} \\ \frac{3}{2} & 1 & \frac{5}{2} \\ \frac{3}{2} & \frac{5}{2} & 1 \end{pmatrix} \tag{50}$$

and

$$\underset{\vee}{A} = \begin{pmatrix} 0 & -\frac{1}{2} & \frac{1}{2} \\ \frac{1}{2} & 0 & \frac{1}{2} \\ -\frac{1}{2} & -\frac{1}{2} & 0 \end{pmatrix}. \tag{51}$$

It is easily verified that $A = \underline{A} + \underset{\vee}{A}$.

(j) Complex and Hermitian Matrices

If A is any matrix of order $(m \times n)$ with complex elements a_{ik}, then the complex conjugate A^* of A is found by taking the complex conjugates of all the elements. For example, if

$$A = \begin{pmatrix} 1+i & 2 & -i \\ 3 & 1-i & 2+i \end{pmatrix} \tag{52}$$

then

$$A^* = \begin{pmatrix} 1-i & 2 & i \\ 3 & 1+i & 2-i \end{pmatrix}. \tag{53}$$

Clearly

$$(A^*)^* = A. \tag{54}$$

A Hermitian matrix is a square matrix which is unchanged by taking the transpose of its complex conjugate; that is, A is Hermitian if

$$(A^*)' = A. \tag{55}$$

For example, if

$$A = \begin{pmatrix} 1 & 1+i \\ 1-i & 3 \end{pmatrix} \tag{56}$$

then

$$A^* = \begin{pmatrix} 1 & 1-i \\ 1+i & 3 \end{pmatrix} \tag{57}$$

and consequently

$$(A^*)' = \begin{pmatrix} 1 & 1+i \\ 1-i & 3 \end{pmatrix} = A. \tag{58}$$

Hence A is Hermitian. It is obvious that the diagonal elements of a Hermitian matrix are always real.

If a square matrix A is such that

$$(A^*)' = -A \tag{59}$$

it is called skew-Hermitian.

(k) The Adjoint and Reciprocal Matrices

If A is the square matrix

$$\begin{pmatrix} a_{11} & a_{12} \dots a_{1n} \\ a_{21} & a_{22} \dots a_{2n} \\ \vdots & \vdots & \vdots \\ a_{n1} & a_{n2} \dots a_{nn} \end{pmatrix} \tag{60}$$

then its adjoint (usually denoted by adj A) is defined as the transpose of the matrix of its cofactors. In other words, if A_{rs} is the cofactor of the element a_{rs} (i.e. the value of the determinant formed by deleting the row and column in which a_{rs} occurs and attaching a plus or minus sign in accordance with Chapter 16, 16.4), then the matrix of the cofactors is the square matrix B (of the same order as A) where

$$B = \begin{pmatrix} A_{11} & A_{12} \dots A_{1n} \\ A_{21} & A_{22} \dots A_{2n} \\ \vdots & \vdots & \vdots \\ A_{n1} & A_{n2} \dots A_{nn} \end{pmatrix}. \tag{61}$$

Consequently

$$\text{adj } A = B' = \begin{pmatrix} A_{11} & A_{21} \dots A_{n1} \\ A_{12} & A_{22} \dots A_{n2} \\ \vdots & \vdots & \vdots \\ A_{1n} & A_{2n} \dots A_{nn} \end{pmatrix} \tag{62}$$

For example, if

$$A = \begin{pmatrix} 1 & 2 & 3 \\ 1 & 3 & 5 \\ 1 & 5 & 12 \end{pmatrix} \tag{63}$$

then the cofactor of a_{11} is $A_{11} = +(3 \cdot 12 - 5 \cdot 5) = 11$ and the cofactor of a_{12} is $A_{12} = -(12 \cdot 1 - 5 \cdot 1) = -7$, and so on. Proceeding in this way we find that the matrix of cofactors is

$$B = \begin{pmatrix} 11 & -7 & 2 \\ -9 & 9 & -3 \\ 1 & -2 & 1 \end{pmatrix}. \tag{64}$$

Hence, by definition,

$$\text{adj } A = B' = \begin{pmatrix} 11 & -9 & 1 \\ -7 & 9 & -2 \\ 2 & -3 & 1 \end{pmatrix}. \tag{65}$$

We now consider the product

$$A(\text{adj } A) = \begin{pmatrix} a_{11} & a_{12} \dots a_{1n} \\ a_{21} & a_{22} \dots a_{2n} \\ \vdots & \vdots & \vdots \\ \vdots & \vdots & \vdots \\ a_{n1} & a_{n2} \dots a_{nn} \end{pmatrix} \begin{pmatrix} A_{11} & A_{21} \dots A_{n1} \\ A_{12} & A_{22} \dots A_{n2} \\ \vdots & \vdots \\ \vdots & \vdots \\ A_{1n} & \dots\dots\dots A_{nn} \end{pmatrix} \tag{66}$$

which becomes, using equations (70)–(73) of Chapter 16, 16.4,

$$\begin{pmatrix} |A| & 0 & 0 & \dots & 0 \\ 0 & |A| & 0 & \dots & 0 \\ 0 & 0 & |A| & & \vdots \\ \vdots & \vdots & \vdots & & \vdots \\ 0 & 0 & 0 & \dots & |A| \end{pmatrix} = |A| \, I, \tag{67}$$

where $|A|$ is the determinant of A, and I is the unit matrix of the same order as A.

It follows that the matrix A^{-1} defined by

$$A^{-1} = \frac{\text{adj } A}{|A|} \tag{68}$$

has the property that

$$AA^{-1} = I, \tag{69}$$

and in this sense it is referred to as the reciprocal (or inverse) matrix of A. It is easily verified that multiplication of matrices and their inverses is commutative in that

$$AA^{-1} = A^{-1}A = I. \tag{70}$$

Clearly only non-singular square matrices have reciprocals, since A^{-1} is undefined when $|A| = 0$.

As an example of a reciprocal matrix we return to the matrix A of (63) and its adjoint matrix (65). Evaluating the determinant of (63) we find $|A| = 3$, and hence by (68)

$$A^{-1} = \frac{\text{adj } A}{|A|} = \begin{pmatrix} \frac{11}{3} & -3 & \frac{1}{3} \\ -\frac{7}{3} & 3 & -\frac{2}{3} \\ \frac{2}{3} & -1 & \frac{1}{3} \end{pmatrix}. \tag{71}$$

It is now easily verified (using (63) and (71)) that (70) is satisfied.

(l) *Orthogonal and Unitary Matrices*

A square matrix A (with real elements) is said to be orthogonal if

$$A' = A^{-1} \tag{72}$$

—that is, if

$$A'A = AA' = I, \tag{73}$$

where I is the unit matrix of the same order as A.

It is easily seen that if A and B are two nth order orthogonal matrices then their product AB is orthogonal. To prove this we note simply that

$$(AB)(AB)' = ABB'A' = I \tag{74}$$

(using (32)). Similarly

$$(AB)'(AB) = B'A'AB = I. \tag{75}$$

Hence from the definition (73) we see that AB is an orthogonal matrix.

If A has complex elements it is said to be unitary if

$$(A^*)' = A^{-1}, \tag{76}$$

which reduces to the definition of an orthogonal matrix when A is real. From (76) we have

$$A^\dagger A = AA^\dagger = I, \tag{77}$$

where $A^\dagger = (A^*)'$.

For example,

$$A = \begin{pmatrix} \dfrac{1}{\sqrt{2}} & \dfrac{i}{\sqrt{2}} \\ -\dfrac{i}{\sqrt{2}} & -\dfrac{1}{\sqrt{2}} \end{pmatrix} \tag{78}$$

is unitary since

$$A^{-1} = \begin{pmatrix} \dfrac{1}{\sqrt{2}} & \dfrac{i}{\sqrt{2}} \\ -\dfrac{i}{\sqrt{2}} & -\dfrac{1}{\sqrt{2}} \end{pmatrix} \quad \text{and} \quad A^\dagger = (A^*)' = \begin{pmatrix} \dfrac{1}{\sqrt{2}} & \dfrac{i}{\sqrt{2}} \\ -\dfrac{i}{\sqrt{2}} & -\dfrac{1}{\sqrt{2}} \end{pmatrix}, \tag{79}$$

whence (77) is satisfied.

Again it may be proved as for orthogonal matrices that the product of two unitary matrices of the same order is a unitary matrix.

17.4 Solution of Linear Equations

Consider the set of n linear equations in the n unknowns $x_1, x_2, ..., x_n$

$$\left.\begin{array}{c} a_{11}x_1 + a_{12}x_2 + ... + a_{1n}x_n = h_1, \\ a_{21}x_1 + a_{22}x_2 + ... + a_{2n}x_n = h_2, \\ \cdot \qquad \cdot \qquad \cdot \\ \cdot \qquad \cdot \qquad \cdot \\ \cdot \qquad \cdot \qquad \cdot \\ a_{n1}x_1 + a_{n2}x_2 + ... a_{nn}x_n = h_n, \end{array}\right\} \qquad (80)$$

where the coefficients a_{ix} and h_i $(i, x = 1, 2, 3, ..., n)$ are known constants. Using the concept of matrix multiplication defined earlier, it is clear that these equations can be written in matrix form as

$$\begin{pmatrix} a_{11} & a_{12} ... a_{1n} \\ a_{21} & a_{22} ... a_{2n} \\ \cdot \\ \cdot \\ \cdot \\ a_{n1} & a_{n2} ... a_{nn} \end{pmatrix} \begin{pmatrix} x_1 \\ x_2 \\ \cdot \\ \cdot \\ \cdot \\ x_n \end{pmatrix} = \begin{pmatrix} h_1 \\ h_2 \\ \cdot \\ \cdot \\ \cdot \\ h_n \end{pmatrix}, \qquad (81)$$

or

$$AX = H, \qquad (82)$$

where

$$A = \begin{pmatrix} a_{11} & a_{12} ... a_{1n} \\ a_{21} & a_{22} ... a_{2n} \\ \cdot \\ \cdot \\ \cdot \\ a_{n1} & a_{n2} ... a_{nn} \end{pmatrix}, \quad X = \begin{pmatrix} x_1 \\ x_2 \\ \cdot \\ \cdot \\ \cdot \\ x_n \end{pmatrix} \qquad (83)$$

and

$$H = \begin{pmatrix} h_1 \\ h_2 \\ \cdot \\ \cdot \\ \cdot \\ h_n \end{pmatrix} \qquad (84)$$

We first assume that $H \neq 0$, and that A has an inverse A^{-1}. Then pre-multiplying (82) by A^{-1} we have

$$A^{-1}AX = A^{-1}H, \tag{85}$$

which, since $A^{-1}A = I$, gives

$$X = A^{-1}H. \tag{86}$$

This is the matrix solution of (82).

Example 1. To solve the equations

$$\begin{aligned} x+y+z &= 6, \\ x+2y+3z &= 14, \\ x+4y+9z &= 36. \end{aligned} \tag{87}$$

Now

$$A = \begin{pmatrix} 1 & 1 & 1 \\ 1 & 2 & 3 \\ 1 & 4 & 9 \end{pmatrix}, \quad X = \begin{pmatrix} x \\ y \\ z \end{pmatrix} \quad \text{and} \quad H = \begin{pmatrix} 6 \\ 14 \\ 36 \end{pmatrix}. \tag{88}$$

Hence

$$A^{-1} = \frac{\text{adj } A}{|A|} = \frac{1}{2} \begin{pmatrix} 6 & -5 & 1 \\ -6 & 8 & -2 \\ 2 & -3 & 1 \end{pmatrix} \tag{89}$$

and consequently, by (86),

$$\begin{pmatrix} x \\ y \\ z \end{pmatrix} = \begin{pmatrix} 3 & -\frac{5}{2} & \frac{1}{2} \\ -3 & 4 & -1 \\ 1 & -\frac{3}{2} & \frac{1}{2} \end{pmatrix} \begin{pmatrix} 6 \\ 14 \\ 36 \end{pmatrix} = \begin{pmatrix} 1 \\ 2 \\ 3 \end{pmatrix} \tag{90}$$

The solutions are therefore $x = 1, y = 2$ and $z = 3$.

The assumption that A^{-1} exists may not always be valid, for we may consider sets of equations for which $|A| = 0$. If $H \neq 0$, we then find that such equations have no finite solutions and are inconsistent. For example,

$$\left. \begin{aligned} 2x-y-z &= -1, \\ x-2y+z &= 2, \\ x+y-2z &= 3, \end{aligned} \right\} \tag{91}$$

for which $|A| = 0$, form an inconsistent set of equations since the sum of the last two equations is clearly inconsistent with the first equation.

If $H = 0$ (i.e. $h_i = 0$, $i = 1, 2, ..., n$) the equations are said to be homogeneous and, by (82), may be written as

$$AX = 0, \qquad (92)$$

where 0 is the zero column matrix of order n. Now if $|A| \neq 0$, then A^{-1} exists and consequently by (86)

$$X = A^{-1}0 = 0 \qquad (93)$$

which is the only solution. This identically zero solution in which the x_i are all zero is called the trivial solution. However, if $|A| = 0$ a different situation arises since A^{-1} does not then exist and we may not deduce (93) from (92). It is clear that under this condition (92) allows non-trivial solutions for X since the product of two matrices may be zero without either matrix itself being zero.

For example, the equations

$$\left. \begin{array}{r} x + 5y + 3z = 0, \\ 5x + y - kz = 0, \\ x + 2y + kz = 0, \end{array} \right\} \qquad (94)$$

where k is a parameter, has the trivial solutions $x = y = z = 0$ for all values of k. However, non-trivial solutions will exist when

$$|A| = \begin{vmatrix} 1 & 5 & 3 \\ 5 & 1 & -k \\ 1 & 2 & k \end{vmatrix} = 27(1 - k) = 0, \qquad (95)$$

that is, when $k = 1$. In this case the equations are not all different, but are linearly dependent since

$$x + 5y + 3z = -\tfrac{1}{3}(5x + y - z) + \tfrac{8}{3}(x + 2y + z). \qquad (96)$$

Consequently there are only two equations for three unknowns. Solving any two of (94) (with $k = 1$) gives an infinity of possible solutions

$$x = -\lambda, \quad y = 2\lambda, \quad z = -3\lambda, \qquad (97)$$

where λ is an arbitrary parameter.

Finally we remark that the calculation of inverses of matrices, which is, as we have seen, an important part of the solution of sets of linear equations, is best done by numerical methods rather than by the adjoint method described here. Many such numerical techniques are available and with the help of computers the inverses of large matrices—say 1000×1000—may now be readily obtained.

17.5 Eigenvalues and Eigenvectors

Consider the matrix equation

$$AX = \lambda X, \tag{98}$$

where A is a square matrix of order n, X is a column vector with n rows, and λ is a parameter. This equation may be written as

$$(A - \lambda I)X = 0, \tag{99}$$

where I is the unit matrix of order n, and forms a homogeneous set of linear equations. Besides the trivial solution $X = 0$, non-trivial solutions will exist (as shown in the previous section) if

$$|A - \lambda I| = 0. \tag{100}$$

This equation is called the characteristic equation of the matrix A, and the n roots $(\lambda_1, \lambda_2, ..., \lambda_n)$ are the characteristic or eigenvalues of A. It is convenient to refer to the left-hand side of (100) as the characteristic polynomial of A and to denote it by $f(\lambda)$.

Example 2. If

$$A = \begin{pmatrix} 1 & 0 & 1 \\ 0 & 2 & 0 \\ 1 & 1 & 3 \end{pmatrix}. \tag{101}$$

then

$$f(\lambda) = |A - \lambda I| = \begin{vmatrix} 1-\lambda & 0 & 1 \\ 0 & 2-\lambda & 0 \\ 1 & 1 & 3-\lambda \end{vmatrix} = 0. \tag{102}$$

Expanding the determinant we have

$$f(\lambda) = (\lambda - 2)(\lambda^2 - 4\lambda + 2) = 0, \tag{103}$$

whence the eigenvalues are

$$\lambda = 2, \quad 2 \pm \sqrt{2}. \tag{104}$$

Now to each eigenvalue there will be a solution X of equation (98). The individual solutions for X are called the eigenvectors of the matrix A. To see how this comes about consider the following example.

Example 3. To find the eigenvalues and eigenvectors of the matrix

$$A = \begin{pmatrix} 4 & 1 \\ 2 & 3 \end{pmatrix}. \tag{105}$$

Now the characteristic equation is

$$|A - \lambda I| = \begin{vmatrix} 4-\lambda & 1 \\ 2 & 3-\lambda \end{vmatrix} = (\lambda - 2)(\lambda - 5) = 0. \tag{106}$$

Hence the two eigenvalues are $\lambda_1 = 2$, $\lambda_2 = 5$. To find the corresponding eigenvectors we take the basic equation $(A - \lambda I)X = 0$ with A given by (105) and solve for X for each λ value. The eigenvector corresponding to the ith eigenvalue λ_i will be denoted by X_i and the elements of X_i by $x_1^{(i)}$, $x_2^{(i)}$. (This notation readily extends to higher order matrices.)

Case $\lambda_1 = 2$.
  We have

$$\left[\begin{pmatrix} 4 & 1 \\ 2 & 3 \end{pmatrix} - 2 \begin{pmatrix} 1 & 0 \\ 0 & 1 \end{pmatrix} \right] \begin{pmatrix} x_1^{(1)} \\ x_1^{(1)} \end{pmatrix} = \begin{pmatrix} 0 \\ 0 \end{pmatrix} \tag{107}$$

whence

$$x_1^{(1)} = -\tfrac{1}{2} x_2^{(1)}. \tag{108}$$

Consequently

$$X_1 = \begin{pmatrix} -\alpha \\ 2\alpha \end{pmatrix}, \quad \text{where } \alpha \text{ is a parameter.} \tag{109}$$

Case $\lambda_2 = 5$.
  Here the equations are

$$\left[\begin{pmatrix} 4 & 1 \\ 2 & 3 \end{pmatrix} - 5 \begin{pmatrix} 1 & 0 \\ 0 & 1 \end{pmatrix} \right] \begin{pmatrix} x_1^{(2)} \\ x_2^{(2)} \end{pmatrix} = \begin{pmatrix} 0 \\ 0 \end{pmatrix}, \tag{110}$$

whence we find

$$x_1^{(2)} = x_2^{(2)}. \tag{111}$$

Consequently

$$X_2 = \begin{pmatrix} \beta \\ \beta \end{pmatrix}, \tag{112}$$

where β is a parameter. We note that in neither case are X_1 or X_2 determined uniquely but contain arbitrary parameters α and β. This is due to the fact that the equations $(A - \lambda I)X = 0$ are satisfied by X or any multiple of X (say αX or βX). To remove this degree of arbitrariness in the eigenvectors it is usual to impose a normalisation condition which requires that their 'lengths' $\sqrt{(X_i^{*'} X_i)}$ are unity ($X_i^{*'}$ is the transpose of X_i^{*}). For example, in this way, using (109), we have

$$(-\alpha \quad 2\alpha) \begin{pmatrix} -\alpha \\ 2\alpha \end{pmatrix} = 1, \tag{113}$$

whence

$$\alpha^2 + 4\alpha^2 = 5\alpha^2 = 1,$$

or

$$\alpha = \pm \frac{1}{\sqrt{5}}. \tag{114}$$

Hence

$$X_1 = \begin{pmatrix} -\dfrac{1}{\sqrt{5}} \\ \dfrac{2}{\sqrt{5}} \end{pmatrix} \quad \text{or} \quad \begin{pmatrix} \dfrac{1}{\sqrt{5}} \\ -\dfrac{2}{\sqrt{5}} \end{pmatrix}. \tag{115}$$

Similarly

$$X_2 = \begin{pmatrix} \dfrac{1}{\sqrt{2}} \\ \dfrac{1}{\sqrt{2}} \end{pmatrix} \quad \text{or} \quad \begin{pmatrix} -\dfrac{1}{\sqrt{2}} \\ -\dfrac{1}{\sqrt{2}} \end{pmatrix}. \tag{116}$$

Either the first expressions for X_i in (115) and (116) must be taken together, or the second expressions, but not the first form for X_1 and the second form for X_2. Consequently the matrix representing the eigenvectors of A (taking the first expressions in (115) and (116) is

$$S = \begin{pmatrix} -\dfrac{1}{\sqrt{5}} & \dfrac{1}{\sqrt{2}} \\[2ex] \dfrac{2}{\sqrt{5}} & \dfrac{1}{\sqrt{2}} \end{pmatrix}. \tag{117}$$

A variety of results exists concerning the eigenvalues and eigenvectors of various types of matrices. For example, the eigenvalues of a real symmetric matrix are real; the eigenvalues of a Hermitian matrix are real. However, we shall not enter into a discussion of these results here.

Finally we note that, if A is a large matrix, the determination of its eigenvalues requires the solution of a polynomial equation of high order. Here again (as with the calculation of matrix inverses) numerical techniques have been devised for the explicit calculation of the eigenvalues. The reader is referred to a more specialised book on the subject for further details.

PROBLEMS 17

1. If $A = \begin{pmatrix} 2 & 1 & 2 \\ 3 & 5 & 7 \\ 1 & 0 & 1 \end{pmatrix}$ and $B = \begin{pmatrix} -3 & 1 & 0 \\ 6 & 2 & 1 \\ 1 & -1 & 2 \end{pmatrix}$

find $A+B$, $A-B$, $B-A$, AB and BA.

2. Given the matrices

$$\gamma_1 = \begin{pmatrix} 0 & 0 & 0 & 1 \\ 0 & 0 & 1 & 0 \\ 0 & 1 & 0 & 0 \\ 1 & 0 & 0 & 0 \end{pmatrix}, \qquad \gamma_2 = \begin{pmatrix} 0 & 0 & 0 & i \\ 0 & 0 & -i & 0 \\ 0 & i & 0 & 0 \\ -i & 0 & 0 & 0 \end{pmatrix}$$

$$\gamma_3 = \begin{pmatrix} 0 & 0 & 1 & 0 \\ 0 & 0 & 0 & -1 \\ 1 & 0 & 0 & 0 \\ 0 & -1 & 0 & 0 \end{pmatrix}, \qquad \gamma_4 = \begin{pmatrix} 1 & 0 & 0 & 0 \\ 0 & 1 & 0 & 0 \\ 0 & 0 & -1 & 0 \\ 0 & 0 & 0 & -1 \end{pmatrix}$$

show that $\gamma_r\gamma_s + \gamma_s\gamma_r = 0$ for $r \neq s$ (where $r, s = 1, 2, 3, 4$), and that

$$\gamma_1^2 = \gamma_2^2 = \gamma_3^2 = \gamma_4^2 = I,$$

where I is the (4×4) unit matrix.

3. Find the symmetric and skew-symmetric parts of the matrix

$$A = \begin{pmatrix} 1 & \frac{3}{2} & -5 \\ \frac{1}{2} & 0 & \frac{3}{4} \\ -1 & \frac{1}{4} & 2 \end{pmatrix}$$

4. Show that if

$$A(\theta) = \begin{pmatrix} \cos\theta & \sin\theta \\ -\sin\theta & \cos\theta \end{pmatrix}$$

then $A(\theta)A(\phi) = A(\theta + \phi)$, and give a geometrical interpretation of this result.

5. If

$$A = \begin{pmatrix} 1 & 0 & 0 & 0 \\ 1 & -1 & 0 & 0 \\ 1 & -2 & 1 & 0 \\ 1 & -3 & 3 & -1 \end{pmatrix}$$

prove that $A^2 = I$, where I is the (4×4) unit matrix. Prove also that if

$$P = AM_1A \quad \text{and} \quad Q = AM_2A,$$

where M_1 and M_2 are arbitrary diagonal matrices of order (4×4), then $PQ = QP$.

6. The (2×2) matrices $\sigma_1, \sigma_2, \sigma_3, I$ are defined by

$$\sigma_1 = \begin{pmatrix} 0 & 1 \\ 1 & 0 \end{pmatrix}, \qquad \sigma_2 = \begin{pmatrix} 0 & -i \\ i & 0 \end{pmatrix},$$

$$\sigma_3 = \begin{pmatrix} 1 & 0 \\ 0 & -1 \end{pmatrix}, \qquad I = \begin{pmatrix} 1 & 0 \\ 0 & 1 \end{pmatrix},$$

where $i^2 = -1$. Prove that

(i) $\sigma_1^2 = \sigma_2^2 = \sigma_3^2 = I$,

(ii) $\sigma_1\sigma_2 = -\sigma_2\sigma_1 = i\sigma_3$,
$\sigma_2\sigma_3 = -\sigma_3\sigma_2 = i\sigma_1$,
$\sigma_3\sigma_1 = -\sigma_1\sigma_3 = i\sigma_2$.

Show also that, if θ is a real number,

$$I + \sum_{n=1}^{\infty} \frac{(i\sigma_1\theta)^n}{n!} = I\cos\theta + i\sigma_1\sin\theta. \qquad \text{(L.U.)}$$

7. Find the reciprocal matrix of

$$A = \begin{pmatrix} 1 & 4 & 0 \\ -1 & 2 & 2 \\ 0 & 0 & 2 \end{pmatrix}$$

and verify that $AA^{-1} = A^{-1}A = I$.

8. Find the inverse matrix of

$$A = \begin{pmatrix} 1 & 2 & 3 & 4 \\ 0 & 1 & 2 & 3 \\ 0 & 0 & 1 & 2 \\ 0 & 0 & 0 & 1 \end{pmatrix}$$

9. Solve the following equations by matrix methods:

(a) $4x - 3y + z = 11$,
$2x + y - 4z = -1$,
$x + 2y - 2z = 1$.

(b) $x + 5y + 3z = 1$,
$5x + y - z = 2$,
$x + 2y + z = 3$.

10. Show that, if A and B are non-singular matrices of the same order,

$$(AB)^{-1} = B^{-1}A^{-1}.$$

319

11. Prove that if A is skew-symmetric (i.e. $A = -A'$) then

$$(I-A)(I+A)^{-1}$$

is orthogonal (assuming that $I+A$ is non-singular).

12. Prove that if A is skew-Hermitian then

$$(I-A)(I+A)^{-1}$$

is unitary (assuming that $I+A$ is non-singular).

13. If

$$S = \begin{pmatrix} \dfrac{1}{\sqrt{2}} & -\dfrac{1}{\sqrt{2}} \\ \dfrac{1}{\sqrt{2}} & \dfrac{1}{\sqrt{2}} \end{pmatrix}$$

show that S is an orthogonal matrix. Hence show that if

$$P = \begin{pmatrix} 1 & 3 \\ 3 & 1 \end{pmatrix}$$

then SPS' is a diagonal matrix.

14. Verify that

$$S = \frac{1}{\sqrt{2}} \begin{pmatrix} 1 & -i \\ -i & 1 \end{pmatrix}$$

is a unitary matrix.

15. If

$$X = \begin{pmatrix} -\dfrac{1}{2} & -\dfrac{\sqrt{3}}{2} & 0 \\ -\dfrac{\sqrt{3}}{2} & \dfrac{1}{2} & 0 \\ 0 & 0 & 2 \end{pmatrix}$$

and

$$P = \begin{pmatrix} \dfrac{1}{2} & \dfrac{\sqrt{3}}{2} & 0 \\ -\dfrac{\sqrt{3}}{2} & \dfrac{1}{2} & 0 \\ 0 & 0 & 1 \end{pmatrix},$$

prove that $P^{-1}XP$ is a diagonal matrix and that X satisfies the equation

$$X^3 - 2X^2 - X + 2I = 0,$$

where I is the unit matrix. (L.U.)

16. Obtain the eigenvalues of the matrix

$$\begin{pmatrix} 1 & 3 \\ 2 & 2 \end{pmatrix}.$$

17. Find the characteristic equation of the matrix

$$A = \begin{pmatrix} 2 & 3 & 1 \\ 3 & 1 & 2 \\ 1 & 2 & 3 \end{pmatrix},$$

and prove that the matrices

$$B = \begin{pmatrix} 3 & 1 & 2 \\ 1 & 2 & 3 \\ 2 & 3 & 1 \end{pmatrix} \quad \text{and} \quad C = \begin{pmatrix} 1 & 2 & 3 \\ 2 & 3 & 1 \\ 3 & 1 & 2 \end{pmatrix}$$

have the same characteristic equations as A.

18. If A and B are square matrices satisfying the equation

$$A^2 = B,$$

show that A and B commute. Hence find matrices A such that

$$A^2 = \begin{pmatrix} 1 & 1 & 0 \\ 0 & 1 & 0 \\ 2 & 0 & 1 \end{pmatrix}.$$

(L.U.)

19. Obtain eigenvalues and eigenvectors normalised to unit length for each of the following matrices:

(a) $\begin{pmatrix} 1 & 4 & 5 \\ 0 & 2 & 6 \\ 0 & 0 & 3 \end{pmatrix}$,   (b) $\begin{pmatrix} 2 & 3 & 1 \\ 0 & 1 & 2 \\ 0 & 0 & 1 \end{pmatrix}$,

(c) $\begin{pmatrix} 2 & 2 & 0 \\ 2 & 2 & 0 \\ 0 & 0 & 1 \end{pmatrix}$,   (d) $\begin{pmatrix} 1 & 1+i \\ 1-i & 2 \end{pmatrix}$.

20. Show that if A has eigenvalues $\lambda_1, \lambda_2, \lambda_3, \ldots, \lambda_n$ then A^m (where m is a positive integer) has eigenvalues $\lambda_1^m, \lambda_2^m, \ldots, \lambda_n^m$.

21. Prove that the eigenvalues of a unitary matrix have absolute value 1.

22. Prove that if A and B are of order n and A is a non-singular matrix then $A^{-1}BA$ and B have the same eigenvalues.

23. (i) If $A = \begin{pmatrix} a & b \\ h & a \end{pmatrix}$ and $B = \begin{pmatrix} p & q \\ r & s \end{pmatrix}$,

show that $AB = BA$ either when $b = 0$ or when $b \neq 0$, $p = s$ and $q = r$.

(ii) Given the matrices

$$\sigma_1 = \begin{pmatrix} 0 & 1 \\ 1 & 0 \end{pmatrix}, \qquad \sigma_2 = \begin{pmatrix} 0 & -1 \\ 1 & 0 \end{pmatrix}, \qquad \sigma_3 = \begin{pmatrix} 1 & 0 \\ 0 & -1 \end{pmatrix}$$

and $I = \begin{pmatrix} 1 & 0 \\ 0 & 1 \end{pmatrix}$, prove that any matrix $\begin{pmatrix} p & q \\ r & s \end{pmatrix}$ may be expressed as a linear combination

$$\begin{pmatrix} p & q \\ r & s \end{pmatrix} = c_1\sigma_1 + c_2\sigma_2 + c_3\sigma_3 + c_4 I,$$

and determine the coefficients c_1, c_2, c_3, c_4. Hence evaluate $\sigma_1\sigma_2 - \sigma_2\sigma_1$ in terms of $\sigma_1, \sigma_2, \sigma_3$ and I.

(L.U.)

24. If

$$S = \begin{pmatrix} \cos \alpha & \sin \alpha \\ -\sin \alpha & \cos \alpha \end{pmatrix}$$

and A is a 2×2 symmetric matrix with elements a_{ik}, show that $B = SAS^T$ is diagonal if

$$\tan 2\alpha = \frac{2a_{12}}{a_{11} - a_{22}},$$

where S^T is the transpose of S. Verify that $TrB = TrA$.

(L.U.)

CHAPTER 18

Groups

18.1 Definitions

In recent years group theory has been applied to many branches of physics and chemistry — notably to problems of atomic and molecular structure. This chapter has been included to give the reader an elementary introduction to the theory of groups, although it is unfortunate that most of the physically interesting applications require more knowledge of the subject than we are able to give here.

We must first define what is meant by a set of elements. Any collection of objects, quantities or operators forms a set, each individual object, quantity or operator being called an element of the set. For example, we might consider a set of students, a set of numbers, a set of points in space, or a set of operators such as $\partial/\partial x_1$, $\partial/\partial x_2$, ... , $\partial/\partial x_n$. If the set contains a finite number of elements it is said to be a finite set, whereas if it contains an infinite number of elements (i.e. the set of all positive integers) it is said to be an infinite set. We shall only be concerned here with finite sets and shall in general denote the elements of an arbitrary set by A, B, C, ... Finally if a set has no elements it is said to be empty (e.g. the set of all integers lying between 0 and 1).

Having defined a set of elements we may now proceed to define a group. A set of elements A, B, C, ... is said to form a group if there exists a rule for combining any two elements to form their ' product ' AB, say, such that

(a) every ' product ' of two elements and the ' square ' of each element are elements of the set,

(b) the associative laws hold such that $A(BC) = (AB)C$,

(c) the set contains a unit (or neutral) element I such that $IA = AI = A$,

(d) each element of the set has an inverse A^{-1} belonging to the set such that $AA^{-1} = A^{-1}A = I$.

The word ' product ' used here is to be understood within the context of the rule of combination. For example, if the elements are to be combined under multiplication then their ' products ' are obtained by multiplying any two elements together. If, however, the rule of combination is addition, the ' product ' of any two elements means their sum $A + B$. (Similar considerations apply to the word ' square ' introduced in (a).) When all multiplication is commutative the group is said to be Abelian. Finally any finite set of elements obeying the group properties (a), (b), (c), (d) are said to form a finite group, the order of the group being equal to the number of elements in it.

18.2 Examples of Groups

Example 1. The set of all integers, positive, negative and zero forms an infinite group under addition. To see that this so we note that (a) is satisfied since the sum of any two integers (and the sum of any integer with itself) is always another integer. Similarly (b) is satisfied since the associative law under addition $A + (B + C) = (A + B) + C$ is true for numbers. Likewise (c) is satisfied since the addition of 0 to any integer A does not alter it; zero is therefore the unit element (or null element as it is sometimes called) of the group since under addition

$$0 + A = A + 0 = A.$$

Finally (d) is satisfied since, if the inverse of any integer is defined as its negative, then

$$A + (-A) = 0.$$

This infinite set of positive and negative integers and zero clearly does not form a group under multiplication since the inverse elements are not then contained in the set as required by (d).

Example 2. Consider now the rotations of a line about the z-axis through angles $\pi/2$, π, $3\pi/2$ and 2π in the xy-plane (see Fig. 18.1).

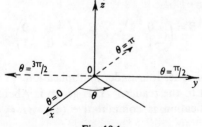

Fig. 18.1

This is a finite set of order four in that it contains four elements, namely the four operations of rotating through $\pi/2$, π, and so on. We may now show that this set of elements forms a group of order four with a rule of combination which is addition. It is clear that if we perform the operation of rotating the line through $\pi/2$ and then follow it by a rotation of π we reach the position corresponding to a rotation of $3\pi/2$. This is true for the combination of any two of the basic rotations. Consequently the group property (a) is satisfied. The group property (b) is also satisfied since the order in which successive rotations are added together is immaterial e.g. $\pi + (\frac{3}{2}\pi - 2\pi) = (\pi + \frac{3}{2}\pi) - 2\pi$. The element 2π corresponds to the unit element (or null element) in that rotating the line through 2π brings it back to its initial position. Hence the group property (c) is satisfied under addition. Finally if the inverse of any rotation is a rotation of the same magnitude but in the opposite direction then (d) is also satisfied (under addition).

Example 3. The set of four numbers 1, i, -1, $-i$, forms a group of order four under a rule of combination which is multiplication. Group property (a) is clearly satisfied since the product of any two elements (and the squares of each element) are also elements of the set (e.g. $1i = i$, $i(-i) = 1$, $i^2 = -1$, $(-i)^2 = -1$, etc.). The associative law (b) also holds for the multiplication of numbers. Similarly (c) holds if we take the unit element I as 1, for then $1i = i$, $1(-1) = -1$, and so on. Finally, the inverse of every element is also a member of the set since $1/1 = 1$, $1/i = -i$, $1/-1 = -1$ and $1/-i = i$. Under multiplication these inverses clearly satisfy the group property (d) with $I = 1$.

Example 4. The set of four 2×2 matrices

$$A = \begin{pmatrix} 1 & 0 \\ 0 & 1 \end{pmatrix}, \quad B = \begin{pmatrix} 0 & 1 \\ -1 & 0 \end{pmatrix}, \quad C = \begin{pmatrix} -1 & 0 \\ 0 & -1 \end{pmatrix}, \quad D = \begin{pmatrix} 0 & -1 \\ 1 & 0 \end{pmatrix}$$

forms a group of order four under multiplication. (The proof of this statement is left to the reader.) This group is Abelian in the sense that all multiplication is commutative (e.g. $AB = BA$, $BC = CB$, etc.).

Example 5. Suppose PQR is an equilateral triangle (see Fig. 18.2). We now consider rotations of PQR in its plane which bring it into

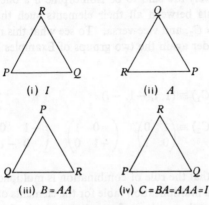

(i) I (ii) A (iii) $B = AA$ (iv) $C = BA = AAA = I$

Fig. 18.2

coincidence with its original position. These rotations may be represented by the following operations:

I ($\equiv 0°$) leaves PQR unchanged (see Fig. 18.2 (i)),

A ($\equiv 2\pi/3$) sends $P \to Q$, $Q \to R$, $R \to P$ (see Fig. 18.2 (ii)),

B ($\equiv 4\pi/3$) sends $P \to R$, $R \to Q$, $Q \to P$ (see Fig. 18.2 (iii)),

C ($\equiv 2\pi$) brings P back to P, Q back to Q, R back to R (see Fig. 18.2 (iv)).

Clearly $C = I$. Hence the possible rotations form a set of three elements represented by I, A and B. This set forms a group of order three under addition since all four group properties are satisfied. Furthermore the operation B is equivalent to performing the operation A twice (since $(4\pi/3) = (2\pi/3) + (2\pi/3)$), and similarly the operation C corresponds to performing A three times. Consequently the elements of the group may be expressed as powers of the single element A as I, A, A^2, $A^3 (= I)$. Such a group is called a cyclic group of order three, and is generated by the element A.

Another typical cyclic group of order n is given by the set of elements 1, A, A^2, ..., A^{n-1}, where $A = e^{2\pi i/n}$ and the rule of combination is multiplication. Again $A^n = e^{2\pi i} = 1$.

18.3 Isomorphism

Two groups G_1 and G_2 with elements A_1, B_1, C_1, ... and A_2, B_2, C_2, ... respectively are said to be isomorphic if a one-to-one correspondence exists between all their elements such that $A_1B_1 = C_1$ implies $A_2B_2 = C_2$, and vice-versa. To see what this means in more detail we consider again the two groups of Examples 3 and 4.

Suppose

$$G_1(I_1, A_1, B_1, C_1) \equiv (1, i, -1, -i).$$

$$G_2(I_2, A_2, B_2, C_2) \equiv \begin{pmatrix} 1 & 0 \\ 0 & 1 \end{pmatrix}, \quad \begin{pmatrix} 0 & 1 \\ -1 & 0 \end{pmatrix}, \quad \begin{pmatrix} -1 & 0 \\ 0 & -1 \end{pmatrix}, \quad \begin{pmatrix} 0 & -1 \\ 1 & 0 \end{pmatrix}$$

Remembering that the rule of combination is multiplication we may now construct a multiplication table for the elements of each of these groups such that the product of any element in the first column with any element in the first row can be obtained by looking at their common position in the table. Proceeding in this way we find

	G_1					G_2			
	I_1	A_1	B_1	C_1		I_2	A_2	B_2	C_2
I_1	I_1	A_1	B_1	C_1	I_2	I_2	A_2	B_2	C_2
A_1	A_1	B_1	C_1	I_1	A_2	A_2	B_2	C_2	I_2
B_1	B_1	C_1	I_1	A_1	B_2	B_2	C_2	I_2	A_2
C_1	C_1	I_1	A_1	B_1	C_2	C_2	I_2	A_2	B_2

Apart from the subscripts 1 and 2 these two tables have exactly the same form. Consequently if $A_1B_1 = C_1$, then $A_2B_2 = C_2$, and so on. G_1 and G_2 are therefore isomorphic.

PROBLEMS 18

1. PQR is an equilateral triangle. If A represents a rotation in the plane of PQR which sends $P \to Q$, $Q \to R$, $R \to P$, and B represents a rotation in space which changes Q and R leaving P unaltered, show that there are six rotations which bring the triangle into coincidence with itself. Verify that these rotations form a group.

2. Verify that the following set of matrices forms a group of order six under multiplication

$$\begin{pmatrix} 1 & 0 \\ 0 & 1 \end{pmatrix}, \begin{pmatrix} -1 & 1 \\ -1 & 0 \end{pmatrix}, \begin{pmatrix} 0 & -1 \\ 1 & -1 \end{pmatrix}, \begin{pmatrix} -1 & 0 \\ 0 & -1 \end{pmatrix}, \begin{pmatrix} 1 & -1 \\ 1 & 0 \end{pmatrix}, \begin{pmatrix} 0 & 1 \\ -1 & 1 \end{pmatrix}.$$

Show that the group is cyclic in that it may be generated by either one of the last two matrices.

3. Show that matrices of the type $\begin{pmatrix} a & 0 \\ 0 & 1 \end{pmatrix}$, where $a \neq 0$, form a group which is isomorphic with the group of real non-zero numbers, the rule of combination being multiplication.

4. Consider three objects numbered 1, 2 and 3, and arranged in a given order. Then there are six ways of permuting these objects (including the permutation which leaves the initial arrangement unchanged).

These permutations are

$$P_1 = \begin{pmatrix} 1 & 2 & 3 \\ 1 & 2 & 3 \end{pmatrix}, \qquad P_2 = \begin{pmatrix} 1 & 2 & 3 \\ 3 & 1 & 2 \end{pmatrix},$$

$$P_3 = \begin{pmatrix} 1 & 2 & 3 \\ 2 & 1 & 3 \end{pmatrix}, \qquad P_4 = \begin{pmatrix} 1 & 2 & 3 \\ 1 & 3 & 2 \end{pmatrix},$$

$$P_5 = \begin{pmatrix} 1 & 2 & 3 \\ 3 & 2 & 1 \end{pmatrix}, \qquad P_6 = \begin{pmatrix} 1 & 2 & 3 \\ 2 & 3 & 1 \end{pmatrix},$$

where, for example, the symbol P_2 means that 1 is replaced by 3, 2 is replaced by 1, and 3 is replaced by 2.

The effect of performing two permutations in succession is to give another permutation of the set. For example, $P_2 P_4$ means first perform P_2 and then P_4. To evaluate this we see that by P_2, 1 is replaced by 3, and by P_4, 3 is replaced by 2. Hence by $P_2 P_4$, 1 is replaced by 2. Similarly by P_2, 2 is replaced by 1, and by P_4, 1 is replaced by 1. Hence by $P_2 P_4$, 2 is replaced by 1. Finally, therefore,

$$P_2 P_4 = \begin{pmatrix} 1 & 2 & 3 \\ 3 & 1 & 2 \end{pmatrix}\begin{pmatrix} 1 & 2 & 3 \\ 1 & 3 & 2 \end{pmatrix} = \begin{pmatrix} 1 & 2 & 3 \\ 2 & 1 & 3 \end{pmatrix} = P_3.$$

Other products may be obtained in the same way. The permutation P_1 is the identity permutation in that the arrangement 1, 2, 3 is left unaltered. The inverse of a given permutation is that permutation which must be performed such that the combined effect of the two permutations is to produce the identity permutation. For example, since

$$P_2P_6 = \begin{pmatrix} 1 & 2 & 3 \\ 3 & 1 & 2 \end{pmatrix}\begin{pmatrix} 1 & 2 & 3 \\ 2 & 3 & 1 \end{pmatrix} = \begin{pmatrix} 1 & 2 & 3 \\ 1 & 2 & 3 \end{pmatrix} = P_1,$$

P_6 is said to be the inverse permutation of P_2.

Verify that this set of six permutations forms a group (the permutation group of order 6), and construct the group multiplication table. Show also that the group is non-commutative in that $P_iP_k \neq P_kP_i$, $(i, k = 2, 3, \ldots 6, i \neq k)$. Verify that this group is isomorphic to the group of six (3×3) matrices

$$S_1 = \begin{pmatrix} 1 & 0 & 0 \\ 0 & 1 & 0 \\ 0 & 0 & 1 \end{pmatrix}, \qquad S_2 = \begin{pmatrix} 0 & 1 & 0 \\ 0 & 0 & 1 \\ 1 & 0 & 0 \end{pmatrix},$$

$$S_3 = \begin{pmatrix} 0 & 1 & 0 \\ 1 & 0 & 0 \\ 0 & 0 & 1 \end{pmatrix}, \qquad S_4 = \begin{pmatrix} 1 & 0 & 0 \\ 0 & 0 & 1 \\ 0 & 1 & 0 \end{pmatrix},$$

$$S_5 = \begin{pmatrix} 0 & 0 & 1 \\ 0 & 1 & 0 \\ 1 & 0 & 0 \end{pmatrix}, \qquad S_6 = \begin{pmatrix} 0 & 0 & 1 \\ 1 & 0 & 0 \\ 0 & 1 & 0 \end{pmatrix}.$$

5. Show that the six functions

$$f_1(x) = x, \qquad f_2(x) = 1-x, \qquad f_3(x) = \frac{x-1}{x},$$

$$f_4(x) = \frac{1}{x}, \qquad f_5(x) = \frac{1}{1-x}, \qquad f_6(x) = \frac{x}{x-1},$$

form a group with the substitution of one function into another as the law of combination.

CHAPTER 19

Vectors

19.1 Introduction

It is well known that many physical quantities (for example, mass, energy and temperature) are unrelated to any direction in space and are defined entirely by a numerical magnitude (in appropriate units). Such quantities are called scalar quantities or, more shortly, scalars, and in virtue of being just numbers obey the fundamental laws of elementary algebra. However, unlike the scalar quantities just mentioned other physical quantities (such as velocity, acceleration, force and electric field) are only completely defined when both a numerical magnitude and a direction in space is specified. These quantities are called vector quantities, or just vectors. A vector obeys the same law of addition as displacement and hence may be represented geometrically by a straight line, say $\overrightarrow{PQ}$, where the arrow indicates the direction of the displacement (i.e. from P to Q). The magnitude of a vector is represented geometrically by the length of $\overrightarrow{PQ}$.

It is usual to denote all vectors by Clarendon type (e.g. **a**, **b**, ... **r**, and so on). The magnitude of a vector, say **a**, is then denoted by $|\,\mathbf{a}\,|$, or by just a, and by definition is always greater than or equal to zero. Vectors with unit magnitude (i.e. $|\,\mathbf{a}\,| = 1$) are called unit vectors and are usually distinguished from other vectors by writing, for example, **â** as the vector of unit magnitude in the direction of **a** Clearly

$$\mathbf{a} = |\,\mathbf{a}\,|\,\mathbf{\hat{a}} = a\mathbf{\hat{a}}. \tag{1}$$

In many cases it is convenient to introduce a vector with zero magnitude; this is denoted by **0** and has no directional property.

Finally two vectors **a** and **b** are said to be equal only when their magnitudes and directions are identical.

19.2 Some Vector Algebra
We now define some simple operations with vectors.

331

(a) Addition and Subtraction of Vectors

Suppose (see Fig. 19.1) the displacements $\overrightarrow{PQ}$, $\overrightarrow{QR}$ represent the vectors **a** and **b** respectively. Since the result of combining these two displacements is to give the displacement $\overrightarrow{PR}$ we define $\overrightarrow{PR}$ as the vector **c** respresenting the sum **a**+**b**. By completing the parallelogram *PQRS* we also have

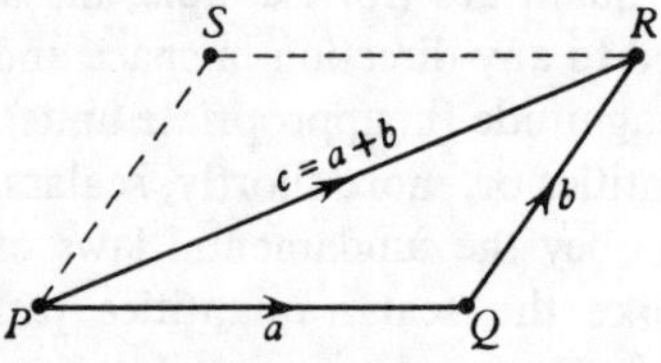

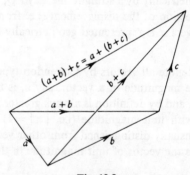

Fig. 19.1

$$\overrightarrow{PR} = \overrightarrow{PQ} + \overrightarrow{QR} = \overrightarrow{PS} + \overrightarrow{SR} \qquad (2)$$

or

$$\mathbf{c} = \mathbf{a} + \mathbf{b} = \mathbf{b} + \mathbf{a}. \qquad (3)$$

The addition of vectors is therefore commutative. Likewise we may show (see Fig. 19.2) that the associative law

$$\cdot \quad \mathbf{a} + (\mathbf{b} + \mathbf{c}) = (\mathbf{a} + \mathbf{b}) + \mathbf{c} \qquad (4)$$

is also true.

Fig. 19.2

Similar results hold for the addition of any finite number of vectors.

In order to define the difference of two vectors we first define the negative of a vector, say **a**, to be that vector which has the same magnitude as **a** but which acts in the opposite direction. This vector is denoted by $-\mathbf{a}$. To define the difference of two vectors **a** and **b** we now write

$$\mathbf{c} = \mathbf{a} - \mathbf{b} = \mathbf{a} + (-\mathbf{b}), \qquad (5)$$

which is illustrated by the geometrical construction shown in Fig. 19.3.

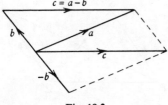

Fig. 19.3

(b) Magnitude of a Vector

Since the magnitude $|\mathbf{a}|$ of a vector $\mathbf{a}$ is just the length of the corresponding displacement, we see immediately (from Fig. 19.1) that

$$|\mathbf{c}| = |\mathbf{a}+\mathbf{b}| \leqq |\mathbf{a}|+|\mathbf{b}|. \qquad (6)$$

This is nothing more than the well-known result that the sum of any two sides of a triangle is greater than the third side (the equality sign occurring when the triangle degenerates into a straight line).

(c) Multiplication of a Vector by a Scalar

If s is a scalar (i.e. a number) then the expression $s\mathbf{a}$ denotes a vector whose magnitude is s times the magnitude of $\mathbf{a}$, and whose direction is the same as $\mathbf{a}$ or is opposite to $\mathbf{a}$ according as s is positive or negative (see Fig. 19.4). From this definition it follows that if s and t are two scalars then

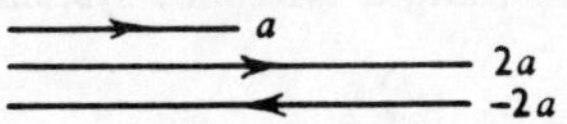

Fig. 19.4

$$s(t\mathbf{a}) = st\mathbf{a} = t(s\mathbf{a}), \qquad (7)$$

$$(s+t)\mathbf{a} = s\mathbf{a}+t\mathbf{a}, \qquad (8)$$

$$s(\mathbf{a}+\mathbf{b}) = s\mathbf{a}+s\mathbf{b}. \qquad (9)$$

To illustrate the results of (a), (b) and (c) we give the following example.

Example 1. To show, using vectors, that the diagonals of a parallelogram bisect each other. Consider the parallelogram $PQRS$ (see Fig. 19.5). Let the midpoint of PR be M, and the mid-point of QS

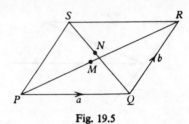

Fig. 19.5

be N. Then writing $\overrightarrow{PQ} = \mathbf{a}$, $\overrightarrow{QR} = \mathbf{b}$ we have

$$\overrightarrow{PN} = \overrightarrow{PQ} + \overrightarrow{QN} = \overrightarrow{PQ} + \tfrac{1}{2}\overrightarrow{QS} \qquad (10)$$

$$= \mathbf{a} + \tfrac{1}{2}(\mathbf{b} - \mathbf{a}) = \tfrac{1}{2}(\mathbf{a} + \mathbf{b}). \qquad (11)$$

Similarly

$$\overrightarrow{PM} = \tfrac{1}{2}\overrightarrow{PR} = \tfrac{1}{2}(\mathbf{a} + \mathbf{b}). \qquad (12)$$

Consequently $\overrightarrow{PM} = \overrightarrow{PN}$; M and N are therefore the same point and the diagonals PR, QS bisect each other.

19.3 Components of a Vector

Consider any three vectors $\mathbf{a}$, $\mathbf{b}$ and $\mathbf{c}$ (not all lying in one plane) forming a reference system with origin O (see Fig. 19.6). Then the

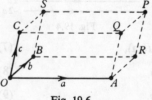

Fig. 19.6

position vector $\mathbf{r}$ of the point P (commonly called the point $\mathbf{r}$ referred to the origin O) is given by

$$\mathbf{r} = \mathbf{a} + \mathbf{b} + \mathbf{c}, \qquad (13)$$

(by completing the parallelepiped $OABCPQRS$). It is often useful to write this in a different form by putting

$$\mathbf{a} = x\hat{\mathbf{a}},$$
$$\mathbf{b} = y\hat{\mathbf{b}}, \qquad (14)$$
$$\mathbf{c} = z\hat{\mathbf{c}},$$

334

where $\hat{\mathbf{a}}$, $\hat{\mathbf{b}}$ and $\hat{\mathbf{c}}$ are unit vectors in the direction of $\mathbf{a}$, $\mathbf{b}$ and $\mathbf{c}$ respectively, and where, by (1), x, y and z are respectively the magnitudes of $\mathbf{a}$, $\mathbf{b}$ and $\mathbf{c}$. It is usual to call x, y and z the components of $\mathbf{r}$ referred to the reference frame $\mathbf{a}$, $\mathbf{b}$, $\mathbf{c}$. Hence

$$\mathbf{r} = x\hat{\mathbf{a}} + y\hat{\mathbf{b}} + z\hat{\mathbf{c}}. \tag{15}$$

This way of writing a vector in terms of its components is clearly unique since only one parallelepiped can be drawn once $\mathbf{a}$, $\mathbf{b}$ and $\mathbf{c}$ have been specified. Consequently if we have two points P_1 and P_2 with position vectors $\mathbf{r}_1$ and $\mathbf{r}_2$ respectively, such that

$$\begin{aligned} \mathbf{r}_1 &= x_1\hat{\mathbf{a}} + y_1\hat{\mathbf{b}} + z_1\hat{\mathbf{c}}, \\ \mathbf{r}_2 &= x_2\hat{\mathbf{a}} + y_2\hat{\mathbf{b}} + z_2\hat{\mathbf{c}}, \end{aligned} \tag{16}$$

then $\mathbf{r}_1 = \mathbf{r}_2$ only when

$$x_1 = x_2, \quad y_1 = y_2, \quad z_1 = z_2. \tag{17}$$

Similarly if $\mathbf{r}_3 = x_3\hat{\mathbf{a}} + y_3\hat{\mathbf{b}} + z_3\hat{\mathbf{c}}$ such that $\mathbf{r}_3 = \mathbf{r}_1 + \mathbf{r}_2$, then

$$x_3\mathbf{a} + y_3\mathbf{b} + z_3\mathbf{c} = (x_1\mathbf{a} + y_1\mathbf{b} + z_1\mathbf{c}) + (x_2\mathbf{a} + y_2\mathbf{b} + z_2\mathbf{c}). \tag{18}$$

Hence

$$x_3 = x_1 + x_2, \quad y_3 = y_1 + y_2, \quad z_3 = z_1 + z_2. \tag{19}$$

Vectors may therefore be added by adding their respective components.

An important and commonly used reference system is the right-handed rectangular Cartesian coordinate frame $Oxyz$ (see Fig. 19.7). (We remark here that the term right-handed as applied to a coordinate frame implies that when looking along the positive Oz direction a clockwise rotation about Oz is needed to take Ox into Oy. Similarly a frame is left-handed (e.g. x and y interchanged in Fig. 19.7) if when looking along the positive Oz direction an anticlockwise rotation is required to take Ox into Oy. The specification of

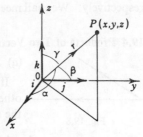

Fig. 19.7

the right-or-left-handedness of a coordinate frame is important in vector algebra as we shall see later on. However, unless otherwise stated we shall always use a right-handed system.)

It is usual when dealing with the decomposition of vectors into components along the three axes Ox, Oy and Oz to define the vectors **i**, **j** and **k** to be unit vectors in these three directions. Consequently the position vector **r** of a point $P(x, y, z)$ is given by

$$\mathbf{r} = x\mathbf{i} + y\mathbf{j} + z\mathbf{k}, \tag{20}$$

where x, y and z are its components in the **i**, **j** and **k** directions respectively.

Similarly, if **a** is any arbitrary vector we have

$$\mathbf{a} = a_x\mathbf{i} + a_y\mathbf{j} + a_z\mathbf{k}, \tag{21}$$

where a_x, a_y, a_z are its components in the **i**, **j** and **k** directions respectively.

In many geometrical applications the three angles α, β and γ which the position vector **r** of P makes with the three axes Ox, Oy and Oz are of interest. For from Fig. 19.7 we see that since

$$\mathbf{r} = \mathbf{i}x + \mathbf{j}y + \mathbf{k}z \tag{22}$$

then

$$x = |\mathbf{r}|\cos\alpha, \quad y = |\mathbf{r}|\cos\beta, \quad z = |\mathbf{r}|\cos\gamma, \tag{23}$$

where $|\mathbf{r}|$ is the length of OP. Consequently

$$\frac{\mathbf{r}}{|\mathbf{r}|} = \hat{\mathbf{r}} = \mathbf{i}\cos\alpha + \mathbf{j}\cos\beta + \mathbf{k}\cos\gamma. \tag{24}$$

The cosines of α, β and γ which appear in (24) are called the direction cosines of the position vector **r**, and are denoted by l, m and n respectively. We shall meet these quantities again in the next section.

19.4 Products of Two Vectors

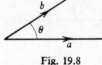

Fig. 19.8

(a) *Scalar Product*

If **a** and **b** are two vectors inclined to each other at an angle θ (see Fig. 19.8) we define the scalar (or dot) product **a . b** as

$$\mathbf{a} \cdot \mathbf{b} = |\mathbf{a}||\mathbf{b}|\cos\theta = ab\cos\theta. \tag{25}$$

Since $a \cos \theta$ is just the magnitude of the projection of **a** on **b** we see that

$$\mathbf{a} \cdot \mathbf{b} = (\text{projection of } \mathbf{a} \text{ on } \mathbf{b})b = (\text{projection of } \mathbf{b} \text{ on } \mathbf{a})a = \mathbf{b} \cdot \mathbf{a}. \quad (26)$$

The scalar product of two vectors is therefore commutative. Furthermore the distributive law

$$\mathbf{a} \cdot (\mathbf{b} + \mathbf{c}) = \mathbf{a} \cdot \mathbf{b} + \mathbf{a} \cdot \mathbf{c} \quad (27)$$

is also seen to hold. Likewise if s and t are arbitrary scalars then

$$(s\mathbf{a}) \cdot (t\mathbf{b}) = st(\mathbf{a} \cdot \mathbf{b}). \quad (28)$$

Unlike the product of two scalars, however, we now see that the scalar product of two vectors can be zero even when **a** and **b** are not themselves zero. For $\mathbf{a} \cdot \mathbf{b} = 0$ when $\theta = \pi/2$; that is, when **a** and **b** are at right angles.

The special case of $\mathbf{b} = \mathbf{a}$ leads to a definition of the square of a vector since

$$\mathbf{a} \cdot \mathbf{a} = |\mathbf{a}| |\mathbf{a}| \cos 0 = |\mathbf{a}|^2 = a^2. \quad (29)$$

Similarly the square of a unit vector $\hat{\mathbf{a}}$ (*say*) is

$$\hat{\mathbf{a}} \cdot \hat{\mathbf{a}} = 1, \quad (30)$$

as required.

In a rectangular Cartesian coordinate system we have

$$\left. \begin{aligned} \mathbf{a} &= \mathbf{i}a_x + \mathbf{j}a_y + \mathbf{k}a_z \\ \mathbf{b} &= \mathbf{i}b_x + \mathbf{j}b_y + \mathbf{k}b_z \end{aligned} \right\}, \quad (31)$$

where a_x, a_y, a_z and b_x, b_y, b_z are the components of **a** and **b** in the x, y and z directions respectively. Hence

$$\mathbf{a} \cdot \mathbf{b} = (\mathbf{i}a_x + \mathbf{j}a_y + \mathbf{k}a_z) \cdot (\mathbf{i}b_x + \mathbf{j}b_y + \mathbf{k}b_z). \quad (32)$$

Now since **i**, **j** and **k** are perpendicular unit vectors we have

$$\mathbf{i} \cdot \mathbf{i} = \mathbf{j} \cdot \mathbf{j} = \mathbf{k} \cdot \mathbf{k} = 1, \quad (33)$$

$$\left. \begin{aligned} \mathbf{i} \cdot \mathbf{j} &= \mathbf{j} \cdot \mathbf{i} = 0, \\ \mathbf{j} \cdot \mathbf{k} &= \mathbf{k} \cdot \mathbf{j} = 0, \\ \mathbf{k} \cdot \mathbf{i} &= \mathbf{i} \cdot \mathbf{k} = 0. \end{aligned} \right\} \quad (34)$$

Hence

$$\mathbf{a} \cdot \mathbf{b} = a_x b_x + a_y b_y + a_z b_z = \mathbf{b} \cdot \mathbf{a}. \qquad (35)$$

Similarly if $\mathbf{r}$ is the position vector of a point $P(x, y, z)$ then since

$$\mathbf{r} = \mathbf{i}x + \mathbf{j}y + \mathbf{k}z \qquad (36)$$

we have

$$r^2 = \mathbf{r} \cdot \mathbf{r} = x^2 + y^2 + z^2 \qquad (37)$$

or

$$r = |\mathbf{r}| = \sqrt{(x^2 + y^2 + z^2)}. \qquad (38)$$

Using (24) we also have

$$\hat{\mathbf{r}} \cdot \hat{\mathbf{r}} = \cos^2 \alpha + \cos^2 \beta + \cos^2 \gamma = l^2 + m^2 + n^2 = 1. \qquad (39)$$

Likewise if $\mathbf{r}_1$ and $\mathbf{r}_2$ are two vectors inclined at an angle θ then

$$\mathbf{r}_1 \cdot \mathbf{r}_2 = |\mathbf{r}_1| \, |\mathbf{r}_2| \cos \theta \qquad (40)$$

or, since $\mathbf{r}_1 = |\mathbf{r}_1| \, \hat{\mathbf{r}}_1, \quad \mathbf{r}_2 = |\mathbf{r}_2| \, \hat{\mathbf{r}}_2,$

$$\cos \theta = \hat{\mathbf{r}}_1 \cdot \hat{\mathbf{r}}_2 = \cos \alpha_1 \cos \alpha_2 + \cos \beta_1 \cos \beta_2 + \cos \gamma_1 \cos \gamma_2$$

$$= l_1 l_2 + m_1 m_2 + n_1 n_2. \qquad (41)$$

Hence $\mathbf{r}_1$ and $\mathbf{r}_2$ are perpendicular if

$$l_1 l_2 + m_1 m_2 + n_1 n_2 = 0, \qquad (42)$$

and parallel if

$$l_1 l_2 + m_1 m_2 + n_1 n_2 = \pm 1. \qquad (43)$$

Example 2. To find the angle between the vectors

$$\mathbf{r}_1 = \mathbf{i} + \mathbf{j} + \mathbf{k}, \quad \mathbf{r}_2 = -\mathbf{i} + 2\mathbf{j} + 3\mathbf{k}. \qquad (44)$$

Here $|\mathbf{r}_1| = \sqrt{3}, \quad |\mathbf{r}_2| = \sqrt{14}.$ Consequently

$$\hat{\mathbf{r}}_1 = \frac{1}{\sqrt{3}}(\mathbf{i} + \mathbf{j} + \mathbf{k}) \quad \text{and} \quad \hat{\mathbf{r}}_2 = \frac{1}{\sqrt{14}}(-\mathbf{i} + 2\mathbf{j} + 3\mathbf{k}) \qquad (45)$$

and

$$\cos \theta = \mathbf{\hat{r}}_1 . \mathbf{\hat{r}}_2 = \frac{1}{\sqrt{42}}(-1+2+3) = \frac{4}{\sqrt{42}}. \tag{46}$$

(b) Vector Product

If **a** and **b** are two vectors in a plane and θ is the angle between them measured from the direction of **a**, then the vector product **a** ∧ **b** is defined as

$$\mathbf{a} \wedge \mathbf{b} = |\mathbf{a}| |\mathbf{b}| \sin \theta \,\mathbf{\hat{n}} = ab \sin \theta \,\mathbf{\hat{n}}, \tag{47}$$

where **n̂** is the unit vector normal to the plane containing **a** and **b** such that **a**, **b**, **n̂** form a right-handed system (see Fig. 19.9). The vector product **a** ∧ **b** is therefore a vector of magnitude $ab \sin \theta$ acting in the direction of **n̂**. Since $ab \sin \theta$ is just the area S of the parallelogram $OACB$ we speak of $\mathbf{a} \wedge \mathbf{b} = S\mathbf{\hat{n}} = \mathbf{S}$ as defining the vector area **S**.

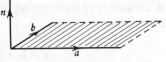

Fig. 19.9

Now since in (47) θ is measured from the direction of **a**, we have

$$\mathbf{b} \wedge \mathbf{a} = |\mathbf{b}| |\mathbf{a}| \sin (-\theta)\mathbf{\hat{n}} = -|\mathbf{a}| |\mathbf{b}| \sin \theta\mathbf{\hat{n}} = -\mathbf{a} \wedge \mathbf{b}. \tag{48}$$

The vector product is therefore not commutative. It is for this reason that we mentioned earlier the difference between a right-handed and a left-handed coordinate frame, for clearly **a** ∧ **b** changes sign in passing from one to the other.

It can be verified from the definition of the vector product (47) that

$$\mathbf{a} \wedge (\mathbf{b}+\mathbf{c}) = \mathbf{a} \wedge \mathbf{b}+\mathbf{a} \wedge \mathbf{c}, \tag{49}$$

and

$$(s\mathbf{a}) \wedge (t\mathbf{b}) = st(\mathbf{a} \wedge \mathbf{b}), \tag{50}$$

where s and t are scalars.

As with the scalar product discussed in (a) the vector product may also be zero without either **a** or **b** being zero. For when $\theta = 0$ (i.e. when **a** and **b** are parallel) we have by (47)

$$\mathbf{a} \wedge \mathbf{b} = 0. \tag{51}$$

339

We now refer the vector product to the right-handed rectangular Cartesian coordinate frame of Fig. 19.7. By (47) the following relations hold between the three perpendicular unit vectors $\mathbf{i}$, $\mathbf{j}$ and $\mathbf{k}$:

$$\mathbf{i} \wedge \mathbf{i} = \mathbf{j} \wedge \mathbf{j} = \mathbf{k} \wedge \mathbf{k} = 0, \tag{52}$$

$$\mathbf{i} \wedge \mathbf{j} = \mathbf{k}, \quad \mathbf{j} \wedge \mathbf{k} = \mathbf{i}, \quad \mathbf{k} \wedge \mathbf{i} = \mathbf{j} \tag{53}$$

(together with the relations $\mathbf{i} \wedge \mathbf{j} = -\mathbf{j} \wedge \mathbf{i}$, etc.). Hence

$$\mathbf{a} \wedge \mathbf{b} = (\mathbf{i}a_x + \mathbf{j}a_y + \mathbf{k}a_z) \wedge (\mathbf{i}b_x + \mathbf{j}b_y + \mathbf{k}b_z)$$
$$= \mathbf{i}(a_y b_z - a_z b_y) + \mathbf{j}(a_z b_x - a_x b_z) + \mathbf{k}(a_x b_y - a_y b_x). \tag{54}$$

A useful way of representing this result is by writing it in determinant form thus

$$\mathbf{a} \wedge \mathbf{b} = \begin{vmatrix} \mathbf{i} & \mathbf{j} & \mathbf{k} \\ a_x & a_y & a_z \\ b_x & b_y & b_z \end{vmatrix} \tag{55}$$

Using the properties of determinants we again deduce that $\mathbf{b} \wedge \mathbf{a} = -\mathbf{a} \wedge \mathbf{b}$, $\mathbf{a} \wedge \mathbf{a} = 0$, and so on.

It is important to note that the vector equation

$$\mathbf{a} \wedge \mathbf{c} = \mathbf{b} \wedge \mathbf{c} \tag{56}$$

does not imply $\mathbf{a} = \mathbf{b}$. For consider

$$\mathbf{a} = \mathbf{b} + s\mathbf{c}, \tag{57}$$

where s is a scalar parameter. Then

$$\mathbf{a} \wedge \mathbf{c} = (\mathbf{b} + s\mathbf{c}) \wedge \mathbf{c} = \mathbf{b} \wedge \mathbf{c}. \tag{58}$$

Consequently (56) is satisfied by any vector with the form (57). In other words $\mathbf{a}$ and $\mathbf{b}$ may differ by any vector parallel to $\mathbf{c}$.

Example 3. To find the angle between the vectors of Example 2 using the vector product.

Here

$$\mathbf{\hat{r}}_1 = \frac{1}{\sqrt{3}}(\mathbf{i}+\mathbf{j}+\mathbf{k}),$$ (59)

$$\mathbf{\hat{r}}_2 = \frac{1}{\sqrt{14}}(-\mathbf{i}+2\mathbf{j}+3\mathbf{k}),$$

whence

$$\mathbf{\hat{r}}_1 \wedge \mathbf{\hat{r}}_2 = \frac{1}{\sqrt{14}}(\mathbf{i}+\mathbf{j}+\mathbf{k}) \wedge \frac{1}{\sqrt{3}}(-\mathbf{i}+2\mathbf{j}+3\mathbf{k})$$ (60)

$$= \frac{1}{\sqrt{42}}(\mathbf{i}-4\mathbf{j}+3\mathbf{k}).$$ (61)

Now since by definition $\mathbf{\hat{r}}_1 \wedge \mathbf{\hat{r}}_2 = \sin\theta\,\mathbf{\hat{n}}$, we have

$$|\mathbf{\hat{r}}_1 \wedge \mathbf{\hat{r}}_2| = |\sin\theta|.$$ (62)

Using (61)

$$|\mathbf{\hat{r}}_1 \wedge \mathbf{\hat{r}}_2|^2 = \frac{1}{42}(\mathbf{i}-4\mathbf{j}+3\mathbf{k}) \cdot (\mathbf{i}-4\mathbf{j}+3\mathbf{k})$$ (63)

$$= \frac{1}{42}(1+16+9) = \frac{26}{42}.$$ (64)

Hence

$$|\sin\theta| = \sqrt{\frac{26}{42}} \quad \text{and} \quad \cos\theta = \frac{4}{\sqrt{42}},$$ (65)

as before.

Example 4. To show that if three vectors **a**, **b**, **c** satisfy the relation $\mathbf{a}+\mathbf{b}+\mathbf{c} = 0$, then $\mathbf{a} \wedge \mathbf{b} = \mathbf{b} \wedge \mathbf{c} = \mathbf{c} \wedge \mathbf{a}$.

This is easily proved by forming the vector product

$$(\mathbf{a}+\mathbf{b}+\mathbf{c}) \wedge \mathbf{b} = \mathbf{a} \wedge \mathbf{b}+\mathbf{c} \wedge \mathbf{b} = 0.$$ (66)

Hence

$$\mathbf{a} \wedge \mathbf{b} = -\mathbf{c} \wedge \mathbf{b} = \mathbf{b} \wedge \mathbf{c}.$$ (67)

The other result follows in a similar way by taking the vector product of $\mathbf{a}+\mathbf{b}+\mathbf{c}$ and **c**.

19.5 Product of Three Vectors

(a) Scalar Triple Product

If **a**, **b** and **c** are three arbitrary vectors the scalar formed by taking the product **a** . (**b** ∧ **c**) is called the scalar triple product. Using the determinantal representation of **b** ∧ **c** given in (55) we have

$$\mathbf{a} \cdot (\mathbf{b} \wedge \mathbf{c}) = (\mathbf{i}a_x + \mathbf{j}a_y + \mathbf{k}a_z) \cdot \begin{vmatrix} \mathbf{i} & \mathbf{j} & \mathbf{k} \\ b_x & b_y & b_z \\ c_x & c_y & c_z \end{vmatrix} \tag{68}$$

$$= \begin{vmatrix} a_x & a_y & a_z \\ b_x & b_y & b_z \\ c_x & c_y & c_z \end{vmatrix}. \tag{69}$$

Hence from the properties of determinants

$$\mathbf{a} \cdot (\mathbf{b} \wedge \mathbf{c}) = \mathbf{b} \cdot (\mathbf{c} \wedge \mathbf{a}) = \mathbf{c} \cdot (\mathbf{a} \wedge \mathbf{b}) \tag{70}$$

and

$$\mathbf{a} \cdot (\mathbf{b} \wedge \mathbf{c}) = -\mathbf{b} \cdot (\mathbf{a} \wedge \mathbf{c}), \quad \text{etc.} \tag{71}$$

Likewise

$$\mathbf{a} \cdot (\mathbf{b} \wedge \mathbf{a}) = 0. \tag{72}$$

From (70) we see that the scalar triple product is unchanged by performing a cyclic permutation of the vectors. However, if the cyclic order is violated (as in (71)) the scalar triple product changes sign. Clearly changing from a right-handed coordinate frame to a left-handed frame also changes the sign of the scalar triple product. In this sense therefore it behaves in a different way from an ordinary scalar which is independent of the handedness of the coordinate frame; for this reason the scalar triple product is often called a pseudoscalar.

The scalar triple product admits a simple geometrical interpretation, for if **a**, **b**, **c** form three sides of a parallelepiped (see Fig. 19.10) then **b** ∧ **c** represents the vector area $S\hat{n}$ of the base, where $\hat{n}$ is the unit vector perpendicular to both **b** and **c**. Hence the scalar triple product

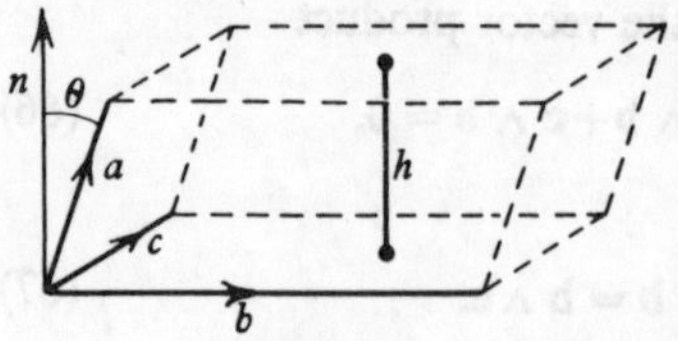

Fig. 19.10

$$\mathbf{a} \cdot (\mathbf{b} \wedge \mathbf{c}) = \mathbf{a} \cdot S\hat{n} = (\mathbf{a} \cdot \mathbf{n})S = S \, | \, a \, | \cos \theta. \tag{73}$$

Since $|a|\cos\theta$ is just the perpendicular height h of the parallelepiped it follows that

$$\mathbf{a}\,.\,(\mathbf{b}\wedge\mathbf{c}) = Sh = V, \tag{74}$$

where V is the volume of the parallelepiped. This volume is zero when $\mathbf{a}$, $\mathbf{b}$ and $\mathbf{c}$ are coplanar and we therefore deduce that the condition for three vectors to be coplanar is

$$\mathbf{a}\,.\,(\mathbf{b}\wedge\mathbf{c}) = 0. \tag{75}$$

Example 5. To solve the equation

$$\mathbf{a}x+\mathbf{b}y+\mathbf{c}z = \mathbf{d}, \tag{76}$$

where $\mathbf{a}$, $\mathbf{b}$, $\mathbf{c}$ and $\mathbf{d}$ are constant vectors.

Consider the scalar triple product

$$(\mathbf{a}x+\mathbf{b}y+\mathbf{c}z)\,.\,(\mathbf{b}\wedge\mathbf{c}) = \mathbf{d}\,.\,(\mathbf{b}\wedge\mathbf{c}). \tag{77}$$

Then since $\mathbf{b}\,.\,(\mathbf{b}\wedge\mathbf{c}) = \mathbf{c}\,.\,(\mathbf{b}\wedge\mathbf{c}) = 0$, we have

$$x\mathbf{a}\,.\,(\mathbf{b}\wedge\mathbf{c}) = \mathbf{d}\,.\,(\mathbf{b}\wedge\mathbf{c}) \tag{78}$$

or

$$x = \frac{\mathbf{d}\,.\,(\mathbf{b}\wedge\mathbf{c})}{\mathbf{a}\,.\,(\mathbf{b}\wedge\mathbf{c})} \tag{79}$$

provided $\mathbf{a}\,.\,(\mathbf{b}\wedge\mathbf{c})\neq 0$.

Similarly

$$(\mathbf{a}x+\mathbf{b}y+\mathbf{c}z)\,.\,(\mathbf{a}\wedge\mathbf{c})\quad\text{and}\quad(\mathbf{a}x+\mathbf{b}y+\mathbf{c}z)\,.\,(\mathbf{a}\wedge\mathbf{b}) = \mathbf{d}\,.\,(\mathbf{a}\wedge\mathbf{b})$$

give expressions for y and z respectively.

(b) Vector Triple Product

If $\mathbf{a}$, $\mathbf{b}$ and $\mathbf{c}$ are three arbitrary vectors, the vectors formed by taking the products $\mathbf{a}\wedge(\mathbf{b}\wedge\mathbf{c})$ and $(\mathbf{a}\wedge\mathbf{b})\wedge\mathbf{c}$ are called vector triple products. The positions of the brackets are important here as we shall see shortly that $\mathbf{a}\wedge(\mathbf{b}\wedge\mathbf{c})$ is not equal to $(\mathbf{a}\wedge\mathbf{b})\wedge\mathbf{c}$.

Consider first the product $\mathbf{a}\wedge(\mathbf{b}\wedge\mathbf{c})$. This represents a vector coplanar with $\mathbf{b}$ and $\mathbf{c}$ since $\mathbf{b}\wedge\mathbf{c}$ is perpendicular to the plane of $\mathbf{b}$

and **c**, and $\mathbf{a} \wedge (\mathbf{b} \wedge \mathbf{c})$ is perpendicular to $\mathbf{b} \wedge \mathbf{c}$ (see Fig. 19.11). To evaluate $\mathbf{a} \wedge (\mathbf{b} \wedge \mathbf{c})$ we now write $\mathbf{b} \wedge \mathbf{c} = \mathbf{f}$. Then by (55) the $x, y\ z$ components of **f** are

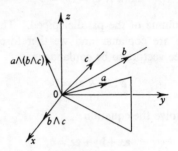

Fig. 19.11

$$f_x = \begin{vmatrix} b_y & b_z \\ c_y & c_z \end{vmatrix}, \quad f_y = \begin{vmatrix} b_z & b_x \\ c_z & c_x \end{vmatrix}, \quad f_z = \begin{vmatrix} b_x & b_y \\ c_x & c_y \end{vmatrix}. \quad (80)$$

Consequently

$$\mathbf{a} \wedge (\mathbf{b} \wedge \mathbf{c}) = \mathbf{a} \wedge \mathbf{f} = \begin{vmatrix} \mathbf{i} & \mathbf{j} & \mathbf{k} \\ a_x & a_y & a_z \\ f_x & f_y & f_z \end{vmatrix} \quad (81)$$

$$= \begin{vmatrix} \mathbf{i} & \mathbf{j} & \mathbf{k} \\ a_x & a_y & a_z \\ \begin{vmatrix} b_y & b_z \\ c_y & c_z \end{vmatrix} & \begin{vmatrix} b_z & b_x \\ c_z & c_x \end{vmatrix} & \begin{vmatrix} b_x & b_y \\ c_x & c_y \end{vmatrix} \end{vmatrix} \quad (82)$$

$$
\begin{aligned}
= &\ \mathbf{i}\{a_y(b_x c_y - b_y c_x) - a_z(b_z c_x - b_x c_z)\} \\
&+ \mathbf{j}\{a_z(b_y c_z - b_z c_y) - a_x(b_x c_y - b_y c_x)\} \\
&+ \mathbf{k}\{a_x(b_z c_x - b_x c_z) - a_y(b_y c_z - b_z c_y)\}
\end{aligned} \quad (83)
$$

$$
\begin{aligned}
= &\ \mathbf{i}\{b_x(a_x c_x + a_y c_y + a_z c_z) - c_x(a_x b_x + a_y b_y + a_z b_z)\} \\
&+ \mathbf{j}\{b_y(a_x c_x + a_y c_y + a_z c_z) - c_y(a_x b_x + a_y b_y + a_z b_z)\} \\
&+ \mathbf{k}\{b_z(a_x c_x + a_y c_y + a_z c_z) - c_z(a_x b_x + a_y b_y + a_z b_z)\}
\end{aligned} \quad (84)
$$

$$= (\mathbf{a} \cdot \mathbf{c})\mathbf{b} - (\mathbf{a} \cdot \mathbf{b})\mathbf{c}. \quad (85)$$

Similarly we find

$$(\mathbf{a} \wedge \mathbf{b}) \wedge \mathbf{c} = (\mathbf{c} \cdot \mathbf{a})\mathbf{b} - (\mathbf{c} \cdot \mathbf{b})\mathbf{a}. \quad (86)$$

344

Accordingly we see that $\mathbf{a} \wedge (\mathbf{b} \wedge \mathbf{c})$ is neither numerically equal to nor parallel to $(\mathbf{a} \wedge \mathbf{b}) \wedge \mathbf{c}$. For example

$$\mathbf{i} \wedge (\mathbf{i} \wedge \mathbf{j}) = \mathbf{i} \wedge \mathbf{k} = -\mathbf{j}, \qquad (87)$$

and

$$(\mathbf{i} \wedge \mathbf{i}) \wedge \mathbf{j} = 0, \quad \text{since} \quad \mathbf{i} \wedge \mathbf{i} = 0. \qquad (88)$$

Example 6. From (85) we have, on putting $c = a$,

$$\mathbf{a} \wedge (\mathbf{b} \wedge \mathbf{a}) = (\mathbf{a} \cdot \mathbf{a})\mathbf{b} - (\mathbf{a} \cdot \mathbf{b})\mathbf{a} = a^2\mathbf{b} - (\mathbf{a} \cdot \mathbf{b})\mathbf{a}. \qquad (89)$$

Hence

$$\mathbf{b} = \frac{(\mathbf{a} \cdot \mathbf{b})\mathbf{a}}{a^2} + \frac{\mathbf{a} \wedge (\mathbf{b} \wedge \mathbf{a})}{a^2}. \qquad (90)$$

In this way the vector $\mathbf{b}$ has been resolved into two component vectors, one parallel to $\mathbf{a}$ (i.e. $(\mathbf{a} \cdot \mathbf{b})\mathbf{a}/a^2$) and the other perpendicular to $\mathbf{a}$ (i.e. $\mathbf{a} \wedge (\mathbf{b} \wedge \mathbf{a})/a^2$).

19.6. Lines and Planes

In the last few sections we have seen how to find the angles between two intersecting straight lines represented by vectors. We now discuss the equation of a straight line in vector form.

Suppose P is an arbitrary point on a straight line, Q is a given point on the same line, and 0 is the origin of some coordinate system (see Fig. 19.12). We suppose that P has a position vector $\mathbf{r}$, that Q has a

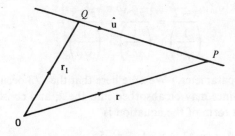

Fig. 19.12

position vector $\mathbf{r}_1$, and that $\hat{\mathbf{u}}$ is a unit vector along the line QP. Then by the composition of vectors we have

$$\mathbf{r} = \mathbf{r}_1 + \lambda\hat{\mathbf{u}}, \qquad (91)$$

where λ is a parameter representing the number of unit vectors $\hat{\mathbf{u}}$ in the distance QP, and which varies as P moves along the line.

Equation (91) is the standard form of the equation of a straight line. If P has coordinates (x, y, z), Q has coordinates (x_1, y_1, z_1), and

$$\hat{\mathbf{u}} = \mathbf{i}l + \mathbf{j}m + \mathbf{k}n,$$

where l, m, n are the direction cosines of the line QP with respect to a Cartesian coordinate system centred at 0, then taking components (91) gives

$$\frac{x - x_1}{l} = \frac{y - y_1}{m} = \frac{z - z_1}{n} = \lambda, \qquad (92)$$

which is the Cartesian equation of a straight line.

Example 7. The line through the point $(2,1,5)$ with direction cosines $(l, m, n) = \left(\frac{1}{\sqrt{3}}, \frac{1}{\sqrt{3}}, \frac{1}{\sqrt{3}} \right)$ is represented by the equation

$$\mathbf{r} = \mathbf{r}_1 + \lambda\hat{\mathbf{u}}, \qquad (93)$$

where

$$\mathbf{r}_1 = 2\mathbf{i} + \mathbf{j} + 5\mathbf{k}, \quad \text{and} \quad \hat{\mathbf{u}} = \frac{1}{\sqrt{3}}\mathbf{i} + \frac{1}{\sqrt{3}}\mathbf{j} + \frac{1}{\sqrt{3}}\mathbf{k}.$$

In Cartesian form we have, by (92),

$$\frac{x-2}{\left(\frac{1}{\sqrt{3}}\right)} = \frac{y-1}{\left(\frac{1}{\sqrt{3}}\right)} = \frac{z-5}{\left(\frac{1}{\sqrt{3}}\right)} = \lambda, \qquad (94)$$

where λ is a parameter. We note here that the $\sqrt{3}$ occurring in the direction cosines may be absorbed into the λ, and consequently an equally valid form of the equation is

$$\frac{x-2}{1} = \frac{y-1}{1} = \frac{z-5}{1} = \lambda' \quad \text{(say)}. \qquad (95)$$

This serves to point to the fact that the numbers l, m, n need not be actual direction cosines but only numbers which are proportional to the cosines. For example, the equation

$$\frac{x-3}{4} = \frac{y+2}{3} = \frac{z-1}{2} = \lambda \tag{96}$$

defines a straight line passing through the point $(3, -2, 1)$ and having direction *ratios* $(4, 3, 2)$. The direction cosines may be obtained from these direction ratios by normalising thus

$$l = \frac{4}{\sqrt{(4^2+3^2+2^2)}}, \quad m = \frac{3}{\sqrt{(4^2+3^2+2^2)}}, \quad n = \frac{2}{\sqrt{(4^2+3^2+2^2)}}, \tag{97}$$

whence $l^2+m^2+n^2 = 1$ as required by (39).

Example 8. To show that the equations

$$\left. \begin{aligned} \frac{x+2}{2} &= \frac{y+8}{5} = \frac{z+5}{3}, \\ \frac{x-2}{4} &= \frac{y-2}{10} = \frac{z-1}{6}, \end{aligned} \right\} \tag{98}$$

represent the same line.

Now the direction ratios of the first line are directly proportional to those of the second line. Hence the direction defined by the first set of direction ratios is the same as that defined by the second set. Accordingly the lines must be either parallel or identical. The point $(-2, -8, -5)$ lies on the first line, and by inserting these (x, y, z) values into the second equation we see that the equation is satisfied. Hence the two lines have a point in common, and act in the same direction. They are therefore the same line.

We now come to the vector equation of a plane. Suppose (see Fig. 19.13) that P is a point in the plane, that O is the origin of the coordinate system, and that Q is a point in the plane such that OQ is perpendicular to the plane. Let P have position vector **r**, and let the direction of OQ be defined by a unit vector **n̂**, as shown.

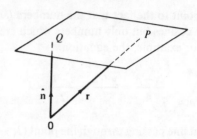

Fig. 19.13

Then the angle OQP is always a right-angle wherever P may be in the plane. Hence if the length of $OQ = p$, we have, using the definition of the scalar product,

$$r \cdot \hat{n} = |r||\hat{n}| \cos \theta = p, \qquad (99)$$

where θ is the angle between OP and OQ. Equation (99) is the vector equation of a plane. Now if we write

$$\hat{n} = il + jm + kn, \qquad (100)$$

where (l, m, n) are the direction cosines of the line OQ, then (99) becomes

$$lx + my + nz = p, \qquad (101)$$

which is the Cartesian form of the equation of a plane.

Example 9. To find the perpendicular distance from the origin to the plane

$$3x + 4y + 5z = 6. \qquad (102)$$

Here the triple of numbers $(3, 4, 5)$ are clearly not the direction cosines of the line OQ (since they are greater than unity). Accordingly we normalise (102) to bring it into the form (101) by dividing by $\sqrt{(3^2 + 4^2 + 5^2)}$ throughout—that is,

$$\frac{3}{\sqrt{(3^2 + 4^2 + 5^2)}} x + \frac{4}{\sqrt{(3^2 + 4^2 + 5^2)}} y + \frac{5}{\sqrt{(3^2 + 4^2 + 5^2)}} z$$

$$= \frac{6}{\sqrt{(3^2 + 4^2 + 5^2)}} . \qquad (103)$$

By comparison with (101) we see that the required perpendicular distance is

$$p = \frac{6}{\sqrt{(3^2 + 4^2 + 5^2)}} = \frac{6}{\sqrt{50}}. \tag{104}$$

It is commonplace to refer to the direction cosines of the normal to the plane as the direction cosines of the plane. Clearly two planes are parallel if they have a common normal, and are identical if, in addition, they have a point in common. If they are neither parallel nor coincident, then they intersect in a straight line. The equation of a straight line may therefore be obtained from the equations of two intersecting planes, as Example 10 shows.

Example 10. Find the standard form of the equation of the straight line defined by the intersection of the planes

$$\left. \begin{array}{l} 3x + y - z = 3, \\ x + 2y + 4z = -4. \end{array} \right\} \tag{105}$$

Let the direction cosines of the required straight line be (l, m, n). Then since this line lies in both planes it must be perpendicular to the normals to each of these planes. Using the condition for two lines to be perpendicular (equation (42)) we therefore have

$$\left. \begin{array}{l} 3l + m - n = 0, \\ l + 2m + 4n = 0, \end{array} \right\} \tag{106}$$

which lead to

$$\frac{l}{6} = -\frac{m}{13} = \frac{n}{5}. \tag{107}$$

The direction *ratios* of the required line are therefore $(6, -13, 5)$. Having established the direction of the line, we now wish to find some point on the line. Such a point will be common to both planes. Suppose we choose a point on the planes for which $x = 0$ (say). Then (105) may be solved to give

$$y = \tfrac{4}{3}, \quad z = -\tfrac{5}{3}. \tag{108}$$

Accordingly the point $(0, \frac{4}{3}, -\frac{5}{3})$ lies on the line defined by the intersection of the two planes. Finally then the equation of the straight line takes the form

$$\frac{x-0}{6} = \frac{y-\frac{4}{3}}{-13} = \frac{z+\frac{5}{3}}{5}. \tag{109}$$

Further results relating to lines and planes may be found in the Problems at the end of the chapter.

19.7 Differentiation of a Vector Function

If for every value of a scalar parameter t in a given interval there exists a vector $\mathbf{a}(t)$, then $\mathbf{a}$ is said to be a vector function of t. In terms of rectangular components we have

$$\mathbf{a}(t) = a_x(t)\mathbf{i} + a_y(t)\mathbf{j} + a_z(t)\mathbf{k}, \tag{110}$$

where $a_x(t)$, $a_y(t)$, $a_z(t)$ are the components in the x, y, z directions respectively.

The derivative of $\mathbf{a}(t)$ with respect to t is defined as

$$\frac{d}{dt}\mathbf{a}(t) = \lim_{\delta t \to 0} \left\{ \frac{\mathbf{a}(t+\delta t) - \mathbf{a}(t)}{\delta t} \right\} = \lim_{\delta t \to 0} \frac{\delta \mathbf{a}(t)}{\delta t}. \tag{111}$$

In terms of rectangular coordinates therefore

$$\frac{d}{dt}\mathbf{a}(t) = \frac{da_x(t)}{dt}\mathbf{i} + \frac{da_y(t)}{dt}\mathbf{j} + \frac{da_z(t)}{dt}\mathbf{k}. \tag{112}$$

From the definition (111) it follows that

$$\frac{d}{dt}(c\mathbf{a}(t)) = c\frac{d\mathbf{a}(t)}{dt}, \tag{113}$$

$$\frac{d}{dt}(\mathbf{a}(t) + \mathbf{b}(t)) = \frac{d\mathbf{a}(t)}{dt} + \frac{d\mathbf{b}(t)}{dt}, \tag{114}$$

$$\frac{d}{dt}(\mathbf{a}(t) \cdot \mathbf{b}(t)) = \mathbf{a}(t) \cdot \frac{d\mathbf{b}(t)}{dt} + \frac{d\mathbf{a}(t)}{dt} \cdot \mathbf{b}(t), \tag{115}$$

$$\frac{d}{dt}(\mathbf{a}(t) \wedge \mathbf{b}(t)) = \mathbf{a}(t) \wedge \frac{d\mathbf{b}(t)}{dt} + \frac{d\mathbf{a}(t)}{dt} \wedge \mathbf{b}(t), \tag{116}$$

where c is a constant, and $\mathbf{a}(t)$, $\mathbf{b}(t)$ are vector functions of t.

Clearly if $\mathbf{a}(t)$ has a constant magnitude but a changing direction with t then

$$\mathbf{a}(t) \cdot \mathbf{a}(t) = a^2 = \text{const.} \tag{117}$$

Hence using (115)

$$2\mathbf{a}(t) \cdot \frac{d\mathbf{a}(t)}{dt} = \frac{da^2}{dt} = 0. \tag{118}$$

Since by assumption $\mathbf{a}(t) \neq 0$, (118) implies that $\mathbf{a}(t)$ and $d\mathbf{a}(t)/dt$ are perpendicular vectors. A special case of this result is when $\mathbf{a}(t)$ is a unit vector.

The operation of vector integration may also be defined. For example, from (115) we have

$$\int \left\{ \mathbf{a}(t) \cdot \frac{d\mathbf{b}(t)}{dt} + \frac{d\mathbf{a}(t)}{dt} \cdot \mathbf{b}(t) \right\} dt = \mathbf{a}(t) \cdot \mathbf{b}(t) + c, \tag{119}$$

where c is a constant of integration.

19.8 Scalar and Vector Fields

In formulating physical problems it is often necessary to associate with every point (x, y, z) of a region R of space some vector $\mathbf{a}(x, y, z)$. It is usual to call $\mathbf{a}(x, y, z)$ a vector function and to say that a vector field exists in R. Similarly if a scalar $\phi(x, y, z)$ is defined at every point (x, y, z) of R then a scalar field is said to exist in R.

We now discuss how certain differential operations may be performed on scalar and vector fields.

Suppose a scalar function $\phi(x, y, z)$ is differentiable in R. Then we define the gradient of $\phi(x, y, z)$ (usually denoted by grad ϕ) as

$$\text{grad } \phi = \mathbf{i} \frac{\partial \phi}{\partial x} + \mathbf{j} \frac{\partial \phi}{\partial y} + \mathbf{k} \frac{\partial \phi}{\partial z}, \tag{120}$$

where $\mathbf{i}$, $\mathbf{j}$ and $\mathbf{k}$ are unit vectors in the Ox, Oy and Oz directions respectively. Clearly grad ϕ is a vector function whose x, y and z

components are the first partial derivatives of $\phi(x, y, z)$ with respect to x, y and z respectively. For example, if

$$\phi(x, y, z) = xyz + 2x \tag{121}$$

then

$$\text{grad } \phi = (yz+2)\mathbf{i} + xz\mathbf{j} + xy\mathbf{k}. \tag{122}$$

It is convenient to write (120) as

$$\text{grad } \phi = \left(\mathbf{i}\,\frac{\partial}{\partial x} + \mathbf{j}\,\frac{\partial}{\partial z} + \mathbf{k}\,\frac{\partial}{\partial y} \right)\phi, \tag{123}$$

where the quantity in brackets is a vector differential operator. This operator is important in much of vector calculus and we accordingly denote it by a special symbol ∇ (pronounced ' del ' or ' nabla ') so that

$$\nabla \equiv \mathbf{i}\,\frac{\partial}{\partial x} + \mathbf{j}\,\frac{\partial}{\partial y} + \mathbf{k}\,\frac{\partial}{\partial z}. \tag{124}$$

Consequently

$$\nabla\phi = \text{grad } \phi = \mathbf{i}\,\frac{\partial\phi}{\partial x} + \mathbf{j}\,\frac{\partial\phi}{\partial y} + \mathbf{k}\,\frac{\partial\phi}{\partial z}. \tag{125}$$

(We note here that ∇ must always be applied to a scalar function since, if $\mathbf{a}$ is a vector function, $\nabla\mathbf{a}$ is undefined.)

Now suppose $d\mathbf{r}$ is an infinitesimal vector displacement such that

$$\frac{d\mathbf{r}}{dt} = \mathbf{i}\,\frac{dx}{dt} + \mathbf{j}\,\frac{dy}{dt} + \mathbf{k}\,\frac{dz}{dt}, \tag{126}$$

where t is some parameter.

Then

$$(\nabla\phi) \cdot \frac{d\mathbf{r}}{dt} = \left(\mathbf{i}\,\frac{\partial\phi}{\partial x} + \mathbf{j}\,\frac{\partial\phi}{\partial y} + \mathbf{k}\,\frac{\partial\phi}{\partial z} \right) \cdot \left(\mathbf{i}\,\frac{dx}{dt} + \mathbf{j}\,\frac{dy}{dt} + \mathbf{k}\,\frac{dz}{dt} \right)$$

$$= \frac{\partial\phi}{\partial x}\frac{dx}{dt} + \frac{\partial\phi}{\partial y}\frac{dy}{dt} + \frac{\partial\phi}{\partial z}\frac{dz}{dt} = \frac{d\phi}{dt}, \tag{127}$$

where $d\phi/dt$ is the total differential coefficient of $\phi(x, y, z)$ with respect to t. Hence considering a surface defined by

$$\phi(x, y, z) = \text{constant} \tag{128}$$

we see from (127) that $\nabla\phi$ must be perpendicular to $d\mathbf{r}/dt$ (since $d\phi/dt = 0$). In other words $\nabla\phi$ is a vector normal to the surface $\phi(x, y, z) = \text{constant}$ at every point. If $\hat{\mathbf{n}}$ is a unit normal to this surface in the direction of increasing $\phi(x, y, z)$ and $\dfrac{\partial\phi}{\partial n}$ is the derivative in this direction then

$$\nabla\phi = \frac{\partial\phi}{\partial n}\,\hat{\mathbf{n}}. \tag{129}$$

Furthermore if ϕ_1 and ϕ_2 are two differentiable scalar functions in R then

$$\nabla(\phi_1 + \phi_2) = \nabla\phi_1 + \nabla\phi_2, \tag{130}$$

$$\nabla(\phi_1\phi_2) = \phi_1\nabla\phi_2 + \phi_2\nabla\phi_1, \tag{131}$$

and

$$\nabla(c\phi_1) = c\nabla\phi_1, \tag{132}$$

where c is an arbitrary constant.

Using the ∇ operator we may now define further quantities. For example, we define $\nabla^2\phi$ (' del-squared ϕ ') as the scalar product

$$\nabla^2\phi = \nabla \cdot \nabla\phi = \left(\mathbf{i}\,\frac{\partial}{\partial x} + \mathbf{j}\,\frac{\partial}{\partial y} + \mathbf{k}\,\frac{\partial}{\partial z}\right) \cdot \left(\mathbf{i}\,\frac{\partial\phi}{\partial x} + \mathbf{j}\,\frac{\partial\phi}{\partial y} + \mathbf{k}\,\frac{\partial\phi}{\partial z}\right) \tag{133}$$

$$= \frac{\partial^2\phi}{\partial x^2} + \frac{\partial^2\phi}{\partial y^2} + \frac{\partial^2\phi}{\partial z^2}, \tag{134}$$

which is just the three-dimensional Laplacian. Laplace's equation in three-dimensions may therefore be written in a convenient and abbreviated form as

$$\nabla^2\phi = 0, \tag{135}$$

where

$$\nabla^2 \equiv \frac{\partial^2}{\partial x^2} + \frac{\partial^2}{\partial y^2} + \frac{\partial^2}{\partial z^2}. \tag{136}$$

Furthermore the operation ∇. may be applied to any vector function $\mathbf{a}(x, y, z)$ to give the divergence of $\mathbf{a}(x, y, z)$ (written as div $\mathbf{a}$). Clearly

$$\nabla \cdot \mathbf{a} = \text{div } \mathbf{a} = \left(\mathbf{i} \frac{\partial}{\partial x} + \mathbf{j} \frac{\partial}{\partial y} + \mathbf{k} \frac{\partial}{\partial z} \right) \cdot (\mathbf{i}a_x + \mathbf{j}a_y + \mathbf{k}a_z) \tag{137}$$

$$= \frac{\partial a_x}{\partial x} + \frac{\partial a_y}{\partial y} + \frac{\partial a_z}{\partial z}, \tag{138}$$

where a_x, a_y, a_z are in general functions of x, y and z. Unlike grad ϕ (which is a vector function) div $\mathbf{a}$ (as can be seen from (138)) is a scalar function. For example, if

$$\mathbf{a} = \mathbf{i}x^2 y + \mathbf{j}y^2 z + \mathbf{k}z^2 x \tag{139}$$

then

$$\nabla \cdot \mathbf{a} = \text{div } \mathbf{a} = 2(xy + yz + zx). \tag{140}$$

We now see that the quantity $\nabla^2 \phi$ defined in (133) and (134) can be written symbolically as

$$\nabla^2 \phi = \nabla \cdot \nabla \phi = \text{div grad } \phi. \tag{141}$$

Besides the operation ∇. as applied to a vector function $\mathbf{a}(x, y, z)$ we may also consider operation $\nabla \wedge$ on $\mathbf{a}(x, y, z)$. The vector product $\nabla \wedge \mathbf{a}$ defines the curl of $\mathbf{a}$ (written as curl $\mathbf{a}$) and is such that

$$\nabla \wedge \mathbf{a} = \text{curl } \mathbf{a} = \left(\mathbf{i} \frac{\partial}{\partial x} + \mathbf{j} \frac{\partial}{\partial y} + \mathbf{k} \frac{\partial}{\partial z} \right) \wedge (\mathbf{i}a_x + \mathbf{j}a_y + \mathbf{k}a_z) \tag{142}$$

$$= \begin{vmatrix} \mathbf{i} & \mathbf{j} & \mathbf{k} \\ \dfrac{\partial}{\partial x} & \dfrac{\partial}{\partial y} & \dfrac{\partial}{\partial z} \\ a_x & a_y & a_z \end{vmatrix} \tag{143}$$

$$= \mathbf{i} \left(\frac{\partial a_z}{\partial y} - \frac{\partial a_y}{\partial z} \right) + \mathbf{j} \left(\frac{\partial a_x}{\partial z} - \frac{\partial a_z}{\partial x} \right) + \mathbf{k} \left(\frac{\partial a_y}{\partial x} - \frac{\partial a_x}{\partial y} \right). \tag{144}$$

Curl **a** is therefore a vector function. For example, taking **a** to be the vector function defined in (139) we have

$$\nabla \wedge \mathbf{a} = \text{curl}\,\mathbf{a} = \mathbf{i}(-y^2) + \mathbf{j}(-z^2) + \mathbf{k}(-x^2). \tag{145}$$

From the definitions of div **a** and curl **a** we now see that if **a** and **b** are two vector functions then

$$\text{div}(\mathbf{a}+\mathbf{b}) = \text{div}\,\mathbf{a} + \text{div}\,\mathbf{b}, \tag{146}$$

or

$$\nabla \cdot (\mathbf{a}+\mathbf{b}) = \nabla \cdot \mathbf{a} + \nabla \cdot \mathbf{b}, \tag{147}$$

and

$$\text{curl}(\mathbf{a}+\mathbf{b}) = \text{curl}\,\mathbf{a} + \text{curl}\,\mathbf{b}, \tag{148}$$

or

$$\nabla \wedge (\mathbf{a}+\mathbf{b}) = \nabla \wedge \mathbf{a} + \nabla \wedge \mathbf{b}. \tag{149}$$

Now since curl **a** is a vector function we may consider the result of taking its divergence.

By (137) and (143) we find

$$\nabla \cdot (\nabla \wedge \mathbf{a}) = \text{div curl}\,\mathbf{a} \tag{150}$$

$$= \left(\mathbf{i}\frac{\partial}{\partial x} + \mathbf{j}\frac{\partial}{\partial y} + \mathbf{k}\frac{\partial}{\partial z}\right) \cdot \begin{vmatrix} \mathbf{i} & \mathbf{j} & \mathbf{k} \\ \dfrac{\partial}{\partial x} & \dfrac{\partial}{\partial y} & \dfrac{\partial}{\partial z} \\ a_x & a_y & a_z \end{vmatrix} \tag{151}$$

$$= \begin{vmatrix} \dfrac{\partial}{\partial x} & \dfrac{\partial}{\partial y} & \dfrac{\partial}{\partial z} \\ \dfrac{\partial}{\partial x} & \dfrac{\partial}{\partial y} & \dfrac{\partial}{\partial z} \\ a_x & a_y & a_z \end{vmatrix} = 0, \tag{152}$$

since two rows of the determinant are identical.

Consequently

$$\nabla \cdot (\nabla \wedge \mathbf{a}) = \text{div curl}\,\mathbf{a} = 0, \tag{153}$$

for all **a**. Conversely if **b** is a vector function such that div **b** = 0, then

355

we may infer that **b** is the curl of some vector function **a**. Vector functions whose divergences are identically zero are said to be solenoidal.

Likewise since grad ϕ is a vector function we may consider the result of taking its curl. By (125) and (143) we have

$$\nabla \wedge \nabla\phi = \text{curl grad } \phi = \begin{vmatrix} \mathbf{i} & \mathbf{j} & \mathbf{k} \\ \dfrac{\partial}{\partial x} & \dfrac{\partial}{\partial y} & \dfrac{\partial}{\partial z} \\ \dfrac{\partial \phi}{\partial x} & \dfrac{\partial \phi}{\partial y} & \dfrac{\partial \phi}{\partial z} \end{vmatrix} = 0. \qquad (154)$$

Hence for all ϕ

$$\nabla \wedge \nabla\phi = \text{curl grad } \phi = 0. \qquad (155)$$

Again conversely we may infer that if **b** is a vector function whose curl is identically zero, then **b** must be the gradient of some scalar function. Vector functions whose curls are identically zero are said to be irrotational. In virtue of (155) we see that if **a** is any irrotational vector function (curl **a** = 0) then another irrotational vector function is

$$\mathbf{a} + \text{grad } \phi, \qquad (156)$$

where ϕ is any scalar function. This follows since

$$\text{curl}(\mathbf{a} + \text{grad } \phi) = \text{curl } \mathbf{a} + \text{curl grad } \phi = 0 + 0 = 0. \qquad (157)$$

Using the definitions given here it is possible to obtain many useful vector differential relations. A few of these are given here and the reader should attempt to verify these from first principles. (We suppose in what follows that **a** and **b** are arbitrary vector functions, and that ϕ is an arbitrary scalar function, the differentiability of all functions being assumed.) Then

$$\nabla . (\phi\mathbf{a}) = \phi(\nabla . \mathbf{a}) + \mathbf{a} . \nabla\phi, \qquad (158)$$

$$\nabla \wedge (\phi\mathbf{a}) = \phi(\nabla \wedge \mathbf{a}) + (\nabla\phi) \wedge \mathbf{a}, \qquad (159)$$

$$\nabla . (\mathbf{a} \wedge \mathbf{b}) = \mathbf{b} . (\nabla \wedge \mathbf{a}) - \mathbf{a} . (\nabla \wedge \mathbf{b}), \qquad (160)$$

$$\nabla \wedge (\mathbf{a} \wedge \mathbf{b}) = \mathbf{a}(\nabla . \mathbf{b}) - \mathbf{b}(\nabla . \mathbf{a}) + (\mathbf{b} . \nabla)\mathbf{a} - (\mathbf{a} . \nabla)\mathbf{b}, \qquad (161)$$

$$\nabla(\mathbf{a.b}) = (\mathbf{a.}\nabla)\,\mathbf{b} + (\mathbf{b.}\nabla)\mathbf{a} + \mathbf{a} \wedge (\nabla \wedge \mathbf{b}) + \mathbf{b} \wedge (\nabla \wedge \mathbf{a}), \qquad (162)$$

$$\nabla \wedge (\nabla \wedge \mathbf{a}) = \nabla(\nabla \cdot \mathbf{a}) - \nabla^2 \mathbf{a}. \qquad (163)$$

As an example, we now prove (161) from first principles. Consider the x component of $\nabla \wedge (\mathbf{a} \wedge \mathbf{b})$. This is

$$\frac{\partial}{\partial y}(\mathbf{a} \wedge \mathbf{b})_z - \frac{\partial}{\partial z}(\mathbf{a} \wedge \mathbf{b})_y \qquad (164)$$

$$= \frac{\partial}{\partial y}(a_x b_y - a_y b_x) - \frac{\partial}{\partial z}(a_z b_x - a_x b_z) \qquad (165)$$

$$= a_x\left(\frac{\partial b_y}{\partial y} + \frac{\partial b_z}{\partial z}\right) - b_x\left(\frac{\partial a_y}{\partial y} + \frac{\partial a_z}{\partial z}\right)$$

$$+ \left(b_y\frac{\partial}{\partial y} + b_z\frac{\partial}{\partial z}\right)a_x - \left(a_y\frac{\partial}{\partial y} + a_z\frac{\partial}{\partial z}\right)b_z. \qquad (166)$$

To this expression we now add $a_x\dfrac{\partial b_x}{\partial x}$ to the first group of terms and subtract it from the fourth group; likewise we add $b_x\dfrac{\partial a_x}{\partial x}$ to the second group of terms and subtract it from the third group. In this way (166) may be written as

$$a_x\left(\frac{\partial b_x}{\partial x} + \frac{\partial b_y}{\partial y} + \frac{\partial b_z}{\partial z}\right) - b_x\left(\frac{\partial a_x}{\partial x} + \frac{\partial a_y}{\partial y} + \frac{\partial a_z}{\partial z}\right)$$

$$+ \left(b_x\frac{\partial}{\partial x} + b_y\frac{\partial}{\partial y} + b_z\frac{\partial}{\partial z}\right)a_x - \left(a_x\frac{\partial}{\partial x} + a_y\frac{\partial}{\partial y} + a_z\frac{\partial}{\partial z}\right)b_x \qquad (167)$$

$$= a_x\,\nabla \cdot \mathbf{b} - b_x\,\nabla \cdot \mathbf{a} + (\mathbf{b} \cdot \nabla)a_x - (\mathbf{a} \cdot \nabla)b_x. \qquad (168)$$

The y and z components of $\nabla \wedge (\mathbf{a} \wedge \mathbf{b})$ may be obtained from (168) by changing the suffix x to y and z respectively. Hence multiplying these three components by $\mathbf{i, j, k}$ respectively and adding, we finally obtain the required result (161).

It may be proved that all these vector relations are independent of the choice of coordinate system.

19.9 Cylindrical and Spherical Polar Coordinates

It is often necessary (especially in solving partial differential equations) to express div A, grad V, curl A, and $\nabla^2 V$ in terms of coordinates other than rectangular Cartesian coordinates (x, y, z). Particularly important are cylindrical polar coordinates and spherical polar coordinates.

(a) Cylindrical Polar Coordinates

As shown in Fig. 19.14, the position of a point P in space is given in terms of three coordinates (r, ϕ, z), where

$$x = r\cos\phi, \quad y = r\sin\phi, \quad z = z \tag{169}$$

express the relationships between the Cartesian coordinates of P and its cylindrical polar coordinates. Now let $\hat{\mathbf{r}}, \hat{\boldsymbol{\phi}}, \hat{\mathbf{k}}$ be unit vectors

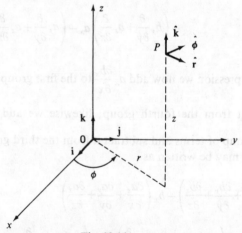

Fig. 19.14

in the directions of increasing r, ϕ, z respectively, as shown. These three unit vectors, like the unit vectors $\mathbf{i}, \mathbf{j}, \mathbf{k}$ along the Cartesian axes, are mutually orthogonal (that is, at right angles). However, unlike $\mathbf{i}, \mathbf{j}, \mathbf{k}$ which are fixed, $\hat{\mathbf{r}}$ and $\hat{\boldsymbol{\phi}}$ vary as P moves relative to the Cartesian axes. By simple geometrical arguments we see that

$$\hat{\mathbf{r}} = \mathbf{i}\cos\phi + \mathbf{j}\sin\phi, \tag{170}$$

$$\hat{\boldsymbol{\phi}} = -\mathbf{i}\sin\phi + \mathbf{j}\cos\phi, \tag{171}$$

$$\hat{\mathbf{k}} = \mathbf{k}. \tag{172}$$

358

Suppose now that we want to find the form of grad V, where V is a scalar function, in terms of the coordinates (r, ϕ, z). From (170) and (171) we easily obtain

$$\left.\begin{aligned} \mathbf{i} &= \hat{\mathbf{r}}\cos\phi - \hat{\boldsymbol{\phi}}\sin\phi \\ \mathbf{j} &= \hat{\mathbf{r}}\sin\phi + \hat{\boldsymbol{\phi}}\cos\phi \end{aligned}\right\}, \tag{173}$$

whence

$$\mathbf{i}\frac{\partial V}{\partial x} \quad \text{becomes} \quad (\hat{\mathbf{r}}\cos\phi - \hat{\boldsymbol{\phi}}\sin\phi)\left(\frac{\partial V}{\partial r}\frac{\partial r}{\partial x} + \frac{\partial V}{\partial \phi}\frac{\partial \phi}{\partial x}\right) \tag{174}$$

$$= (\hat{\mathbf{r}}\cos\phi - \hat{\boldsymbol{\phi}}\sin\phi)\left(\cos\phi\frac{\partial V}{\partial r} - \frac{\sin\phi}{r}\frac{\partial V}{\partial \phi}\right). \tag{175}$$

Likewise

$$\mathbf{j}\frac{\partial V}{\partial y} \quad \text{becomes} \quad (\hat{\mathbf{r}}\sin\phi + \hat{\boldsymbol{\phi}}\cos\phi)\left(\frac{\partial V}{\partial r}\frac{\partial r}{\partial y} + \frac{\partial V}{\partial \phi}\frac{\partial \phi}{\partial y}\right) \tag{176}$$

$$= (\hat{\mathbf{r}}\sin\phi + \hat{\boldsymbol{\phi}}\cos\phi)\left(\sin\phi\frac{\partial V}{\partial r} + \frac{\cos\phi}{r}\frac{\partial V}{\partial \phi}\right). \tag{177}$$

Finally,

$$\mathbf{k}\frac{\partial V}{\partial z} = \hat{\mathbf{k}}\frac{\partial V}{\partial z} \tag{178}$$

since the z coordinate is common to both coordinate systems. Adding (175), (172) and (178) we find that

$$\nabla V \quad \text{becomes} \quad \hat{\mathbf{r}}\frac{\partial V}{\partial r} + \frac{\hat{\boldsymbol{\phi}}}{r}\frac{\partial V}{\partial \phi} + \mathbf{k}\frac{\partial V}{\partial z}, \tag{179}$$

which is the required result.

To find the form of div $\mathbf{A}$, where $\mathbf{A}$ is a vector function, we proceed in a similar fashion by writing

$$\nabla \cdot \mathbf{A} = \nabla \cdot (\hat{\mathbf{r}}A_r + \hat{\boldsymbol{\phi}}A_\phi + \hat{\mathbf{k}}A_z), \tag{180}$$

where A_r, A_ϕ, A_z are the components of $\mathbf{A}$ in the r, ϕ, z directions.

Hence

$$\nabla \cdot \mathbf{A} = \nabla \cdot (\hat{\mathbf{r}}A_r) + \nabla \cdot (\hat{\boldsymbol{\phi}}A_\phi) + \frac{\partial A_z}{\partial z}. \tag{181}$$

We note that $\hat{\mathbf{r}}$ and $\hat{\boldsymbol{\phi}}$ cannot be taken outside the ∇ operator sign since, as we have seen, they are not constant vectors. To proceed we have to use the basic result (158) on the first two terms of (181) to get

$$\nabla \cdot \mathbf{A} = (A_r \nabla \cdot \hat{\mathbf{r}} + \hat{\mathbf{r}} \cdot \nabla A_r) + (A_\phi \nabla \cdot \hat{\boldsymbol{\phi}} + \hat{\boldsymbol{\phi}} \cdot \nabla A_\phi) + \frac{\partial A_z}{\partial z}. \qquad (182)$$

Now

$$\nabla \cdot \hat{\mathbf{r}} = \left(\mathbf{i} \frac{\partial}{\partial x} + \mathbf{j} \frac{\partial}{\partial y} \right) \cdot (\mathbf{i} \cos \phi + \mathbf{j} \sin \phi) \qquad (183)$$

(using (170))

$$= \frac{\partial}{\partial x} \left(\frac{x}{\sqrt{(x^2 + y^2)}} \right) + \frac{\partial}{\partial y} \left(\frac{y}{\sqrt{(x^2 + y^2)}} \right) \qquad (184)$$

$$= \frac{1}{\sqrt{(x^2 + y^2)}} = \frac{1}{r}. \qquad (185)$$

Similarly

$$\nabla \cdot \hat{\boldsymbol{\phi}} = \left(\mathbf{i} \frac{\partial}{\partial x} + \mathbf{j} \frac{\partial}{\partial y} \right) \cdot (-\mathbf{i} \sin \phi + \mathbf{j} \cos \phi) \qquad (186)$$

(using (171))

$$= -\frac{\partial}{\partial x} (\sin \phi) + \frac{\partial}{\partial y} (\cos \phi) = 0. \qquad (187)$$

Finally using the form of ∇V established earlier in this section to obtain ∇V_r and ∇A_ϕ, we have

$$\hat{\mathbf{r}} \cdot \nabla A_r = \frac{\partial A_r}{\partial r}, \quad \hat{\boldsymbol{\phi}} \cdot \nabla A_\phi = \frac{1}{r} \frac{\partial A_\phi}{\partial \phi}. \qquad (188)$$

Inserting (185), (187) and (188) into (182) we now obtain

$$\nabla \cdot \mathbf{A} = \frac{\partial A_r}{\partial r} + \frac{A_r}{r} + \frac{1}{r} \frac{\partial A_\phi}{\partial \phi} + \frac{\partial A_z}{\partial z}, \qquad (189)$$

which is the form of div **A** in cylindrical polar coordinates. Similarly it may be shown that

$$\text{curl}\,\mathbf{A} = \nabla \wedge \mathbf{A} = \begin{vmatrix} \dfrac{\hat{\mathbf{r}}}{r} & \hat{\boldsymbol{\phi}} & \dfrac{\hat{\mathbf{k}}}{r} \\[2ex] \dfrac{\partial}{\partial r} & \dfrac{\partial}{\partial \phi} & \dfrac{\partial}{\partial z} \\[2ex] A_r & rA_\phi & A_z \end{vmatrix}, \tag{190}$$

and

$$\text{div}\,\text{grad}\,V = \nabla^2 V = \frac{1}{r}\frac{\partial}{\partial r}\left(r\frac{\partial V}{\partial r}\right) + \frac{1}{r^2}\frac{\partial^2 V}{\partial \phi^2} + \frac{\partial^2 V}{\partial z^2}. \tag{191}$$

(b) Spherical Polar Coordinates

By reference to Fig. 19.15, the position of a point P in space can be described in terms of three coordinates (r, θ, ϕ), where

$$x = r\cos\phi\sin\theta, \quad y = r\sin\phi\sin\theta, \quad z = r\cos\theta, \tag{192}$$

Q being the projection of P on the xOy plane. If $\hat{\mathbf{r}}, \hat{\boldsymbol{\theta}}, \hat{\boldsymbol{\phi}}$ are unit

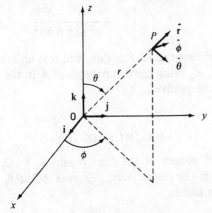

Fig. 19.15

vectors in the directions of increasing r, θ, ϕ respectively, then in terms of the fixed unit vectors **i**, **j**, **k**

$$\left.\begin{aligned} \hat{\mathbf{r}} &= \mathbf{i}\sin\theta\cos\phi + \mathbf{j}\sin\theta\sin\phi + \mathbf{k}\cos\theta, \\ \hat{\boldsymbol{\theta}} &= \mathbf{i}\cos\theta\cos\phi + \mathbf{j}\cos\theta\sin\phi - \mathbf{k}\sin\theta, \\ \hat{\boldsymbol{\phi}} &= -\mathbf{i}\sin\phi + \mathbf{j}\cos\phi. \end{aligned}\right\} \tag{193}$$

By the methods used in the section on cylindrical polar coordinates, we may obtain the corresponding results which we give here without proof—

$$\text{grad } V = \nabla V = \hat{\mathbf{r}} \frac{\partial V}{\partial r} + \frac{\hat{\boldsymbol{\theta}}}{r} \frac{\partial V}{\partial \theta} + \frac{\hat{\boldsymbol{\phi}}}{r \sin \theta} \frac{\partial V}{\partial \phi}, \tag{194}$$

$$\text{div } \mathbf{A} = \nabla . \mathbf{A} = \frac{1}{r^2} \frac{\partial}{\partial r}(r^2 A_r) + \frac{1}{r \sin \theta} \frac{\partial}{\partial \theta}(A_\theta \sin \theta) + \frac{1}{r \sin \theta} \frac{\partial A_\phi}{\partial \phi}, \tag{195}$$

$$\text{curl } \mathbf{A} = \nabla \wedge \mathbf{A} = \begin{vmatrix} \dfrac{\hat{\mathbf{r}}}{r^2 \sin \theta} & \dfrac{\hat{\boldsymbol{\theta}}}{r \sin \theta} & \dfrac{\hat{\boldsymbol{\phi}}}{r} \\[2mm] \dfrac{\partial}{\partial r} & \dfrac{\partial}{\partial \theta} & \dfrac{\partial}{\partial \phi} \\[2mm] A_r & rA_\theta & r \sin \theta A_\phi \end{vmatrix}, \tag{196}$$

and

$$\text{div grad } V = \nabla^2 V = \frac{1}{r^2} \frac{\partial}{\partial r}\left(r^2 \frac{\partial V}{\partial r}\right) + \frac{1}{r^2 \sin \theta} \frac{\partial}{\partial \theta}\left(\sin \theta \frac{\partial V}{\partial \theta}\right)$$
$$+ \frac{1}{r^2 \sin^2 \theta} \frac{\partial^2 V}{\partial \phi^2}, \tag{197}$$

where V is an arbitrary scalar function, and $\mathbf{A}$ is an arbitrary vector function, A_r, A_θ, A_ϕ being the components of $\mathbf{A}$ in the directions of increasing r, θ, ϕ respectively.

PROBLEMS 19

1. The position vectors of the four points A, B, C, D are $\mathbf{a}$, $\mathbf{b}$, $2\mathbf{a}+3\mathbf{b}$ and $\mathbf{a}-2\mathbf{b}$ respectively. Express AC, DB, BC and CD in terms of $\mathbf{a}$ and $\mathbf{b}$.

2. Find the scalar product $\mathbf{a} . \mathbf{b}$ and the vector product $\mathbf{a} \wedge \mathbf{b}$ when

    (i) $\mathbf{a} = \mathbf{i}+2\mathbf{j}-\mathbf{k}$, $\mathbf{b} = 2\mathbf{i}+3\mathbf{j}+\mathbf{k}$,
    (ii) $\mathbf{a} = \mathbf{i}-3\mathbf{j}+\mathbf{k}$, $\mathbf{b} = 2\mathbf{i}+\mathbf{j}+\mathbf{k}$.

3. Find the angles and sides of the triangle whose vertices have position vectors $\mathbf{i}+\mathbf{j}-\mathbf{k}$, $2\mathbf{i}-\mathbf{j}+\mathbf{k}$, $-\mathbf{i}+\mathbf{j}+\mathbf{k}$.

4. The position vectors of the foci of an ellipse are $\mathbf{c}$ and $-\mathbf{c}$, and the length of the major axis is $2a$. Prove that the equation of the ellipse can be written in the form

$$a^4 - a^2(\mathbf{r}^2 + \mathbf{c}^2) + (\mathbf{r} \cdot \mathbf{c})^2 = 0.$$

5. Form the scalar and vector products of the vectors

$$\mathbf{A} = \mathbf{i} \cos \theta + \mathbf{j} \sin \theta,$$
$$\mathbf{B} = \mathbf{i} \cos \phi + \mathbf{j} \sin \phi,$$

and hence prove that

$$\cos(\theta - \phi) = \cos \theta \cos \phi + \sin \theta \sin \phi,$$
$$\sin(\theta - \phi) = \sin \theta \cos \phi - \cos \theta \sin \phi.$$

6. Show that the equation $\mathbf{r} = \mathbf{r}_0 + \lambda \mathbf{s} + \mu \mathbf{t}$ represents a plane, where $\mathbf{r}_0, \mathbf{s}, \mathbf{t}$ are constant vectors, and λ and μ are scalar parameters.

7. If $\mathbf{p}, \mathbf{q}, \mathbf{r}$ are any vectors, prove that $\mathbf{a} = \mathbf{q} + \lambda \mathbf{r}$, $\mathbf{b} = \mathbf{r} + \mu \mathbf{p}$, $\mathbf{c} = \mathbf{p} + \nu \mathbf{q}$ are coplanar provided $\lambda \mu \nu = -1$, where λ, μ, ν are scalars. Show that this condition is satisfied when $\mathbf{a}$ is perpendicular to $\mathbf{p}$, $\mathbf{b}$ to $\mathbf{q}$, and $\mathbf{c}$ to $\mathbf{r}$.

8. Prove that the length of the shortest distance from the point $\mathbf{a}$ to the line joining the points $\mathbf{b}$ and $\mathbf{c}$ is given by

$$\frac{|\mathbf{a} \wedge \mathbf{b} + \mathbf{b} \wedge \mathbf{c} + \mathbf{c} \wedge \mathbf{a}|}{|\mathbf{b} - \mathbf{c}|}.$$

9. Show that the two straight lines $\mathbf{r} = \mathbf{a} + k\mathbf{u}$ and $\mathbf{r} = \mathbf{b} + h\mathbf{v}$, where h and k are scalar parameters, intersect if

$$\mathbf{v} \cdot \mathbf{b} \wedge \mathbf{u} = \mathbf{v} \cdot \mathbf{a} \wedge \mathbf{u},$$

and that the point of intersection is

$$\mathbf{a} + \left(\frac{\mathbf{a} \cdot \mathbf{b} \wedge \mathbf{v}}{\mathbf{v} \cdot \mathbf{a} \wedge \mathbf{u}} \right) \mathbf{u} \quad \text{or} \quad \mathbf{b} + \left(\frac{\mathbf{a} \cdot \mathbf{b} \wedge \mathbf{u}}{\mathbf{v} \cdot \mathbf{b} \wedge \mathbf{u}} \right) \mathbf{v}.$$

10. Resolve the vector **r** into a component along a given vector **a** and component perpendicular to **a**, showing that the decomposition is

$$r = \left(\frac{a \cdot r}{a^2}\right)a + \frac{a \wedge (r \wedge a)}{a^2}.$$

11. Show that the equation of the perpendicular from the point **b** to the line $r = a + ku$ is

$$r = b + hu \wedge \{(a - b) \wedge u\}.$$

12. If $u \wedge b = c \wedge b$ and $u \cdot a = 0$, prove that

$$u = c - \left(\frac{a \cdot c}{a \cdot b}\right)b$$

provided $a \cdot b \neq 0$.

13. Prove that four vectors along the outward normals to the faces of a tetrahedron and proportional to the areas of the corresponding faces have vector sum zero.

14. Prove the identities

(i) $(a \wedge b) \cdot (c \wedge d) = \begin{vmatrix} a \cdot c & a \cdot d \\ b \cdot c & b \cdot d \end{vmatrix}$,

(ii) $a \wedge (b \wedge c) + b \wedge (c \wedge a) + c \wedge (a \wedge b) = 0.$

(iii) $(a \wedge b) \wedge (c \wedge d) = \{d \cdot (a \wedge b)\}c - \{c \cdot (a \wedge b)\}d$
$= \{a \cdot (c \wedge d)\}b - \{b \cdot (c \wedge d)\}a.$

15. The set of vectors b_1, b_2, b_3 reciprocal to the set a_1, a_2, a_3 is defined by

$$b_1 = \frac{a_2 \wedge a_3}{L}, \quad b_2 = \frac{a_3 \wedge a_1}{L}, \quad b_3 = \frac{a_1 \wedge a_2}{L}$$

where $L = a_1 \cdot a_2 \wedge a_3$.

Show that

$$b_1 \cdot b_2 \wedge b_3 = \frac{1}{a_1 \cdot a_2 \wedge a_3}$$

16. Given

$$r' = r + \left\{ \frac{(\beta - 1)\mathbf{v} \cdot \mathbf{r}}{v^2} - \beta t \right\} \mathbf{v},$$

$$t' = \beta \left(t - \frac{\mathbf{v} \cdot \mathbf{r}}{c^2} \right),$$

where $\beta = \dfrac{1}{\sqrt{\{1 - (v^2/c^2)\}}}$, show that

$$r = r' + \left\{ \frac{(\beta - 1)\mathbf{v} \cdot \mathbf{r}'}{v^2} + \beta t' \right\} \mathbf{v},$$

$$t = \beta \left(t' + \frac{\mathbf{v} \cdot \mathbf{r}'}{c^2} \right).$$

17. If $\mathbf{u}$ is a unit vector, $\mathbf{v}$ is an arbitrary vector and $\mathbf{w}$ is the reflection of $\mathbf{v}$ in a line in the direction $\mathbf{u}$, show that

$$\mathbf{w} = 2(\mathbf{u} \cdot \mathbf{v})\mathbf{u} - \mathbf{v}.$$

Writing $\mathbf{w} = R\mathbf{v}$, and taking the Cartesian components of $\mathbf{u}$ and $\mathbf{v}$ to be (u_1, u_2) and (v_1, v_2) find the components of the matrix R. Verify that $R^2 = I$, where I is the unit matrix.   (C.U.)

18. Consider a triangle ABC on the surface of a sphere of unit radius and centre O, where BC, CA, AB are arcs of great circles; such a triangle is called a spherical triangle. The angle which the arc BC subtends at O is called the side a of the triangle, and the angle between the planes AOB and AOC is the angle A of the triangle. The quantities b, B and c, C are similarly defined.

Let $\mathbf{i}, \mathbf{j}, \mathbf{k}$ be the three unit vectors OA, OB, OC (not mutually perpendicular in general). Consider the vectors $\mathbf{i} \wedge \mathbf{j}$ and $\mathbf{i} \wedge \mathbf{k}$, and show that

$$(\mathbf{i} \wedge \mathbf{j}) \cdot (\mathbf{i} \wedge \mathbf{k}) = \mathbf{j} \cdot \mathbf{k} - (\mathbf{i} \cdot \mathbf{k})(\mathbf{i} \cdot \mathbf{j}).$$

Hence deduce the fundamental formula of spherical trigonometry

$$\cos a = \cos b \cos c + \sin b \sin c \cos A.$$

365

19. If ω, **a**, **b** are constants, show that

$$\mathbf{r} = \mathbf{a} \cos \omega t + \mathbf{b} \sin \omega t$$

is the equation of an ellipse, and verify that

$$\frac{d^2\mathbf{r}}{dt^2} + \omega^2 \mathbf{r} = 0$$

and

$$\mathbf{r} \wedge \frac{d\mathbf{r}}{dt} = \omega \mathbf{a} \wedge \mathbf{b}.$$

20. Obtain the gradients of the following scalar functions

(a) x, (b) r^n, (c) $\mathbf{r} \cdot \nabla(x+y+z)$, (d) $\mathbf{A} \cdot \mathbf{r}$ (**A** a constant vector),

where $\mathbf{r} = \mathbf{i}x + \mathbf{j}y + \mathbf{k}z$.

21. Evaluate $\oint_C \mathbf{A} \cdot d\mathbf{s}$ for

$$\mathbf{A} = \frac{\mathbf{i}x + \mathbf{j}y}{\sqrt{(x^2 + y^2)}}$$

around the square whose sides are $x = 0$, $x = a$, $y = 0$, $y = a$.

22. Obtain the divergences of the following vectors

(a) $\mathbf{r}$, (b) $xyz(\mathbf{i}+\mathbf{j}+\mathbf{k})$, (c) $\mathbf{r}/r^3$.

23. Evaluate $\oint_C \mathbf{r} \wedge d\mathbf{r}$ around the circle $x^2 + y^2 = a^2$, $z = 0$.

24. Obtain the curls of the following vectors

(a) $x\mathbf{i}$, (b) $\mathbf{r}$, (c) $(x\mathbf{i} - y\mathbf{j})/(x+y)$, (d) $\mathbf{i} \sin y + \mathbf{j}x(1 + \cos y)$.

25. If curl $\mathbf{A} = 0$, where $\mathbf{A} = (xyz)^m(x^n\mathbf{i} + y^n\mathbf{j} + z^n\mathbf{k})$ show that either $m = 0$ or $n = -1$.

26. If $\mathbf{v} = \mathbf{r}(\mathbf{a} \cdot \mathbf{r})$, where **a** is a constant vector show that

$$\text{div } \mathbf{v} = 4(\mathbf{a} \cdot \mathbf{r}),$$
$$\text{curl } \mathbf{v} = \mathbf{a} \wedge \mathbf{r},$$

and

$$\text{curl } (\mathbf{a} \wedge \mathbf{r}) = 2\mathbf{a}.$$

366

27. Given Maxwell's equations

$$\operatorname{div} \mathbf{E} = 0, \qquad \operatorname{curl} \mathbf{E} = -\frac{1}{c}\frac{\partial \mathbf{H}}{\partial t},$$

$$\operatorname{div} \mathbf{H} = 0, \qquad \operatorname{curl} \mathbf{H} = \frac{1}{c}\frac{\partial \mathbf{E}}{\partial t},$$

where c is a constant, prove that

$$\nabla^2 \mathbf{E} = \frac{1}{c^2}\frac{\partial^2 \mathbf{E}}{\partial t^2}, \qquad \nabla^2 \mathbf{H} = \frac{1}{c^2}\frac{\partial^2 \mathbf{H}}{\partial t^2}.$$

28. If $\mathbf{a}$ is a constant unit vector, show that

$$\mathbf{a} \cdot \{\nabla(\mathbf{v} \cdot \mathbf{a}) - \nabla \wedge (\mathbf{v} \wedge \mathbf{a})\} = \nabla \cdot \mathbf{v}.$$

29. Show that

   (a) $\operatorname{curl}\operatorname{curl}(r^2 \mathbf{a} \wedge \mathbf{r}) = -10\mathbf{a} \wedge \mathbf{r}$,

   (b) $\operatorname{curl}\operatorname{curl}\operatorname{curl}(r^2 \mathbf{a} \wedge \mathbf{r}) = -20\mathbf{a}$,

   where $\mathbf{a}$ is a constant vector, and $\mathbf{r}$ is the position vector with respect to a fixed origin.

367

CHAPTER 20

Solution of Algebraic and Transcendental Equations

20.1 Algebraic Equations

In the first part of this chapter we are mainly concerned with the properties and solution of algebraic equations of the type

$$f(x) \equiv a_0 x^n + a_1 x^{n-1} + \dots + a_{n-1} x + a_n = 0, \qquad (1)$$

where the coefficients $a_0, a_1, \dots, a_n$ are real constants and n (the degree of the equation) is a positive integer. It can be proved that (1) may be written as the product of n factors such that

$$a_0 x^n + a_1 x^{n-1} + \dots + a_{n-1} x + a_n = a_0(x - \alpha_1)(x - \alpha_2) \dots (x - \alpha_n). \quad (2)$$

This is the fundamental theorem of algebra.

The equation $f(x) = 0$ is now satisfied when x takes any of the values $\alpha_1, \alpha_2, \alpha_3, \dots, \alpha_n$. These n values are called the roots of the equation. If it happens that $k \, (\leq n)$ of these roots are identical the equation is said to have a repeated root of multiplicity k; clearly every algebraic equation of degree n has exactly n roots provided any repeated root of multiplicity k is counted as k roots. To illustrate these ideas consider the quartic equation (degree 4)

$$x^4 + x^3 - 7x^2 - x + 6 = 0. \qquad (3)$$

Since this factorises at sight into

$$(x - 1)(x - 2)(x + 1)(x + 3) = 0, \qquad (4)$$

we see that the four roots are $x = 1, 2, -1, -3$. Similarly the cubic equation (degree 3)

$$x^3 - 3x + 2 = 0 \qquad (5)$$

factorises into

$$(x - 1)^2 (x + 2) = 0, \qquad (6)$$

368

and hence has a repeated root of multiplicity 2 at $x = 1$, and a third root at $x = -2$.

Many algebraic equations have complex roots. For example, the roots of the quadratic equation

$$x^2 - 2x + 2 = 0 \tag{7}$$

are easily found to be $x = 1 + i$, $1 - i$. These roots form a complex conjugate pair. Indeed the complex roots of any algebraic equation with real coefficients must always occur in conjugate pairs for otherwise the coefficients a_0, a_1, ..., a_n would themselves be complex. For example, suppose a quadratic equation has complex roots $x = z = a + ib$, $x = w = c + id$, where a, b, c and d are real. Then by (2) the equation is

$$(x - z)(x - w) = x^2 + a_1 x + a_2 = 0, \tag{8}$$

where

$$a_1 = (a + c) + i(b + d), \tag{9}$$

$$a_2 = (ac - bd) + i(bc + ad). \tag{10}$$

Hence in general (8) has complex coefficients given by (9) and (10). If, however, the two roots form a complex conjugate pair, then $w = \bar{z}$, $c = a$, $d = -b$, and the coefficients a_1 and a_2 become real.

Immediate consequences of this result are that

(a) every algebraic equation with real coefficients can have only an even number (or zero) of complex roots which occur in conjugate pairs,

(b) every algebraic equation with real coefficients of odd degree must have at least one real root.

In general, as is well known, the roots of a quadratic equation (whether they be real or complex) may be found exactly by the use of a simple formula. Similar, but more complicated, formulae also exist for the roots of the general cubic and quartic. However, it has been shown by Galois that such formulae cannot exist for the solution of general algebraic equations of degree higher than four. The real roots of these equations, of necessity therefore, have to be found by some suitable numerical approximation method. Such approximation methods are usually best even for the solution of cubic and quartic equations, since the formulae giving their exact solutions are too cumbersome to justify their use (unless, of course, exact solutions are required). A few of the various numerical methods of solving algebraic equations (and transcendental equations such as $e^x = 2+x$) will be discussed in 20.5.

20.2 Relations between Roots and Coefficients

Returning now to the quadratic equation

$$a_0x^2 + a_1x + a_2 = 0 \tag{11}$$

we see that if α_1 and α_2 are the roots then (by (2))

$$a_0x^2 + a_1x + a_2 = a_0(x-\alpha_1)(x-\alpha_2)$$
$$= a_0x^2 - a_0(\alpha_1+\alpha_2)x + a_0\alpha_1\alpha_2. \tag{12}$$

Hence equating coefficients of corresponding powers of x on each side of (12) we find

$$\alpha_1 + \alpha_2 = -\frac{a_1}{a_0}, \tag{13}$$

$$\alpha_1\alpha_2 = \frac{a_2}{a_0}. \tag{14}$$

Equations (13) and (14) are simple expressions, symmetrical in α_1 and α_2, which give respectively the sum and the product of the roots in terms of the coefficients of the equation. Likewise in the case of the cubic

$$a_0x^3 + a_1x^2 + a_2x + a_3 = 0, \tag{15}$$

if the roots are α_1, α_2 and α_3, then

$$\alpha_1 + \alpha_2 + \alpha_3 = -\frac{a_1}{a_0}, \tag{16}$$

$$\alpha_1\alpha_2 + \alpha_2\alpha_3 + \alpha_3\alpha_1 = \frac{a_2}{a_0}, \tag{17}$$

$$\alpha_1\alpha_2\alpha_3 = -\frac{a_3}{a_0}. \tag{18}$$

In the same way, if α_1, α_2, α_3, α_3 are the roots of the quartic equation

$$a_0 x^4 + a_1 x^3 + a_2 x^2 + a_3 x + a_4 = 0, \qquad (19)$$

we find

$$\alpha_1 + \alpha_2 + \alpha_3 + \alpha_4 = -\frac{a_1}{a_0}, \qquad (20)$$

$$\alpha_1 \alpha_2 + \alpha_2 \alpha_3 + \alpha_3 \alpha_4 + \alpha_4 \alpha_1 + \alpha_1 \alpha_3 + \alpha_2 \alpha_4 = \frac{a_2}{a_0}, \qquad (21)$$

$$\alpha_1 \alpha_2 \alpha_3 + \alpha_2 \alpha_3 \alpha_4 + \alpha_3 \alpha_4 \alpha_1 + \alpha_4 \alpha_1 \alpha_2 = -\frac{a_3}{a_0}, \qquad (22)$$

$$\alpha_1 \alpha_2 \alpha_3 \alpha_4 = \frac{a_4}{a_0}. \qquad (23)$$

Similar relations may easily be found for the general algebraic equation of the nth degree, but they will not interest us here. The following examples show how the relations between roots and coefficients may be used in solving equations.

Example 1. Solve the equation

$$x^3 + x^2 - 10x + 8 = 0 \qquad (24)$$

given that the sum of two of its roots is equal to 3. Suppose the roots are α_1, α_2, α_3. Let

$$\alpha_1 + \alpha_2 = 3. \qquad (25)$$

Then since, by (17),

$$\alpha_1 + \alpha_2 + \alpha_3 = -1 \qquad (26)$$

we have $\alpha_3 = -4$. To find α_1 and α_2 we now use (18) which gives the product of the roots

$$\alpha_1 \alpha_2 \alpha_3 = -8. \qquad (27)$$

Consequently $\alpha_1 \alpha_2 = 2$ which with (25) gives $\alpha_1 = 1$, $\alpha_2 = 2$ (or $\alpha_1 = 2$, $\alpha_2 = 1$, which is the same thing). Hence the three roots are $x = 1, 2, -4$.

Example 2. Solve the equation

$$x^3 - 7x^2 + 14x - 8 = 0 \qquad (28)$$

given that the roots are in geometrical progression. Let the three roots be α_1, $\alpha_1 k$, $\alpha_1 k^2$, where k is the common ratio of the progression.

Then using (16) and (18) we have

$$\alpha_1 + \alpha_1 k + \alpha_1 k^2 = 7, \qquad (29)$$

and

$$\alpha_1^3 k^3 = 8 \quad \text{or} \quad \alpha_1 k = 2. \qquad (30)$$

Eliminating α_1 from (29) and (30) we find $k = 2$ or $\frac{1}{2}$, which immediately leads to the three roots $x = 1, 2, 4$.

20.3 Reduced Form of a Cubic Equation

The general cubic equation

$$a_0 x^3 + a_1 x^2 + a_2 x + a_3 = 0 \qquad (31)$$

may be transformed into its reduced form

$$y^3 + Ay + B = 0, \qquad (32)$$

(where A and B are expressible in terms of a_0, a_1, a_2, a_3) by the substitution

$$x = y - \frac{1}{3} \frac{a_1}{a_0}. \qquad (33)$$

This reduced equation has no y^2 term and is a useful starting point for the trignometric method of finding real roots discussed in the next section. To illustrate the transformation of a cubic to reduced form we consider the equation

$$x^3 - 3x^2 + 2x - 1 = 0. \qquad (34)$$

From (33) we find

$$x = y + 1. \qquad (35)$$

Consequently (34) becomes

$$y^3 - y - 1 = 0. \qquad (36)$$

372

Cubic equations in which $a_2 = 0$ can best be put into reduced form by the substitution $x = 1/y$. For example,

$$x^3 + 2x^2 + 2 = 0 \qquad (37)$$

becomes

$$2y^3 + 2y + 1 = 0. \qquad (38)$$

20.4 Trignometric Solution of a Cubic Equation

The reduced form of a cubic is particularly useful in finding its real roots as shown by the following example.

Example 3. To solve the equation

$$8x^3 - 6x - 1 = 0 \qquad (39)$$

we first put $x = \lambda \cos \theta$ and compare the resulting equation

$$8\lambda^3 \cos^3 \theta - 6\lambda \cos \theta = 1 \qquad (40)$$

with the trignometric identity

$$4 \cos^3 \theta - 3 \cos \theta = \cos 3\theta. \qquad (41)$$

If these two equations are to be the same then we must have

$$\frac{8\lambda^3}{4} = \frac{6\lambda}{3} = \frac{1}{\cos 3\theta}. \qquad (42)$$

Hence

$$\lambda = 0, \pm 1 \qquad (43)$$

and

$$\cos 3\theta = \frac{1}{2\lambda}. \qquad (44)$$

Clearly the case of $\lambda = 0$ is to be excluded since this requires $\cos 3\theta$ to exceed unity. With $\lambda = +1$ we find

$$\cos 3\theta = \tfrac{1}{2}. \qquad (45)$$

Therefore

$$3\theta = 2\pi k \pm \frac{\pi}{3}, \quad (k = 0, 1, 2 \ldots) \qquad (46)$$

$$\theta = \frac{2\pi k}{3} \pm \frac{\pi}{9}, \quad (k = 0, 1, 2 \ldots) \qquad (47)$$

$$= \pm \frac{\pi}{9}, \frac{5\pi}{9}, \frac{7\pi}{9}, \frac{11\pi}{9}, \frac{13\pi}{9}. \tag{48}$$

Since $x = \lambda \cos \theta$ and $\lambda = 1$ we obtain from (48) only three different roots, namely

$$x_1 = \cos \frac{\pi}{9}, \quad x_2 = \cos \frac{5\pi}{9}, \quad x_3 = \cos \frac{7\pi}{9}. \tag{49}$$

It is easily verified that $\lambda = -1$ just reproduces these three roots.

20.5 Numerical Methods of Solution

Most numerical methods depend on (*a*) finding a rough approximation to the root of the equation and (*b*) improving this approximation to a required accuracy. Usually the location of a root of an equation, say $f(x) = 0$, may be made either by trial and error or by drawing a rough graph of the curve $y = f(x)$ and noting the point (or points) at which the curve intersects the x-axis. For example, a rough graph of $y = \frac{3}{8} x^4 + 10x - 1$ shows that the two real roots of the equation

$$\tfrac{3}{8} x^4 + 10x - 1 = 0 \tag{50}$$

occur near $x = 0 \cdot 1$ and $x = -3$ (see Fig. 20.1). Similarly to find an approximation to the non-zero root of the equation

$$e^x - 2x - 1 = 0 \tag{51}$$

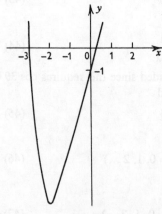

Fig. 20.1

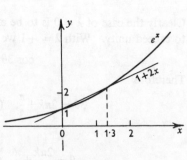

Fig. 20.2

we may draw rough graphs of the functions

$$y_1(x) = e^x \quad \text{and} \quad y_2(x) = 1 + 2x.$$

The roots of (51) then occur where $y_1(x) = y_2(x)$; that is, at the points of intersection of the two curves (see Fig. 20.2). In this way a fair approximation to the required root is found to be $x = 1\cdot3$.

The location of roots of $f(x) = 0$ is often made easier by evaluating $f(x)$ for various different values of x. For if $f(a) > 0$ and $f(b) < 0$ (or $f(a) < 0$ and $f(b) > 0$) then clearly $f(x)$ must become zero somewhere in the interval $a < x < b$ (assuming here that $b > a$). This is illustrated in Fig. 20.3. There may, in fact, be any odd number of

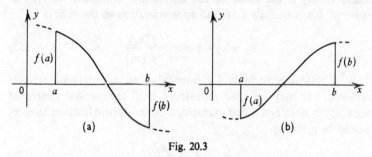

Fig. 20.3

roots in this interval as shown by Fig. 20.4. In the case of (50), for example, where

$$f(x) = \tfrac{3}{8}x^4 + 10x - 1 \tag{52}$$

we find $f(0) = -1$ and $f(1) = 9\tfrac{3}{8}$. Hence since the function changes sign for $0 < x < 1$ there must be a root of $f(x) = 0$ somewhere in this interval.

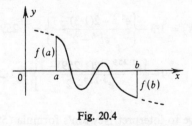

Fig. 20.4

Having shown how to obtain rough approximations to roots of equations, we now discuss Newton's method of improving them.

Suppose $x = x_1$ is a known rough approximation to a root of $f(x) = 0$. Let the exact root be at $x = x_1 + h$ so that $f(x_1 + h) = 0$, h being a small quantity. Then by Taylor's expansion

$$f(x_1 + h) = 0 = f(x_1) + hf'(x_1) + \frac{h^2}{2!}f''(x_1) + \dots . \qquad (53)$$

Hence, neglecting terms in h^2 and higher, we have

$$h \simeq -\frac{f(x_1)}{f'(x_1)}, \qquad (54)$$

where $f'(x_1)$ is the value of the differential coefficient of $f(x)$ at $x = x_1$. Consequently a second approximation to the root is

$$x_2 = x_1 + h = x_1 - \frac{f(x_1)}{f'(x_1)}. \qquad (55)$$

This is Newton's formula for improving an approximate root of $f(x) = 0$. It may be used repeatedly until the desired degree of accuracy is obtained. For example, a better approximation than x_2 would be given by

$$x_3 = x_2 - \frac{f(x_2)}{f'(x_2)}, \qquad (56)$$

and so on.

Example 4. Since $x_1 = 1 \cdot 3$ is a rough approximation to the non-zero root of the equation

$$f(x) = e^x - 2x - 1 = 0, \qquad (57)$$

we have, using Newton's method, the improved approximations

$$x_2 = 1 \cdot 3 - \left\{ \frac{e^{1 \cdot 3} - 2(1 \cdot 3) - 1}{e^{1 \cdot 3} - 2} \right\} = 1 \cdot 259, \qquad (58)$$

$$x_3 = 1 \cdot 259 - \left\{ \frac{e^{1 \cdot 259} - 2(1 \cdot 259) - 1}{e^{1 \cdot 259} - 2} \right\} = 1 \cdot 257, \qquad (59)$$

and so on.

It is instructive to interpret Newton's formula (55) geometrically. Suppose (see Fig. 20.5) the curve $y = f(x)$ cuts the x-axis at a point A. Then $OA(= x_0)$ is an exact root of $f(x) = 0$. If now $P(x_1, y_1)$ is a point on the curve close to A such that $OB(=x_1)$ is a first approximation to x_0, then a better approximation to x_0 is given by $OC(=x_2)$,

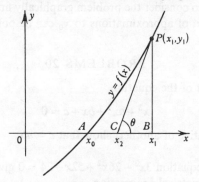

Fig. 20.5

where C is the point such that PC is the tangent to $y = f(x)$ at $x = x_1$. Clearly

$$x_2 = OC = OB - CB = x_1 - PB \cot \theta. \tag{60}$$

Now PB is the value of the function $y = f(x)$ at $x = x_1$ and is therefore $f(x_1)$. Also $\tan \theta$ is the gradient of $y = f(x)$ at $x = x_1$, which is $f'(x_1)$. Hence (60) becomes

$$x_2 = x_1 - \frac{f(x_1)}{f'(x_1)}, \tag{61}$$

which is Newton's formula (55).

Using this geometrical interpretation we now see that in certain circumstances Newton's method may fail. For example (see Fig. 20.6), if there is a maximum in the neighbourhood of the root

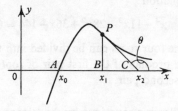

Fig. 20.6

$x_0(=OA)$, and OB is taken as the first approximation x_1, then the second approximation x_2 is given by OC. Since $OC > OB$, x_2 is a worse approximation to x_0 than x_1. A similar situation arises if a minimum occurs near x_0. In applying Newton's method it is therefore advisable to consider the problem graphically first to ensure that a convergent set of approximations to x_0 can be obtained.

PROBLEMS 20

1. If one root of the equation

$$x^3 + ax^2 + bx + c = 0$$

is the negative of another, show that $c = ab$.

2. Solve the equation $3x^3 - 26x^2 + 52x - 24 = 0$ given that its roots are in geometrical progression.

3. Show that there is only one value of k for which the sum of two roots of the equation

$$x^4 - 4x^3 + x^2 + kx + 2 = 0$$

is equal to the sum of the other two roots. Find the value of k and all the roots of the equation. (L.U.)

4. If α_1, α_2, α_3 are the roots of the equation

$$a_0 x^3 + a_1 x^2 + a_2 x + a_3 = 0,$$

show that (i) $\alpha_1^2 + \alpha_2^2 + \alpha_3^2 = (a_1^2 - 2a_0 a_2)/a_0^2$,

(ii) $\alpha_1^2 \alpha_2^2 + \alpha_2^2 \alpha_3^2 + \alpha_3^2 \alpha_1^2 = (a_2^2 - 2a_1 a_3)/a_0^2$.

5. Solve the equation

$$x^4 - 11x^3 + 28x^2 + 36x - 144 = 0$$

given that the four roots can be divided into two pairs in such a way that the product of the first pair of roots is minus the product of the second pair. (C.U.)

6. Solve the following equations by the trigonometric method

    (a) $x^3 - 6x - 4 = 0$,

    (b) $x^3 + 3x^2 - 1 = 0$.

7. Solve the equation

$$x^4 - 3x^3 + 4x^2 - 3x + 1 = 0$$

by putting $x + (1/x) = y$.

8. Show that $x^3 + 3x^2 + 6x - 3 = 0$ has only one real root. Prove that this lies between 0 and 1, and find it correct to one place of decimals.

9. Show that the equation $e^{2x} - x = 2$ has two real roots, and find the positive one correct to three places of decimals.

10. Show graphically that the equation

$$\sin \theta = 0.25 + 0.35\theta$$

has three real roots. Verify that the negative root is near $\theta = -2.45$, and find its value correct to three significant figures.

(L.U.)

11. If $(1 + x^2)y = \log_e (1 + x)$, show that a stationary value of y corresponds with a root of the equation

$$2x(x + 1) \log_e (1 + x) = 1 + x^2.$$

There is one real root of this equation: show that it lies between 0.5 and 1, and determine its value correct to three places of decimals.

(L.U.)

12. Show that the equation

$$1000x^3 - x - 1 = 0$$

has only one real root which is close to zero, and determine it to three decimal places.

13. Bessel's function $J_0(x)$ is given by the infinite series

$$J_0(x) = 1 - \frac{x^2}{2^2} + \frac{x^4}{2^2 . 4^2} - \frac{x^6}{2^2 . 4^2 . 6^2} + \dots .$$

Assuming $J_0(x) = 0$ has a real root near $x = 2.5$ find this root to two decimal places taking four terms of the above series.

CHAPTER 21

Ordinary Differential Equations

21.1 Introduction

Any equation containing differential coefficients is called a differential equation. Such equations can be divided into two main types — ordinary and partial, ordinary differential equations involving only one independent variable (and therefore only ordinary differential coefficients), and partial differential equations involving two or more independent variables (and therefore partial differential coefficients). In general, therefore, any function of x, y and the derivatives of y up to any order such that

$$f\left(x, y, \frac{dy}{dx}, \frac{d^2y}{dx^2} \dots \right) = 0 \qquad (1)$$

defines an ordinary differential equation for y (the dependent variable) in terms of x (the independent variable). For example, the equations

$$\frac{dy}{dx} = 3y, \qquad (2)$$

$$\frac{d^2y}{dx^2} + 2\frac{dy}{dx} + y = \sin x, \qquad (3)$$

$$\left(\frac{dy}{dx}\right)^3 + y^2 = x, \qquad (4)$$

$$\frac{d^3y}{dx^3} + 6\sqrt{\left\{\left(\frac{dy}{dx}\right)^2 + y^2\right\}} = 0, \qquad (5)$$

are all ordinary differential equations, whereas the equations

$$yx\frac{\partial u}{\partial x} + y^2\frac{\partial u}{\partial y} = u, \quad \frac{\partial^2 u}{\partial x^2} + \frac{\partial^2 u}{\partial y^2} = 0 \qquad (6)$$

are partial differential equations for u (the dependent variable) in terms of the independent variables x and y. In this chapter (and the

380

one that follows it) we shall consider only the solution of certain types of ordinary differential equations, the solution of partial differential equations being dealt with separately in Chapter 24.

We must now define two important terms — the order and the degree — which are applied to ordinary differential equations. The order of a differential equation is the order of the highest differential coefficient contained in it. For example, equations (2), (3), (4) and (5) are of the first, second, first and third order respectively. The degree of a differential equation is the power to which the highest order differential coefficient is raised when the equation is rationalised (i.e. fractional powers removed). For example, equations (2), (3) and (4) are of the first, first and third degree respectively. Equation (5), however, is of second degree since when written in rationalised form it becomes

$$\left(\frac{d^3y}{dx^3}\right)^2 = 36\left\{\left(\frac{dy}{dx}\right)^2 + y^2\right\}, \tag{7}$$

the highest differential coefficient d^3y/dx^3 now appearing to the second power.

Any differential equation of order n is said to be linear if it is linear in the dependent variable y and the derivatives

$$\frac{dy}{dx}, \frac{d^2y}{dx^2}, ..., \frac{d^ny}{dx^n}$$

(for example, (2) and (3)). When this condition is not satisfied the equation is said to be non-linear (for example, (4) and (5)). The solution of non-linear equations usually presents a difficult problem and for the most part therefore we shall only discuss the solution of linear equations. However, as we shall see in 21.9 (d) certain types of non-linear equations may either be transformed into linear equations by a suitable change of variable, or else be solved in terms of elliptic functions (see Problem 21).

21.2 Formation of Ordinary Differential Equations

In general if y is expressed as a function of x which contains n arbitrary constants, then n differentiations are just sufficient to eliminate the constants and reduce the relation between y and x to an ordinary differential equation of order n. For example, if

$$y^2 = 4a(x+a), \quad (a = \text{constant}) \tag{8}$$

then differentiating once we have

$$2y\frac{dy}{dx} = 4a. \tag{9}$$

Consequently eliminating a from (8) using (9) we find the first order (non-linear) equation

$$y\left(\frac{dy}{dx}\right)^2 + 2x\frac{dy}{dx} - y = 0. \tag{10}$$

Similarly, if

$$y = Ae^{-x} + Be^{-3x} \tag{11}$$

(A, B constants) then

$$\frac{dy}{dx} = -Ae^{-x} - 3Be^{-3x}, \tag{12}$$

and

$$\frac{d^2y}{dx^2} = Ae^{-x} + 9Be^{-3x}. \tag{13}$$

Hence solving for A and B from (12) and (13) and substituting in (11) we find the second order (linear) equation

$$\frac{d^2y}{dx^2} - 4\frac{dy}{dx} + 3y = 0. \tag{14}$$

Clearly the relations (8) and (11) may be regarded as solutions of the differential equations (10) and (14) which have been formed from them.

Following these arguments in reverse it is now reasonable to define the general solution of an ordinary differential equation of order n as that solution containing n arbitrary constants. However, this definition is not always quite sufficient (as we s hall see in 1.4) and it is usually better to state conversely that (without exception) any solution of an nth order equation not containing n arbitrary constants cannot be the general solution.

In many physical problems the solution of a differential equation has to satisfy certain specified conditions. These conditions, usually known as initial or boundary conditions, determine the values of the arbitrary constants in the solution. For example, if (11) (which is the general solution of (14)) is to satisfy the boundary conditions

$$y = 1 \text{ at } x = 0, \text{ and } \frac{dy}{dx} = 3 \text{ at } x = 0 \qquad (15)$$

then we find $A + B = 1$, and $-A - 3B = 3$, and hence $A = 3$, $B = -2$.

21.3 First Order Equations

Equations of this type can, in general, be written as

$$\frac{dy}{dx} = F(x, y), \qquad (16)$$

where $F(x, y)$ is a given function. However, despite the apparent simplicity of this equation analytic solutions can usually only be found when $F(x, y)$ has particularly simple forms. Four such forms are discussed below.

(a) Variables Separable

If

$$F(x, y) = f(x)g(y), \qquad (17)$$

where $f(x)$ and $g(y)$ are respectively functions of x only and y only, then (16) becomes

$$\frac{dy}{dx} = f(x)g(y). \qquad (18)$$

Since the variables x and y are now separate we have, integrating (18),

$$\int \frac{dy}{g(y)} = \int f(x) \, dx, \qquad (19)$$

which expresses y implicitly in terms of x.

Example 1. Solve the equation

$$\frac{dy}{dx} = \frac{y+1}{x-1} \tag{20}$$

given the boundary condition $y = 1$ at $x = 0$.

Writing (20) such that all terms in y are to one side and all terms in x to the other and integrating, we have

$$\int \frac{dy}{y+1} = \int \frac{dx}{x-1}, \tag{21}$$

or

$$\log_e (y+1) = \log_e (x-1) + \log_e C, \tag{22}$$

where C is an arbitrary constant of integration. Hence

$$\frac{y+1}{x-1} = C. \tag{23}$$

Now if $y = 1$ at $x = 0$, then from (23) we find

$$C = -2. \tag{24}$$

Substituting this value of C into (23) the required solution of (20) satisfying the given boundary condition is therefore

$$y = 2(1-x)-1. \tag{25}$$

(b) The Homogeneous Equation

If

$$F(x, y) = \frac{f(x, y)}{g(x, y)}, \tag{26}$$

where $f(x, y)$ and $g(x, y)$ are homogeneous functions of x and y of the same degree, say n (see Chapter 9, 9.8), then (16) becomes

$$\frac{dy}{dx} = \frac{f(x, y)}{g(x, y)} = \frac{x^n f\left(\frac{y}{x}\right)}{x^n g\left(\frac{y}{x}\right)} = \phi\left(\frac{y}{x}\right), \tag{27}$$

where $\bar{f}$, $\bar{g}$ and ϕ are functions of y/x. This equation is usually called the homogeneous first order differential equation, and may always be reduced to a variable separable equation by the substitution

$$y = vx, \tag{28}$$

where v is a function of x. For differentiating (28) we have

$$\frac{dy}{dx} = v + x\frac{dv}{dx} \tag{29}$$

which, with (28), enables (27) to be written as

$$v + x\frac{dv}{dx} = \phi(v). \tag{30}$$

This equation is clearly of the variable separable type and may be integrated directly to give

$$\int \frac{dv}{\phi(v) - v} = \int \frac{dx}{x}, \tag{31}$$

from which v (and hence y) may be found in terms of x.

Example 2. Solve the equation

$$2xy\frac{dy}{dx} = x^2 + y^2 \tag{32}$$

given $y = 0$ at $x = 1$.

Writing (32) as

$$\frac{dy}{dx} = \frac{x^2 + y^2}{2xy} = \frac{1}{2}\left(\frac{x}{y} + \frac{y}{x}\right), \tag{33}$$

and using (28) we have

$$v + x\frac{dv}{dx} = \frac{1}{2}\left(\frac{1}{v} + v\right). \tag{34}$$

Hence

$$2x\frac{dv}{dx} = \frac{1 - v^2}{v} \tag{35}$$

385

and therefore

$$\int \frac{2v}{1-v^2}\,dv = \int \frac{dx}{x}. \tag{36}$$

Consequently

$$-\log_e(1-v^2) = \log_e x + \log_e C, \tag{37}$$

where C is an arbitrary constant of integration. Rewriting (37) without logarithms we finally get

$$y^2 - x^2 = -\frac{x}{C}. \tag{38}$$

Now since $y = 0$ at $x = 1$, we must have

$$C = 1. \tag{39}$$

The solution of (32) satisfying the given boundary condition is therefore

$$x^2 - y^2 = x. \tag{40}$$

Some non-homogeneous equations can be put into homogeneous form by a simple change of variables. For example, the equation

$$\frac{dy}{dx} = \frac{ax+by+c}{fx+gy+h}, \tag{41}$$

where a, b, c, f, g and h are given constants becomes on putting

$$\left.\begin{array}{l} x = x_0 + X, \\ y = y_0 + Y, \end{array}\right\} \tag{42}$$

(where x_0 and y_0 are constants)

$$\frac{dY}{dX} = \frac{aX+bY+ax_0+by_0+c}{fX+gY+fx_0+gy_0+h}. \tag{43}$$

If x_0 and y_0 are now chosen so that

$$\left.\begin{array}{l} ax_0 + by_0 + c = 0, \\ fx_0 + gy_0 + h = 0, \end{array}\right\} \tag{44}$$

then

$$\frac{dY}{dX} = \frac{aX+bY}{fX+gY} = \frac{a+b(Y/X)}{f+g(Y/X)}, \quad (45)$$

which is clearly of homogeneous form.

However, (44) only determines x_0 and y_0 when $ag-bf \neq 0$ (see Chapter 16). If $ag-bf = 0$ then $f/a = g/b = k$ (say), and (41) becomes

$$\frac{dy}{dx} = \frac{(ax+by)+c}{k(ax+by)+h}. \quad (46)$$

Hence putting $u = ax+by$, (46) becomes

$$\frac{du}{dx} = a+b\left(\frac{u+c}{ku+h}\right), \quad (47)$$

which is now of variable separable type.

Example 3. To solve the equation

$$\frac{dy}{dx} = \frac{x+y+3}{x-y-5} \quad (48)$$

we use (42) to obtain

$$\frac{dY}{dX} = \frac{X+Y+x_0+y_0+3}{X-Y+x_0-y_0-5}. \quad (49)$$

Hence putting

$$\left.\begin{array}{l} x_0+y_0+3 = 0, \\ x_0-y_0-5 = 0, \end{array}\right\} \quad (50)$$

(49) reduces to the homogeneous form

$$\frac{dY}{dX} = \frac{X+Y}{X-Y} \quad (51)$$

from which Y may be found in terms of X. Finally to express y in terms of x we note from (50) that $x_0 = 1$, $y_0 = -4$ and hence by (42) $x = 1+X$, $y = -4+Y$.

Example 4. The equation

$$\frac{dy}{dx} = \frac{x+y}{1-x-y} \tag{52}$$

is such that when $x = x_0 + X$, $y = y_0 + Y$, x_0 and y_0 (by (44)) do not exist. Hence putting $x + y = u$ (say) we have

$$\frac{dy}{dx} = \frac{du}{dx} - 1 = \frac{u}{1-u} \tag{53}$$

or

$$\frac{du}{dx} = \frac{1}{1-u}. \tag{54}$$

This equation is of variable separable type and can be integrated directly to give

$$\int (1-u) \, du = \int dx \tag{55}$$

or

$$u - \frac{u^2}{2} = x + C, \tag{56}$$

where C is an arbitrary constant of integration. Hence since $u = x + y$, (56) becomes

$$y - \frac{(x+y)^2}{2} = C, \tag{57}$$

which is therefore the solution of (52).

(c) The Exact Equation

$$F(x, y) = -\frac{P(x, y)}{Q(x, y)}, \tag{58}$$

where P and Q are given functions of x and y, then the general first order equation (16) becomes

$$Q(x, y) \frac{dy}{dx} + P(x, y) = 0. \tag{59}$$

388

Now depending on the forms of $P(x, y)$ and $Q(x, y)$ it may be possible to write the left-hand side of this equation as the total differential coefficient of some function, say $u(x, y)$. From Chapter 9, 9.6 we have

$$\frac{du}{dx} = \left(\frac{\partial u}{\partial y}\right)\frac{dy}{dx} + \frac{\partial u}{\partial x}. \tag{60}$$

Consequently (comparing (59) and (60)) we see that if

$$\left.\begin{array}{l} P(x, y) = \dfrac{\partial u}{\partial x}, \\[2mm] Q(x, y) = \dfrac{\partial u}{\partial y}, \end{array}\right\} \tag{61}$$

then (59) may be written as

$$\frac{du}{dx} = 0, \tag{62}$$

the solution of which is

$$u(x, y) = \text{constant.} \tag{63}$$

Differentiating (61) (and assuming P and Q have continuous first derivatives) we have

$$\frac{\partial P(x, y)}{\partial y} = \frac{\partial^2 u}{\partial y\, \partial x} = \frac{\partial^2 u}{\partial x\, \partial y} = \frac{\partial Q(x, y)}{\partial x}. \tag{64}$$

This condition is a necessary and sufficient condition for (59) to be expressible as an exact or total differential coefficient. We illustrate the solution of exact equations by the following examples.

Example 5. The equation

$$(8y - x^2 y)\frac{dy}{dx} + (x - xy^2) = 0 \tag{65}$$

is exact since $P(x, y) = x - xy^2$, $Q(x, y) = 8y - x^2 y$ and

$$\frac{\partial P(x, y)}{\partial y} = -2xy = \frac{\partial Q(x, y)}{\partial x}. \tag{66}$$

To find $u(x, y)$ we therefore use (61) which gives

$$\frac{\partial u}{\partial x} = P(x, y) = x - xy^2. \tag{67}$$

Accordingly, integrating with respect to x, we have

$$u(x, y) = \frac{x^2}{2}(1 - y^2) + \phi(y), \tag{68}$$

where $\phi(y)$ is an arbitrary function of y. Now since by (61)

$$\frac{\partial u}{\partial y} = Q(x, y) = 8y - x^2 y \tag{69}$$

we see by differentiating (68) with respect to y and comparing with (69) that

$$8y - x^2 y = -x^2 y + \frac{d\phi}{dy}. \tag{70}$$

Hence

$$\phi = 4y^2 + c, \tag{71}$$

where c is an arbitrary constant of integration. Finally (68) and (71) together give

$$u(x, y) = \frac{x^2}{2}(1 - y^2) + 4y^2 + c \tag{72}$$

and the solution of (65) is therefore (by (63))

$$\frac{x^2}{2}(1 - y^2) + 4y^2 = \text{constant}, \tag{73}$$

(the constant c being absorbed in the constant term on the right).

Example 6. The equation

$$2x \log_e x \frac{dy}{dx} + y = 0 \tag{74}$$

is not exact since $P(x, y) = y$, $Q(x, y) = 2x \log_e x$ and

$$\frac{\partial P(x, y)}{\partial y} \neq \frac{\partial Q(x, y)}{\partial x}.$$

390

However, if (74) is multiplied through by the factor y/x we find

$$2y \log_e x \frac{dy}{dx} + \frac{y^2}{x} = 0, \tag{75}$$

which, since

$$\frac{\partial}{\partial y}\left(\frac{y^2}{x}\right) = \frac{\partial}{\partial x}(2y \log_e x), \tag{76}$$

is exact. Consequently following the method of the last example the solution of (75) (and therefore of (74)) is found to be

$$u(x, y) = y^2 \log_e x = \text{constant}. \tag{77}$$

The factor y/x used here to make (74) exact is called an integrating factor.

In general if the equation

$$Q(x, y)\frac{dy}{dx} + P(x, y) = 0 \tag{78}$$

is not exact, then there exists an integrating factor $\mu(x, y)$ which makes it exact, although the form of μ may be difficult to find. Suppose (78) is multiplied through by $\mu(x, y)$ to give

$$\mu(x, y)Q(x, y)\frac{dy}{dx} + \mu(x, y)P(x, y) = 0. \tag{79}$$

Then (writing P, Q and μ for $P(x, y)$, $Q(x, y)$ and $\mu(x, y)$, respectively) (79) is exact provided μ satisfies the partial differential equation

$$\frac{\partial}{\partial x}(\mu Q) = \frac{\partial}{\partial y}(\mu P) \tag{80}$$

or

$$\mu\left(\frac{\partial Q}{\partial x} - \frac{\partial P}{\partial y}\right) + Q\frac{\partial \mu}{\partial x} - P\frac{\partial \mu}{\partial y} = 0. \tag{81}$$

Except when P and Q have exceptionally simple forms (81) is usually difficult to solve. The determination of integrating factors is at this stage therefore largely a matter of trial and error.

Example 7. The equation

$$x \frac{dy}{dx} - y = 0 \tag{82}$$

is not exact. Suppose there exists an integrating factor μ such that

$$\mu x \frac{dy}{dx} - \mu y = 0 \tag{83}$$

is exact. Then

$$\frac{\partial}{\partial x}(\mu x) = \frac{\partial}{\partial y}(-\mu y) \tag{84}$$

or

$$x \frac{\partial \mu}{\partial x} + y \frac{\partial \mu}{\partial y} + 2\mu = 0. \tag{85}$$

Any solution of this partial differential equation will be a suitable integrating factor. It is easily verified that

$$\mu = \frac{1}{x^2}, \frac{1}{y^2}, \frac{1}{xy}, \frac{1}{x^2 + y^2}, \frac{1}{x^2 - y^2} \tag{86}$$

are all solutions of (85) and therefore all qualify as integrating factors. For example, taking $\mu = 1/x^2$, (83) becomes

$$\frac{1}{x} \frac{dy}{dx} - \frac{y}{x^2} = 0, \tag{87}$$

which is exact. Since (87) can now be written as $\dfrac{d}{dx}\left(\dfrac{y}{x}\right) = 0$, the solution of (82) is therefore

$$y/x = \text{constant.} \tag{88}$$

Similarly, taking $\mu = 1/(x^2 - y^2)$, (83) becomes

$$\frac{x}{x^2 - y^2} \frac{dy}{dx} - \frac{y}{x^2 - y^2} = 0, \tag{89}$$

which again is exact. Since (89) may be written as

$$\frac{d}{dx} \log_e (x^2 - y^2) = 0$$

we again find $y/x = $ constant as in (88).

(*d*) *The Linear Equation*

If

$$F(x, y) = Q(x) - P(x)y, \tag{90}$$

where $P(x)$ and $Q(x)$ are given functions of x (or constants), then (16) becomes

$$\frac{dy}{dx} + P(x)y = Q(x). \tag{91}$$

This equation (known as the general linear first order equation) may be solved with the help of an integrating factor. For multiplying (91) through by an arbitrary function $R(x)$ (the integrating factor) we have

$$R(x)\frac{dy}{dx} + R(x)P(x)y = Q(x)R(x). \tag{92}$$

If $R(x)$ is now chosen such that the left-hand side of (92) is equal to $\frac{d}{dx}\{R(x)y\}$ then

$$\frac{d}{dx}\{R(x)y\} = R(x)\frac{dy}{dx} + y\frac{dR(x)}{dx} = R(x)\frac{dy}{dx} + R(x)P(x)y. \tag{93}$$

Hence comparing terms in (93) we have

$$y\frac{dR(x)}{dx} = R(x)P(x)y, \tag{94}$$

which gives on integration (assuming $y \neq 0$)

$$R(x) = e^{\int P(x)\,dx}. \tag{95}$$

Since $P(x)$ is known, the form of the integrating factor is uniquely determined by (95).

Finally from (92) and (93)

$$\frac{d}{dx}\{R(x)y\} = Q(x)R(x), \tag{96}$$

and hence

$$R(x)y = \int Q(x)R(x)\,dx. \tag{97}$$

With $R(x)$ given by (95), (97) therefore represents the general solution of (91).

Example 8. To solve

$$\frac{dy}{dx} + \frac{3}{x}y = x^2 \tag{98}$$

given $y = 1/6$ when $x = 1$.

Here the integrating factor R is given by

$$R(x) = e^{\int (3/x)dx} = e^{3\{\log_e x\}} = e^{\log_e x^3} = x^3. \tag{99}$$

Hence by (97) the solution of (98) is

$$x^3 y = \int x^2 \cdot x^3 \, dx = \frac{x^6}{6} + c, \tag{100}$$

or

$$y = \frac{x^3}{6} + \frac{c}{x^3}. \tag{101}$$

With $y = 1/6$ when $x = 1$, we find $c = 0$, and hence the required solution of (98) is

$$y = \frac{x^3}{6}. \tag{102}$$

21.4 Linear Equations

The linear first order equation discussed in the last section is a special case of the general linear equation of order n

$$a_0(x)\frac{d^n y}{dx^n} + a_1(x)\frac{d^{n-1}y}{dx^{n-1}} + \ldots + a_{n-1}(x)\frac{dy}{dx} + a_n(x)y = f(x), \tag{103}$$

where $a_0(x)$, $a_1(x)$ ... $a_n(x)$ and $f(x)$ are given functions of x or are constants. This equation is said to be homogeneous when $f(x) = 0$ and to be inhomogeneous when $f(x) \neq 0$. As we shall see later it is useful in dealing with inhomogeneous equations of the type (103) to consider the corresponding homogeneous (or reduced) equation obtained by putting $f(x) = 0$. For example, the reduced equation corresponding to

$$x \frac{d^2 y}{dx^2} + 2 \frac{dy}{dx} + 3y = \sin x \tag{104}$$

is

$$x \frac{d^2 y}{dx^2} + 2 \frac{dy}{dx} + 3y = 0. \tag{105}$$

Suppose now y_1 and y_2 are two independent (see Chapter 16, 16.8) solutions of the reduced equation of (103), namely,

$$a_0(x) \frac{d^n y}{dx^n} + a_1(x) \frac{d^{n-1} y}{dx^{n-1}} + \ldots + a_{n-1}(x) \frac{dy}{dx} + a_n(x)y = 0. \tag{106}$$

Then clearly the linear combination

$$y = c_1 y_1 + c_2 y_2, \tag{107}$$

where c_1 and c_2 are arbitrary constants, is also a solution since substituting (107) into the left-hand side of (106) we have

$$\left(a_0(x) \frac{d^n y_1}{dx^n} + a_1(x) \frac{d^{n-1} y_1}{dx^{n-1}} + \ldots + a_{n-1}(x) \frac{dy_1}{dx} + a_n(x)y_1 \right)$$

$$+ \left(a_0(x) \frac{d^n y_2}{dx^n} + a_1(x) \frac{d^{n-1} y_2}{dx^{n-1}} + \ldots + a_{n-1}(x) \frac{dy_2}{dx} + a_n(x)y_2 \right). \tag{108}$$

This expression vanishes identically, since each bracket is zero in virtue of y_1 and y_2 being solutions of (106). Similarly we may show that if $y_1, y_2 \ldots y_n$ are n independent solutions of (106) then the linear combination

$$y = c_1 y_1 + c_2 y_2 + \ldots + c_n y_n, \tag{109}$$

where $c_1, c_2, \ldots, c_n$ are arbitrary constants, is also a solution. Since (109) contains n constants it is reasonable (following the ideas of

21.2) to take this as the general solution of the nth order equation (106).

We now define the general solution of the inhomogeneous equation (103) as the sum of the general solution of the reduced equation and any particular solution of the inhomogeneous equation. In other words if Y is a particular solution of (103) then the general solution is

$$y = c_1 y_1 + c_2 y_2 + \ldots + c_n y_n + Y. \tag{110}$$

Clearly the phrase 'general solution' is overworked here and to avoid confusion it is usual when discussing inhomogeneous equations to call the general solution of the reduced (homogeneous) equation the 'complementary function' and the particular solution Y of the inhomogeneous equation a 'particular integral'. Hence for inhomogeneous equations

General solution = Complementary Function + Particular Integral.

21.5 Linear Homogeneous Equations with Constant Coefficients

The general linear equation discussed in the last section is usually difficult to solve and requires special techniques (see Chapter 22, and also 21.9 (a)). However, an important and special case occurs when the coefficients $a_0(x)$, $a_1(x)$, ... $a_n(x)$ are constants, the equation then being called a constant coefficient equation. In this section we shall discuss the solution of the homogeneous constant coefficient equation

$$a_0 \frac{d^n y}{dx^n} + a_1 \frac{d^{n-1} y}{dx^{n-1}} + \ldots + a_{n-1} \frac{dy}{dx} + a_n y = 0, \tag{111}$$

dealing in particular with the second order equation ($n = 2$)

$$a_0 \frac{d^2 y}{dx^2} + a_1 \frac{dy}{dx} + a_2 y = 0. \tag{112}$$

Suppose we try a solution of (112) of the form

$$y = e^{mx}. \tag{113}$$

Then

$$(a_0 m^2 + a_1 m + a_2) e^{mx} = 0. \tag{114}$$

Hence (113) is a solution of (112) when m is a root of

$$a_0 m^2 + a_1 m + a_2 = 0. \tag{115}$$

If m_1 and m_2 are the two roots of this equation (which is usually known as the auxiliary equation) then

$$y_1 = e^{m_1 x} \quad \text{and} \quad y_2 = e^{m_2 x} \tag{116}$$

are both solutions of (112). Hence the general solution is

$$y = A_1 e^{m_1 x} + A_2 e^{m_2 x}, \tag{117}$$

where A_1 and A_2 are arbitrary constants. A slight difficulty arises, however, when the auxiliary equation has equal roots such that $m_1 = m_2 = m$(say). For in this case the two solutions in (116) are not independent, and (117) becomes

$$y = (A_1 + A_2) e^{mx} = C e^{mx}, \tag{118}$$

where C is a new arbitrary constant. This solution does not now qualify as the general solution of (112) since it only contains one arbitrary constant. To obtain another solution we therefore write

$$y = u e^{mx}, \tag{119}$$

where u is a function of x.

Substituting in (112) we now get

$$a_0 \frac{d^2 u}{dx^2} + (2ma_0 + a_1) \frac{du}{dx} + (a_0 m^2 + a_1 m + a_2) u = 0. \tag{120}$$

In virtue of (115) the last bracket of (120) vanishes; likewise since both roots of (115) are equal to m we have their sum

$$m + m = -\frac{a_1}{a_0} \tag{121}$$

or

$$2ma_0 + a_1 = 0. \tag{122}$$

Hence the second term of (120) also vanishes. The form of u is therefore determined by the remaining equation

$$\frac{d^2u}{dx^2} = 0, \tag{123}$$

which gives

$$u = A_1 + A_2 x, \tag{124}$$

where A_1 and A_2 are arbitrary constants.

The general solution of (112) when the auxiliary equation has two equal roots $m_1 = m_2 = m$ is therefore

$$y = (A_1 + A_2 x)e^{mx}. \tag{125}$$

Similarly if the nth order constant coefficient equation (111) is such that the auxiliary equation has roots $m_1, m_2, \ldots$ of multiplicity $k_1, k_2, \ldots$ then the appropriate terms in the general solution are

$$\begin{aligned}
(A_1 + A_2 x + A_3 x^2 + \ldots + A_{k_1} x^{k_1 - 1})e^{m_1 x}, \\
(B_1 + B_2 x + B_3 x^2 + \ldots + B_{k_2} x^{k_2 - 1})e^{m_2 x},
\end{aligned} \tag{126}$$

and so on.

We now illustrate these results by the following examples.

Example 9. To solve the equation

$$\frac{d^2y}{dx^2} + 3\frac{dy}{dx} + 2y = 0. \tag{127}$$

Here putting $y = e^{mx}$ we have the auxiliary equation

$$m^2 + 3m + 2 = 0, \tag{128}$$

which has roots $m = -1, -2$. Hence the general solution of (127) is

$$y = Ae^{-x} + Be^{-2x}, \tag{129}$$

A and B being arbitrary constants.

Example 10. To solve the equation

$$\frac{d^2y}{dx^2} + \frac{dy}{dx} + y = 0. \tag{130}$$

Here the auxiliary equation is

$$m^2 + m + 1 = 0, \tag{131}$$

which has roots $m = -\tfrac{1}{2} \pm i(\tfrac{1}{2}\sqrt{3})$.

Consequently the general solution is

$$y = Ae^{\left[-\frac{1}{2}+\frac{i}{2}\sqrt{3}\right]x} + Be^{\left[-\frac{1}{2}-\frac{i}{2}\sqrt{3}\right]x} \tag{132}$$

When, as here, the roots of the auxiliary equation are complex the general solution may be written in a different form using the relation (see Chapter 7, equation (42))

$$e^{ix} = \cos x + i \sin x, \quad (x \text{ real}). \tag{133}$$

In this way (132) becomes

$$y = e^{-\frac{1}{2}x}\left(E \cos \frac{\sqrt{3}}{2}x + F \sin \frac{\sqrt{3}}{2}x \right), \tag{134}$$

where E and F are arbitrary constants.

Example 11. To solve the equation

$$\frac{d^2y}{dx^2} - 6\frac{dy}{dx} + 9y = 0. \tag{135}$$

Here the auxiliary equation

$$m^2 - 6m + 9 = 0 \tag{136}$$

has two equal roots $m = 3$ (twice).

Consequently the general solution is

$$y = (A + Bx)e^{3x}, \tag{137}$$

A and B being arbitrary constants as before.

Example 12. To solve

$$\frac{d^3y}{dx^3} - \frac{d^2y}{dx^2} - \frac{dy}{dx} + y = 0. \tag{138}$$

399

Here the auxiliary equation

$$m^3 - m^2 - m + 1 = 0$$

(obtained by putting $y = e^{mx}$ in (138)) factorises to give

$$(m-1)^2(m+1) = 0,$$

or $m = 1$ (twice) and $m = -1$. The general solution is therefore

$$y = (A + Bx)e^x + Ce^{-x}, \tag{139}$$

A, B and C being arbitrary constants.

21.6 Linear Inhomogeneous Constant Coefficient Equations

We now consider equations of the general type

$$a_0 \frac{d^n y}{dx^n} + a_1 \frac{d^{n-1}y}{dx^{n-1}} + \ldots + a_{n-1}\frac{dy}{dx} + a_n y = f(x), \tag{140}$$

where $f(x)$ is a given function of x, and $a_0, a_1, \ldots a_n$ are constants. As we have already seen the general solution of this equation is the sum of the general solution of its reduced equation (obtained by putting $f(x) = 0$) and a particular integral of (140). However, since the reduced (homogeneous) equation may be solved by the method of the last section, the only remaining problem is to find a particular integral in a given case. This may sometimes be done by inspection. For example, the equation

$$\frac{d^2y}{dx^2} + y = 3x \tag{141}$$

is clearly satisfied by $y = 3x$, which is therefore a particular integral. The general solution of (141) is therefore the sum of this particular solution and the general solution (the complementary function) of

$$\frac{d^2y}{dx^2} + y = 0. \tag{142}$$

Solving (142) we find $m^2 + 1 = 0$, and hence $m = \pm i$. The complementary function is therefore of the form

$$y = Ae^{ix} + Be^{-ix} = E \cos x + F \sin x, \tag{143}$$

where A, B, E, F are constants. Consequently the general solution of (141) is

$$y = E \cos x + F \sin x + 3x. \qquad (144)$$

However, in all but very simple cases it is virtually impossible to find particular integrals by inspection and, in general, other methods have to be used. Two such methods will be discussed here; a third method based on the properties of a linear operator (the D-operator) will be discussed in 21.8. We remark here that in Chapter 23 we shall meet an important method of solving ordinary linear differential equations with constant coefficients which does not involve finding the complementary function and particular integral separately.

(a) Method of Undetermined Coefficients

This method is based on assuming a trial form for the particular integral $Y(x)$ of (140) which is dependent on the form of the function $f(x)$ and contains a finite number of arbitrary constants. The trial function is then substituted into the equation and the constants so chosen that it is a solution. The following rules (which can be justified rigorously by other methods) are useful in fixing the forms of $Y(x)$ for particular forms of $f(x)$.

(i) if $f(x) = ae^{bx}$, where a and b are constants, and if the auxiliary equation of (140) has $m = b$ as a root of multiplicity k (i.e. root occurs k times), then we take

$$Y(x) = Ax^k e^{bx},$$

where A is a constant to be determined. (If $m = b$ is not a root, then $k = 0$.)

(ii) if $f(x) = a \sin bx$ or $a \cos bx$, where a and b are constants, and if $(m^2 + b^2)$ is a factor of the auxiliary equation of multiplicity k, then we take

$$Y(x) = x^k(A \sin bx + B \cos bx),$$

where A and B are constants to be determined. (If $m^2 + b^2$ is not a factor, then $k = 0$.)

(iii) if $f(x) = ax^s$, where a and s are constants, and if the auxiliary equation has $m = 0$ as a root of multiplicity k, then we take

$$Y(x) = x^k(Ax^s + Bx^{s-1} + \dots + Px + Q)$$

where $A, B, ..., P, Q$ are constants to be determined. (Again if $m = 0$ is not a root, then $k = 0$.)

If $f(x)$ is the sum of any two or all of these special forms, the particular integral is then the appropriate sum of the individual particular integrals. The following examples illustrate the use of these rules.

Example 13. To find a particular integral of

$$\frac{d^2y}{dx^2} - 3\frac{dy}{dx} + 2y = e^{2x}. \tag{145}$$

The auxiliary equation

$$m^2 - 3m + 2 = 0 \tag{146}$$

has roots $m = 1$ and $m = 2$. Therefore, following (i), since one of these roots is equal to the value of $b(= 2)$, we must take $k = 1$ giving

$$Y(x) = Axe^{2x}. \tag{147}$$

Substituting (147) into (145) we find $A = 1$ and hence the particular integral is

$$Y(x) = xe^{2x}. \tag{148}$$

The general solution of (145) is therefore

$$y = Ce^x + De^{2x} + xe^{2x}, \tag{149}$$

where C and D are arbitrary constants.

Example 14. To find a particular integral of

$$\frac{d^2y}{dx^2} + 4y = 3\sin 2x. \tag{150}$$

Here the auxiliary equation is

$$(m^2 + 4) = 0. \tag{151}$$

Now since (in the notation of (ii)) $b = 2$, we see that $m^2 + b^2$ coincides with the factor $m^2 + 4$ of (151). Hence $k = 1$ and accordingly

$$Y(x) = x(A\sin 2x + B\cos 2x), \tag{152}$$

where A and B are constants to be found. Substituting (152) into (150) we find $A = 0$, $B = -\frac{3}{4}$. The particular integral is therefore

$$Y(x) = -\tfrac{3}{4}x \cos 2x. \tag{153}$$

Consequently the general solution (using (151) is

$$\begin{aligned} y &= Ce^{2ix} + De^{-2ix} - \tfrac{3}{4}x \cos 2x, \\ &= E \cos 2x + F \sin 2x - \tfrac{3}{4}x \cos 2x, \end{aligned} \tag{154}$$

where C, D, E and F are arbitrary constants.

Example 15. To find a particular integral of

$$\frac{d^3y}{dx^3} - 2\frac{d^2y}{dx^2} - \frac{dy}{dx} + 2y = 6x + \sin x. \tag{155}$$

Here the auxiliary equation

$$m^3 - 2m^2 - m + 2 = 0 \tag{156}$$

has roots $m = 1$, -1 and 2. Hence by (iii) and (ii) the forms of the particular integrals appropriate to the $6x$ term and the $\sin x$ term are respectively $Ax + B$ and $C \sin x + D \cos x$, where A, B, C and D are constants. The form of $Y(x)$ for (155) is therefore

$$Y(x) = Ax + B + C \sin x + D \cos x. \tag{157}$$

Substituting (157) into (155) we find $A = 3$, $B = 3/2$, $C = 1/5$ and $D = 1/10$, which give the particular integral as

$$Y(x) = 3(x + \tfrac{1}{2}) + \tfrac{1}{10}(2 \sin x + \cos x). \tag{158}$$

(b) Method of Variation of Constants

To illustrate this technique (which is sometimes known as the method of variation of parameters) we consider the second order constant coefficient equation

$$a_0 \frac{d^2y}{dx^2} + a_1 \frac{dy}{dx} + a_2 y = f(x). \tag{159}$$

Suppose the complementary function of this equation is

$$y = A_1 y_1 + A_2 y_2, \tag{160}$$

where y_1 and y_2 are independent solutions of the reduced equation

$$a_0 \frac{d^2y}{dx^2} + a_1 \frac{dy}{dx} + a_2 y = 0,$$ (161)

and A_1, A_2 are constants. We now replace these constants by functions $v_1(x)$ and $v_2(x)$ and define a new function $Y(x)$ as

$$Y(x) = v_1(x)y_1 + v_2(x)y_2.$$ (162)

The functions $v_1(x)$ and $v_2(x)$ are now to be found such that $Y(x)$ is a solution of (159). Now substituting (162) into (159) only leads to one condition on these two functions; they are therefore not determined by the requirement that $Y(x)$ be a solution of (159). Consequently we may impose one further condition on $v_1(x)$ and $v_2(x)$ which in conjunction with the first condition will determine them uniquely. It is convenient to take this condition to be

$$v_1'(x)y_1 + v_2'(x)y_2 = 0, \quad \text{for all } x,$$ (163)

(dashes denoting derivatives with respect to x) since the expression for $dY(x)/dx$ which will be needed in what follows then simplifies to

$$\frac{dY(x)}{dx} = v_1(x)y_1' + v_2(x)y_2'.$$ (164)

Differentiating (164) we find

$$\frac{d^2Y(x)}{dx^2} = v_1(x)y_1'' + v_2(x)y_2'' + v_1'(x)y_1' + v_2'(x)y_2'.$$ (165)

Hence substituting (162), (164) and (165) into (159) we have

$$a_0[v_1(x)y_1'' + v_2(x)y_2'' + v_1'(x)y_1' + v_2'(x)y_2']$$
$$+ a_1[v_1(x)y_1' + v_2(x)y_2'] + a_2[v_1(x)y_1 + v_2(x)y_2] = f(x), \quad (166)$$

or

$$v_1(x)(a_0y_1'' + a_1y_1' + a_2y_1) + v_2(x)(a_0y_2'' + a_1y_2' + a_2y_2)$$
$$+ a_0[v_1'(x)y_1' + v_2'(x)y_2'] = f(x). \quad (167)$$

Now since by definition y_1 and y_2 are solutions of (161) the first two brackets in (167) vanish and consequently

$$v_1'(x)y_1' + v_2'(x)y_2' = \frac{f(x)}{a_0}, \quad (a_0 \neq 0).$$ (168)

Hence solving (168) and (163) simultaneously we find

$$v_1'(x) = -\frac{y_2 f(x)}{a_0(y_1 y_2' - y_1' y_2)} \tag{169}$$

and

$$v_2'(x) = \frac{y_1 f(x)}{a_0(y_1 y_2' - y_1' y_2)}, \tag{170}$$

from which

$$v_1(x) = -\frac{1}{a_0} \int \frac{y_2 f(x)}{y_1 y_2' - y_1' y_2} \, dx \tag{171}$$

and

$$v_2(x) = \frac{1}{a_0} \int \frac{y_1 f(x)}{y_1 y_2' - y_1' y_2} \, dx. \tag{172}$$

We note here that since y_1 and y_2 are, by assumption, independent solutions, the Wronskian $y_1 y_2' - y_2 y_1'$ is non-zero (see Chapter 16, 16.8).

With these forms of $v_1(x)$ and $v_2(x)$, (162) is now a particular integral of (159). The general solution is therefore the sum of (160) and (162), namely

$$y = [A_1 + v_1(x)]y_1 + [A_2 + v_2(x)]y_2. \tag{173}$$

Example 16. To solve

$$\frac{d^2 y}{dx^2} - y = e^x. \tag{174}$$

Here the two solutions of the reduced equation are $y_1 = e^x, y_2 = e^{-x}$. The complementary function is therefore

$$y = A_1 e^x + A_2 e^{-x}. \tag{175}$$

Consequently the particular integral $Y(x)$ is

$$Y(x) = v_1(x)e^x + v_2(x)e^{-x}, \tag{176}$$

where, by (171) and (172),

405

$$v_1(x) = -\int \frac{e^{-x} \cdot e^x}{(-e^x \cdot e^{-x} - e^x \cdot e^{-x})} \, dx = \frac{x}{2}, \tag{177}$$

$$v_2(x) = \int \frac{e^x \cdot e^x}{(-e^x \cdot e^{-x} - e^x \cdot e^{-x})} \, dx = -\frac{e^{2x}}{4}. \tag{178}$$

By (173) the general solution of (174) is therefore

$$y = \left(A_1 + \frac{x}{2}\right)e^x + \left(A_2 - \frac{e^{2x}}{4}\right)e^{-x}, \tag{179}$$

$$= A_1' e^x + A_2 e^{-x} + \frac{x}{2}e^x, \tag{180}$$

where $A_1'(= A_1 - \frac{1}{4})$ and A_2 are arbitrary constants. (We note here that integration constants need not be added in (177) and (178) since they are already contained in A_1 and A_2 as shown by (173).)

21.7 The D-Operator

In discussing linear differential equations with constant coefficients it is often useful to let the symbol D represent the differential operator d/dx. Hence

$$Dy = \frac{dy}{dx}, \quad D^2y = D(Dy) = \frac{d^2y}{dx^2}, \quad D^3y = D(D^2y) = \frac{d^3y}{dx^3}, \tag{181}$$

and so on. From the rules of differentiation we now see

(i) $D[f(x) + g(x)] = Df(x) + Dg(x)$,

where $f(x)$ and $g(x)$ are differentiable functions;

(ii) $D[cf(x)] = cDf(x)$, where c is a constant;

and

(iii) $D^m[D^n f(x)] = D^n[D^m f(x)] = D^{m+n} f(x)$, where m and n are positive integers.

$$\left.\vphantom{\begin{array}{c}1\\1\\1\\1\\1\\1\end{array}}\right\} \tag{182}$$

Consequently we see that D satisfies three of the rules of elementary algebra and in this sense may be treated like an ordinary algebraic quantity (i.e. a number) despite the fact that it is an operator and has no actual numerical value.

The general nth order linear differential equation with constant coefficients

$$a_0 \frac{d^n y}{dx^n} + a_1 \frac{d^{n-1} y}{dx^{n-1}} + \ldots + a_{n-1} \frac{dy}{dx} + a_n y = f(x) \qquad (183)$$

may now be written in terms of the D-operator as

$$(a_0 D^n + a_1 D^{n-1} + \ldots + a_{n-1} D + a_n) y = f(x) \qquad (184)$$

or, in symbolic form, as

$$F(D)y = f(x), \qquad (185)$$

where

$$F(D) = a_0 D^n + a_1 D^{n-1} + \ldots + a_{n-1} D + a_n \qquad (186)$$

is a polynomial operator in D of degree n. Now since D behaves as an algebraic quantity so also must $F(D)$ (which is only a linear combination of powers of D). Hence $F(D)$ may be factorised in the same way as algebraic expressions. For example, $(D^2 + 4D + 3)y$ may be written as $(D+3)(D+1)y$, where $(D+3)$ now operates on $(D+1)y$. The order in which the factors are written down is irrelevant since it is easily seen that $(D+1)(D+3)y = (D+3)(D+1)y$. We note here that the order of writing the factors of (186) is important when $a_0, a_1, \ldots a_n$ are not constants but functions of x. For example

$$(D+2x)(D+1)y = D^2 y + 2xDy + Dy + 2xy, \qquad (187)$$

whereas

$$(D+1)(D+2x)y = D^2 y + D(2xy) + Dy + 2xy. \qquad (188)$$

Hence $(D+2x)(D+1)y \neq (D+1)(D+2x)y$. To further illustrate the factorisation of $F(D)$ we give the following examples:

$$\left.\begin{array}{l} (D^2-1)y = (D+1)(D-1)y = (D-1)(D+1)y, \\ (D^2+2D+1)y = (D+1)(D+1)y = (D+1)^2 y, \\ (D^3-3D-2)y = (D+1)(D+1)(D-2)y = (D+1)^2(D-2)y \\ \qquad = (D-2)(D+1)^2 y, \\ (D^2+1)y = (D+i)(D-i)y = (D-i)(D+i)y. \end{array}\right\} \qquad (189)$$

We now prove three useful theorems on the polynomial operator $F(D)$ defined by (186).

Theorem 1. If k is a constant

$$F(D)e^{kx} = F(k)e^{kx}. \qquad (190)$$

To prove this we note that $De^{kx} = ke^{kx}$, $D^2e^{kx} = k^2e^{kx}$, and so on. Consequently

$$\begin{aligned}
F(D)e^{kx} &= (a_0D^n + a_1D^{n-1} + \dots + a_{n-1}D + a_n)e^{kx} \\
&= (a_0k^n + a_1k^{n-1} + \dots + a_{n-1}k + a_n)e^{kx} \\
&= F(k)e^{kx}.
\end{aligned}$$

For example,

$$(6D^2 + 3D + 2)e^{3x} = (6.3^2 + 3.3 + 2)e^{3x} = 65e^{3x}. \qquad (191)$$

Theorem 2. If k is a constant and $V(x)$ is an arbitrary function of x

$$F(D)\{e^{kx}V(x)\} = e^{kx}F(D+k)V(x). \qquad (192)$$

To prove this we note that

$$D\{e^{kx}V(x)\} = e^{kx}DV(x) + ke^{kx}V(x) = e^{kx}(D+k)V(x),$$
$$\begin{aligned}
D^2\{e^{kx}V(x)\} &= D\{e^{kx}(D+k)V(x)\} = e^{kx}\{(D+k)(D+k)V(x)\} \\
&= e^{kx}(D+k)^2V(x),
\end{aligned}$$

and, by Leibnitz's theorem, for arbitrary integral n

$$\begin{aligned}
D^n\{e^{kx}V(x)\} &= e^{kx}D^nV(x) + {}^nc_1D(e^{kx})D^{n-1}V(x) + \dots \\
&\qquad + {}^nc_{n-1}D^{n-1}(e^{kx})DV(x) + {}^nc_nD^n(e^{kx})V(x) \\
&= e^{kx}D^nV(x) + kne^{kx}D^{n-1}V(x) + \dots \\
&\qquad + nk^{n-1}e^{kx}DV(x) + k^ne^{kx}V(x) \\
&= e^{kx}(D+k)^nV(x).
\end{aligned}$$

Hence

$$F(D)\{e^{kx}V(x)\} = e^{kx}F(D+k)V(x).$$

For example,

$$\begin{aligned}
(D^2 - 4D + 1)\{e^{2x}V(x)\} &= e^{2x}\{(D+2)^2 - 4(D+2) + 1\}V(x) \\
&= e^{2x}(D^2 - 3)V(x). \qquad (193)
\end{aligned}$$

Theorem 3. If k is a constant,

$$F(D^2) \sin kx = F(-k^2) \sin kx, \Big\}$$
and
$$F(D^2) \cos kx = F(-k^2) \cos kx, \Big\} \qquad (194)$$

where $F(D^2)$ is the polynomial operator (186) with D everywhere replaced by D^2.

To prove this we note that $D^2 \sin kx = -k^2 \sin kx$, $D^4 \sin kx = k^4 \sin kx = (-k^2)^2 \sin kx$, and so on.

Hence

$$F(D^2) \sin kx = (a_0 D^{2n} + a_1 D^{2(n-1)} + \ldots + a_{n-1} D^2 + a_n) \sin kx$$
$$= \{a_0(-k^2)^n + a_1(-k^2)^{n-1} + \ldots + a_{n-1}(-k^2) + a_n\} \sin kx$$
$$= F(-k^2) \sin kx.$$

A similar proof exists for $F(D^2) \cos kx$.

We see that Theorem 3 can in fact be deduced from Theorem 1 using the relation

$$e^{ikx} = \cos kx + i \sin kx. \qquad (195)$$

For we now have

$$\cos kx = \mathrm{R}\, e^{ikx},$$
$$\sin kx = \mathrm{I}\, e^{ikx}, \qquad (196)$$

where R and I denote the real and imaginary parts respectively of the function that follows them. Consequently

$$F(D^2) \sin kx = F(D^2) \mathrm{I}\, e^{ikx},$$

which by Theorem 1 is equal to $F(-k^2) \mathrm{I}\, e^{ikx}$, or $F(-k^2) \sin kx$. Hence we have Theorem 3.

A similar argument applies to $F(D^2) \cos kx$. To illustrate Theorem 3 we give the following simple examples

$$(D^4 + 3D^2 - 1) \sin 2x = \{(-2^2)^2 + 3(-2^2) - 1\} \sin 2x = 3 \sin 2x, \qquad (197)$$

$$(D^4 - 2D^2) \cos 2x = \{(-2^2)^2 - 2(-2^2)\} \cos 2x = 24 \cos 2x. \qquad (198)$$

In Chapter 4 we defined the operation of indefinite integration as the inverse operation to differentiation so that

$$\frac{d}{dx}\int^x y(u)du = y(x). \tag{199}$$

Consequently corresponding to the operator D we now define an inverse operator D^{-1} such that

$$D^{-1} \equiv \frac{1}{D} \equiv \int, \tag{200}$$

and

$$D(D^{-1}y) = y. \tag{201}$$

Hence just as D^n denotes n operations of differentiation, so $1/D^n \equiv D^{-n}$ denotes n operations of integration. We must remember when writing inverse operations in the form $1/D^n$ that since an operator only acts on functions which appear to its right, the expression $f(x)/D^n$ is ambiguous. It is not clear whether this implies that $1/D^n$ operates on $f(x)$ to give the function $(1/D^n)f(x)$, or whether it simply defines the operator $f(x)(1/D^n)$. To avoid this ambiguity we therefore write all functions either to the left or the right of $1/D^n$ as required, thus keeping $1/D^n$ as a separate symbol.

21.8 D-Operator Method for Particular Integrals

Consider the linear first order constant coefficient equation

$$(D-k)y = f(x), \tag{202}$$

where k is a constant and $f(x)$ a given function. The complementary function of this equation is

$$y = Ae^{kx}, \tag{203}$$

where A is an arbitrary constant. To find a particular integral $Y(x)$ we now assume the form

$$Y(x) = V(x)e^{kx}. \tag{204}$$

Then by Theorem 2

$$(D-k)Y(x) = (D-k)\{e^{kx}V(x)\} = e^{kx}DV(x) = f(x). \tag{205}$$

Hence

$$DV(x) = e^{-kx}f(x), \qquad (206)$$

and therefore

$$V(x) = \frac{1}{D}\{e^{-kx}f(x)\} = \int^{x} e^{-k\xi}f(\xi)\,d\xi, \qquad (207)$$

where ξ is an arbitrary variable of integration. The particular integral (by (204)) is therefore

$$Y(x) = e^{kx}\int^{x} e^{-k\xi}f(\xi)\,d\xi, \qquad (208)$$

whence the general solution of (202) is

$$y = Ae^{kx} + e^{kx}\int^{x} e^{-k\xi}f(\xi)\,d\xi. \qquad (209)$$

(We note that no integration constant need be added in (207) since if it were it would merely become absorbed in the arbitrary constant A in (209).)

Consider now the second order equation

$$(D^2 - 2kD + k^2)y = (D-k)^2 y = f(x). \qquad (210)$$

Here the complementary function has the form

$$y = (Ax + B)e^{kx} \qquad (211)$$

in virtue of both roots of the auxiliary equation being equal (A and B being arbitrary constants). We now assume a particular integral of the form

$$Y(x) = V(x)e^{kx}. \qquad (212)$$

Hence by Theorem 2

$$(D-k)^2 Y(x) = (D-k)^2\{e^{kx}V(x)\} = e^{kx}D^2V(x) = f(x), \quad (213)$$

which gives

$$D^2V(x) = e^{-kx}f(x). \qquad (214)$$

Consequently

$$V(x) = \frac{1}{D^2}\{e^{-kx}f(x)\} = \int^x d\eta \int^\eta e^{-k\xi}f(\xi)\,d\xi, \tag{215}$$

where ξ and η are arbitrary variables of integration. The particular integral is therefore

$$Y(x) = e^{kx}\int^x d\eta \int^\eta e^{-k\xi}f(\xi)\,d\xi \tag{216}$$

and the general solution

$$y = (Ax+B)e^{kx} + e^{kx}\int^x d\eta \int^\eta e^{-k\xi}f(\xi)\,d\xi. \tag{217}$$

(Again no integration constant is required in (216) for the same reasons as given above.)

Finally we consider the solution of the equation

$$(D-k_1)(D-k_2)y = f(x), \tag{218}$$

where k_1 and k_2 are constants ($k_1 \neq k_2$).

Here the complementary function is

$$y = Ae^{k_1 x} + Be^{k_2 x}, \tag{219}$$

where A and B are arbitrary constants.

We now assume a particular integral of the form

$$Y(x) = V(x)e^{k_1 x}. \tag{220}$$

Then

$$(D-k_1)(D-k_2)\{e^{k_1 x}V(x)\} = e^{k_1 x}\{(D+k_1)-k_1\}\{(D+k_1)-k_2\}V(x)$$
$$= e^{k_1 x}D(D+k_1-k_2)V(x) = f(x). \tag{221}$$

Hence

$$D(D+k_1-k_2)V(x) = e^{-k_1 x}f(x)$$

and consequently

$$(D+k_1-k_2)V(x) = \frac{1}{D}\{e^{-k_1 x}f(x)\} = \int^x e^{-k_1 \xi}f(\xi)\,d\xi. \tag{222}$$

This equation has the same intrinsic form as (202) and therefore has the solution

$$V(x) = e^{(k_2-k_1)x} \int^x e^{(k_1-k_2)\eta} \, d\eta \int^\eta e^{-k_1\xi} f(\xi) \, d\xi, \qquad (223)$$

whence, by (220),

$$Y(x) = e^{k_2 x} \int^x e^{(k_1-k_2)\eta} \, d\eta \int^\eta e^{-k_1\xi} f(\xi) \, d\xi. \qquad (224)$$

This form of the particular integral may be simplified to some extent by integrating (224) by parts (i.e. integrating the $e^{(k_1-k_2)x}$ term and differentiating

$$\int^\eta e^{-k_1\xi} f(\xi) \, d\xi).$$

In this way we find

$$Y(x) = \frac{e^{k_1 x}}{k_1-k_2} \int^x e^{-k_1\xi} f(\xi) \, d\xi - \frac{e^{k_2 x}}{k_1-k_2} \int^x e^{-k_2\xi} f(\xi) \, d\xi. \qquad (225)$$

Hence the general solution of (218) is

$$y = Ae^{k_1 x} + Be^{k_2 x} + \frac{e^{k_1 x}}{k_1-k_2} \int^x e^{-k_1\xi} f(\xi) \, d\xi - \frac{e^{k_2 x}}{k_1-k_2} \int^x e^{-k_2\xi} f(\xi) \, d\xi. \qquad (226)$$

Similar results can be found for higher order equations and we leave their derivation to the reader. The following example illustrates the method of this section.

Example 17. To solve the equation

$$(D^2+4)y = (D+2i)(D-2i)y = e^{3x}. \qquad (227)$$

Here $k_1 = -2i$, $k_2 = 2i$ and hence by (225)

$$Y(x) = \frac{e^{-2ix}}{-4i} \int^x e^{2i\xi} e^{3\xi} \, d\xi - \frac{e^{2ix}}{-4i} \int^x e^{-2i\xi} e^{3\xi} \, d\xi \qquad (228)$$

$$= -\frac{e^{-2ix}}{4i} \cdot \frac{e^{(3+2i)x}}{3+2i} + \frac{e^{2ix}}{4i} \cdot \frac{e^{(3-2i)x}}{3-2i} \qquad (229)$$

$$= e^{3x}/13. \qquad (230)$$

413

Since the complementary function of (227) is $y = A \cos 2x + B \sin 2x$, where A and B are arbitrary constants, the general solution is therefore

$$y = A \cos 2x + B \sin 2x + \frac{e^{3x}}{13}. \qquad (231)$$

21.9 Euler's Equation

The equation

$$a_0 x^n \frac{d^n y}{dx^n} + a_1 x^{n-1} \frac{d^{n-1} y}{dx^{n-1}} + \ldots + a_{n-1} x \frac{dy}{dx} + a_n y = f(x), \qquad (232)$$

where $a_0, a_1, \ldots, a_{n-1}, a_n$ are given constants and $f(x)$ is a given function of x, is usually referred to as Euler's equation. The general solution of this equation may always be found by first reducing (232) to a constant coefficient equation by means of the substitution

$$x = e^t, \qquad (233)$$

and then solving the constant coefficient equation for y as a function of t in the usual way. To see how this comes about we note that (233) implies

$$\frac{dy}{dx} = \frac{dy}{dt}\frac{dt}{dx} = \frac{1}{x}\frac{dy}{dt} \qquad (234)$$

or

$$x\frac{dy}{dx} = \frac{dy}{dt}. \qquad (235)$$

Likewise we find

$$x^2 \frac{d^2 y}{dx^2} = \frac{d^2 y}{dt^2} - \frac{dy}{dt} \qquad (236)$$

and

$$x^n \frac{d^n y}{dx^n} = D_t(D_t - 1)(D_t - 2)\ldots(D_t - n + 1)y, \qquad (237)$$

where D_t is the operator $\dfrac{d}{dt}$.

For example, using these results the equation

$$x^2 \frac{d^2y}{dx^2} + 2x\frac{dy}{dx} + 2y = x^2 \tag{238}$$

becomes

$$\frac{d^2y}{dt^2} + \frac{dy}{dt} + 2y = e^{2t}, \tag{239}$$

which may be solved using one of the earlier methods.

Example 18. To solve the equation

$$x^2 \frac{d^2y}{dx^2} - x\frac{dy}{dx} + 4y = \cos(\log_e x). \tag{240}$$

Writing $x = e^t$, (240) becomes

$$\frac{d^2y}{dt^2} - 2\frac{dy}{dt} + 4y = \cos t. \tag{241}$$

The complementary function (that is, the solution of the equation (241) with *zero* on the right-hand side) is easily found to be

$$y = e^t(A\cos\sqrt{3}t + B\sin\sqrt{3}t). \tag{242}$$

A particular integral of (241) may be obtained using D-operator methods to give

$$y = \frac{1}{D^2 - 2D + 4}\cdot\cos t = \frac{1}{(-1)^2 - 2D + 4}\cos t \tag{243}$$

$$= \frac{1}{3 - 2D}\cos t \tag{244}$$

$$= \frac{3 + 2D}{9 - 4D^2}\cos t \tag{245}$$

$$= \frac{1}{13}(3 + 2D)\cos t \tag{246}$$

$$= \frac{1}{13}(3\cos t - 2\sin t). \tag{247}$$

The general solution of (240) is finally obtained by adding (242) and (247) and substituting $t = \log_e x$. In this way, we find

$$y = x(A\cos(\sqrt{3}\log_e x) + B\sin(\sqrt{3}\log_e x))$$
$$+ \frac{1}{13}(3\cos(\log_e x) - 2\sin(\log_e x)). \quad (248)$$

21.10 The General Linear Second Order Equation

The Euler equation discussed in the last section is a special case of the general linear second order equation

$$\frac{d^2y}{dx^2} + p(x)\frac{dy}{dx} + q(x)y = f(x), \quad (249)$$

where $p(x), q(x)$ and $f(x)$ are given functions of x. Usually this type of equation has to be solved by series approximation methods (see next chapter), but in certain instances we may proceed either by means of a substitution or by knowing one solution. The following examples illustrate these approaches.

Example 19. The substitution $x = z^{1/2}$ transforms the equation

$$\frac{d^2y}{dx^2} + \left(4x - \frac{1}{x}\right)\frac{dy}{dx} + 4x^2y = 0 \quad (250)$$

into one with constant coefficients, namely

$$\frac{d^2y}{dz^2} + 2\frac{dy}{dz} + y = 0, \quad (251)$$

which has the solution

$$y = (A + Bz)e^{-z}. \quad (252)$$

Hence the general solution of (250) is

$$y = (A + Bx^2)e^{-x^2}. \quad (253)$$

Example 20. If one solution is assumed to be known, we may obtain the second, and hence the general, solution in the following way: Suppose $y = v(x)$ is a known solution of the equation

$$x\frac{d^2y}{dx^2} + \frac{dy}{dx} + xy = 0. \quad (254)$$

416

Then a second solution may be found by writing

$$y = u(x)v(x) \tag{255}$$

and solving for $u(x)$. In this way we find

$$\frac{1}{w}\frac{dw}{dx} + \frac{2}{v}\frac{dv}{dx} + \frac{1}{x} = 0, \tag{256}$$

where $w = \dfrac{du}{dx}$, whence finally on integration

$$\frac{du}{dx} = \frac{A}{xv^2}, \tag{257}$$

A being a constant of integration.

Hence

$$u(x) = A \int \frac{dx}{xv^2} + B, \tag{258}$$

where B is an integration constant. The required second solution is therefore

$$y = Av \int \frac{dx}{xv^2} + Bv. \tag{259}$$

Example 21. One solution of the equation

$$x^2 \frac{d^2y}{dx^2} - x(x+2)\frac{dy}{dx} + (x+2)y = 0 \tag{260}$$

is $y_1 = x$. We may find a second solution by writing $y_2 = xu(x)$ and solving for $u(x)$. In this way (260) leads to

$$\frac{d^2u}{dx^2} - \frac{du}{dx} = 0 \tag{261}$$

which on integration gives

$$u = Ae^x + B, \tag{262}$$

where A and B are arbitrary constants. The general solution of (260) is therefore

$$\begin{aligned} y &= x(Ae^x + B) + Cx, \\ &= x(Ae^x + B'), \end{aligned} \tag{263}$$

where B' is an arbitrary constant.

21.11. Simultaneous Equations

The general solution of simultaneous equations with constant co-efficients in two or more dependent variables may be found by solving for each dependent variable separately. The following examples illustrate the method.

Example 22. To solve the equations

$$\left.\begin{array}{c} \dfrac{dx}{dt}+2y+3x = 0, \\[2mm] 3x+\dfrac{dy}{dt}-2y = 0, \end{array}\right\} \tag{264}$$

we first write the operator $\dfrac{d}{dt}$ as D to give

$$\left.\begin{array}{c} (D+3)x+2y = 0, \\ 3x+(D-2)y = 0. \end{array}\right\} \tag{265}$$

Now to eliminate y (say) from this pair of equations we operate on the first equation with $D-2$, and multiply the second by a factor 2. Hence

$$\left.\begin{array}{c} (D-2)(D+3)x+2(D-2)y = 0, \\ 6x+2(D-2)y = 0. \end{array}\right\} \tag{266}$$

Subtracting the first equation from the second leads directly to the constant coefficient equation

$$(D^2+D-6)x-6x = 0 \tag{267}$$

or

$$(D^2+D-12)x = 0. \tag{268}$$

The solution of (268) is readily found to be

$$x(t) = Ae^{3t}+Be^{-4t}. \tag{269}$$

Having found x it is a simple matter to insert (269) back into the original equations to find y. In this way, we have

$$y(t) = -3Ae^{3t}+\tfrac{1}{2}Be^{-4t}. \tag{270}$$

418

Example 23. To solve

$$\left.\begin{array}{c} \dfrac{dx}{dt}+y = t^3, \\[2mm] \dfrac{dy}{dt}-x = t, \end{array}\right\} \tag{271}$$

we may differentiate the first equation with respect to t to obtain

$$\frac{dy}{dt} = 3t^2 - \frac{d^2x}{dt^2}. \tag{272}$$

Inserting this expression into the second equation of (271) we find

$$\frac{d^2x}{dt^2}+x = 3t^2 - t \tag{273}$$

as an equation for x. This equation may be solved by the methods already discussed and the final solutions are

$$\left.\begin{array}{l} x = A\cos t + B\sin t + 3t^2 - t - 6, \\[1mm] y = A\sin t - B\cos t + t^3 - 6t + 1, \end{array}\right\} \tag{274}$$

as the reader may verify.

Similar techniques may be applied to higher order linear simultaneous equations, but the complexity may be considerable in the process of eliminating one or more of the dependent variables. Other methods of solution may sometimes be used with profit. One of these depends on the Laplace transform (see Chapter 23) and will be discussed later, whilst the other is based on writing the equations in matrix form. The following example illustrates the basic idea.

Example 24. Consider the following three equations

$$\left.\begin{array}{l} \dfrac{dx_1}{dt} = a_{11}x_1 + a_{12}x_2 + a_{13}x_3, \\[2mm] \dfrac{dx_2}{dt} = a_{21}x_1 + a_{22}x_2 + a_{23}x_3, \\[2mm] \dfrac{dx_3}{dt} = a_{31}x_1 + a_{32}x_2 + a_{33}x_3, \end{array}\right\} \tag{275}$$

where the coefficients a_{ik} $(i, k = 1, 2, 3)$ are constants and x_1, x_2, x_3 are to be found.

We may write these simultaneous equations in matrix form as

$$\frac{dX}{dt} = AX, \tag{276}$$

where

$$X = \begin{pmatrix} x_1 \\ x_2 \\ x_3 \end{pmatrix}_. \quad \text{and} \quad A = \begin{pmatrix} a_{11} & a_{12} & a_{13} \\ a_{21} & a_{22} & a_{23} \\ a_{31} & a_{32} & a_{33} \end{pmatrix} \tag{277}$$

The solution of (276) is easily verified to be

$$X = e^{At}X(0), \tag{278}$$

where $X(0)$ is the column matrix of the values of x_1, x_2 and x_3 at $t = 0$ (assumed given), and

$$e^{At} = I + A + \frac{A^2}{2!}t^2 + \frac{A^3}{3!}t^3 + \dots, \tag{279}$$

I being the unit matrix of order 3. Hence we see that the solution of the set of equations (275) depends on the evaluation of the matrix (279). Unless A is of low order the evaluation of e^{At} involves a great deal of numerical work (except in the case where many of the elements happen to be zero). However, modern computing methods enable this task to be carried out, and the matrix approach is much to be recommended when the number of dependent variables is increased.

21.12. Simple Non-Linear Equations

In a few simple cases, non-linear equations may be reduced to linear form. For example, Bernoulli's equation

$$\frac{dy}{dx} + P(x)y = Q(x)y^n, \tag{280}$$

where $P(x)$ and $Q(x)$ are given functions of x, reduces to the linear equation

$$\frac{1}{(1-n)}\frac{dz}{dx} + P(x)z = Q(x) \tag{281}$$

by putting $y^{1-n} = z$.

Other non-linear equations may sometimes be solved by writing

$$\frac{dy}{dx} = p, \tag{282}$$

whence

$$\frac{d^2y}{dx^2} = \frac{dp}{dx} = p\frac{dp}{dy}. \tag{283}$$

For example, the equation

$$\frac{d^2y}{dx^2} - 2y = 2y^3 \tag{284}$$

may be written (using (283)) as

$$p\frac{dp}{dy} = 2y + 2y^3, \tag{285}$$

which is now a variable-separable equation for p in terms of y. A further integration then gives y in terms of x.

21.13 Other Methods of Solving Differential Equations

We have concentrated in this chapter on the solution of ordinary linear equations (mainly with constant coefficients) and those equations which can be reduced to such. A further useful method for constant coefficient equations involves the idea of the Laplace transformation and this is discussed in Chapter 23.

Linear equations with variable coefficients are, in general, more difficult to solve, and as we shall see in the next chapter a series approximation technique is the basic method of solution.

Non-linear equations are usually only soluble by some numerical method and for a full discussion of the numerical solution of differential equations (both linear and non-linear) the reader should consult an appropriate text.

PROBLEMS 21

1. Solve the following first order equations

(a) $\dfrac{dy}{dx} = \dfrac{y+1}{x+1}$; given that $y = 1$ at $x = 0$,

(b) $x\dfrac{dy}{dx}+3y=2,$ given that $y=2$ at $x=1,$

(c) $x^2\dfrac{dy}{dx}+xy=y^2,$

(d) $\dfrac{dy}{dx}=\dfrac{x+2y+3}{3x+6y+7},$

(e) $\dfrac{dy}{dx}+\dfrac{y}{x}=\sin x,$ given $y=0$ at $x=\pi.$

2. (i) Solve the differential equation

$$(x+1)\frac{dy}{dx}-3y=(x+1)^5$$

given that $y=\frac{3}{2}$ when $x=0.$

(ii) Find the general solution of

$$\frac{d^2y}{dx^2}+4\frac{dy}{dx}+4y=2\cos^2 x. \qquad \text{(C.U.)}$$

3. Solve the equations

(i) $x(x+1)\dfrac{dy}{dx}-(x+2)y=x^3(2x-3),$

(ii) $\dfrac{d^2y}{dx^2}+3\dfrac{dy}{dx}+2y=xe^{-x}.$ \qquad (C.U.)

4. Find the general solution of each of the differential equations:

(i) $\dfrac{d^2y}{dx^2}-6\dfrac{dy}{dx}+10y=20-e^{2x},$

(ii) $(x^2+3x+2)\dfrac{dy}{dx}+xy=x(x+1).$ \qquad (C.U.)

5. Solve

(i) $\dfrac{dy}{dx}+y\log_e x=e^{-x\log_e x},$

(ii) $\dfrac{d^2y}{dx^2} - n^2y = e^{nx}$, $\quad(n \text{ constant})$. $\hfill$ (C.U.)

6. Test the following equations for 'exactness', and solve those found to be exact. The others may be made exact by some transformation. In each case find this transformation and hence solve the equation.

  (a) $e^y \sin x \dfrac{dy}{dx} + (1 + e^y) \cos x = 0$,

  (b) $x \dfrac{dy}{dx} + y(x^2 + \log_e y) = 0$,

  (c) $(y^2 - x^2)\dfrac{dy}{dx} + 2xy = 0$,

  (d) $(y^4 + 2y^2 - x^3 + 5x^2y - 21xy^2)\dfrac{dy}{dx}$

$$+ (x^3 - 3x^2y + 5xy^2 - 7y^3) = 0.$$

7. Obtain the solution of the equation

$$\frac{d^2y}{dx^2} + 6\frac{dy}{dx} + 8y = \cosh 2x$$

  for which $y = 0$, $\dfrac{dy}{dx} = 1$ at $x = 0$. $\hfill$ (C.U.)

8. Solve the equations

  (i) $(D^2 - D - 2)y = \sin x$,

  (ii) $(D^2 + 2D + 1)y = 4 \sinh x$,

  (iii) $(D^2 - 2D + 5)y = e^x \sin x$,

  (iv) $(D^2 - 4D + 5)y = x^2 e^{2x} \sin x$.

9. Find the solution of

$$\frac{d^3y}{dx^3} - 2\frac{d^2y}{dx^2} + \frac{dy}{dx} - 2y = 12 \sin 2x - 4x$$

  which satisfies $\dfrac{d^2y}{dx^2} = -4$, $\dfrac{dy}{dx} = 5$, $y = 2$ at $x = 0$. $\hfill$ (C.U.)

10. By means of the substitution $x = \sin \theta$ transform the equation

$$\cos \theta \frac{d^2y}{d\theta^2} + \sin \theta \frac{dy}{d\theta} + 4y \cos^3 \theta = 8 \cos^5 \theta$$

into one with x as independent variable. Hence solve the equation.

11. Change the independent variable in the differential equation

$$x^2 \frac{d^2y}{dx^2} + x \frac{dy}{dx} + y = 0$$

from x to z, where $z = \log_e x$, and hence obtain the general solution. (C.U.)

12. Find n such that the substitution $y = zx^n$ transforms the differential equation

$$x^2 \frac{d^2y}{dx^2} + 2x(x+2)\frac{dy}{dx} + 2(x+1)^2 y = e^{-x} \cos x$$

into one with constant coefficients. Hence solve the original equation and show that in all solutions y is small when x is large and positive. (L.U.)

13. The vector $\mathbf{r} = (x, y, z)$ satisfies the vector equation

$$m \frac{d^2\mathbf{r}}{dt^2} = e\mathbf{E} + \frac{e}{c}\left(\frac{d\mathbf{r}}{dt} \wedge \mathbf{H}\right),$$

where $\mathbf{E} = (0, \mathcal{E}, 0)$, $\mathbf{H} = (0, 0, H)$, and e, m, c, E, H are constants.

Write the equation in component form and show by solving the equation that

$$x = \frac{cEt}{H} - \frac{mc^2E}{eH^2} \sin\left(\frac{eH}{mc}t\right),$$

$$y = \frac{mc^2E}{eH^2}\left\{1 - \cos\left(\frac{eH}{mc}t\right)\right\},$$

$$z = 0,$$

given that $\mathbf{r} = 0, \dot{\mathbf{r}} = 0$ at $t = 0$. (C.U.)

424

14. Using the transformation $x = e^t$, solve the equations

    (i) $x^2 \dfrac{d^2y}{dx^2} + x \dfrac{dy}{dx} - y = 0$,

    (ii) $\dfrac{d^2y}{dx^2} + \dfrac{1}{x}\dfrac{dy}{dx} - \dfrac{4}{x^2}y = 1 + x^2$,

    (iii) $x^3 \dfrac{d^3y}{dx^3} + 3x^2 \dfrac{d^2y}{dx^2} + x \dfrac{dy}{dx} + 8y = 65 \cos(\log_e x)$.

15. Find the general solutions of the following simultaneous differential equations

    (i) $\dfrac{dx}{dt} = x - 2y$, $\quad \dfrac{dy}{dt} = y - 2x$,

    (ii) $\dfrac{d^2x}{dt^2} = x - y$, $\quad \dfrac{d^2y}{dt^2} = y - x$.

16. Obtain the solution of the simultaneous equations

    $$\frac{d^2x}{dt^2} + 3\frac{dx}{dt} - 2x + \frac{dy}{dt} - 3y = 2e^{-t},$$

    $$2\frac{dx}{dt} - x + \frac{dy}{dt} - 2y = 0,$$

    that satisfies the conditions $x = 0$, $dx/dt = 0$, $y = 4$ when $t = 0$.

    (C.U.)

17. Solve the simultaneous equations

    $$\frac{d^2x}{dt^2} + 2a\frac{dy}{dt} + a^2x = \cos at,$$

    $$\frac{d^2y}{dt^2} + 2a\frac{dx}{dt} + a^2y = 0,$$

    where a is a constant.

    (L.U.)

425

18. Solve the simultaneous equations

$$\frac{dx}{dt} + k(y+z) = 0,$$

$$\frac{dy}{dt} + k(z+x) = 0,$$

$$\frac{dz}{dt} + k(x+y) = 0,$$

where k is a constant.

19. Solve the following equations by first transforming them to linear form

(a) $\dfrac{dy}{dx} = \sin^2 (x - 2y),$

(b) $x \sec^2 y \dfrac{dy}{dx} + \tan y = x^2.$

20. Solve the non-linear equations

(a) $\dfrac{d^2y}{dx^2} + 2\left(\dfrac{dy}{dx}\right)^2 = y^2,$ given $\dfrac{dy}{dx} = \dfrac{1}{4}$ when $y = 0,$

(b) $\dfrac{d^2y}{dx^2} - 2y = 2y^3,$ given $\dfrac{dy}{dx} = 1$ when $y = 0,$

(c) $x\dfrac{d^2y}{dx^2} = (1+y)\dfrac{dy}{dx},$ given $y = 0, \dfrac{dy}{dx} = 0$ when $x = 0,$

(d) $y\dfrac{d^2y}{dx^2} = 1 + \left(\dfrac{dy}{dx}\right)^2,$ given $\dfrac{dy}{dx} = 0$ when $y = 1.$

21. Show that the solution of the non-linear equation

$$\frac{d^2y}{dx^2} + y + \frac{y^3}{b^2} = 0$$

satisfying the conditions $y = a$, $dy/dx = 0$ at $x = 0$ (where a and b are constants) is

$$x = \left(1 + \frac{a^2}{b^2}\right)^{-\frac{1}{2}} F(k, \phi),$$

where $k^2 = \dfrac{a^2}{2(a^2 + b^2)}$, $\phi = \cos^{-1}\left(\dfrac{y}{a}\right)$, and $F(k, \phi)$ is the elliptic integral of the first kind (see Chapter 11).

22. If two families of curves cross at right angles everywhere they are called orthogonal families: each family is then a set of orthogonal trajectories for the other. Find the orthogonal trajectories of the following families of curves

   (a) $y = x^2 + c$, (b) $y = ce^x$, (c) $x^2 + y^2 = 2cx$, where c is an arbitrary parameter.

23. Show that if $P(x) + Q(x) + 1 = 0$, then $y = e^x$ is a solution of the equation

$$\frac{d^2y}{dx^2} + P(x)\frac{dy}{dx} + Q(x)y = 0.$$

Hence show that the general solution of the equation

$$x\frac{d^2y}{dx^2} - (2x+1)\frac{dy}{dx} + (x+1)y = 0$$

is $\qquad\qquad y = (Ax^2 + B)e^x.$

24. Show that

$$\frac{d^2y}{dx^2} + \omega^2 y = 0$$

(ω constant) has a non-zero solution satisfying the conditions $y = 0$ when $x = 0$, and $y = 0$ when $x = l$ (l constant) only if $\omega = \dfrac{n\pi}{l}$, where n is an integer. (This infinite set of discrete values of ω is called the set of eigenvalues, and the corresponding solutions the eigenfunctions).

25. Show that the equation

$$\frac{d^4y}{dx^4}+(\lambda^2-\mu^2)\frac{d^2y}{dx^2}-\lambda^2\mu^2y = 0$$

has a non-zero solution satisfying the conditions

$$y = 0, \quad \frac{dy}{dx}=0 \quad \text{at} \quad x = 0,$$

$$y = 0, \quad \frac{d^2y}{dx^2}=0 \quad \text{at} \quad x = l,$$

where λ, μ, l are given constants, only if

$$\frac{\tan \lambda l}{\lambda}=\frac{\tanh \mu l}{\mu}.$$

26. Using the transformation $x = t^3$, or otherwise, show that the equation

$$9x^2\frac{d^2y}{dx^2}+6x\frac{dy}{dx}+\lambda x^{2/3}y = 0,$$

where λ is a constant, has a non-zero solution satisfying the conditions $y = 0$ when $x = 0$ and when $x = 1$ only if $\lambda = n^2\pi^2$ $(n = 1, 2, 3 \ldots)$. (L.U.)

27. Determine the matrix A such that the three simultaneous differential equations

$$\frac{dy_1}{dt} = y_1+y_2+3y_3,$$

$$\frac{dy_2}{dt} = 5y_1+2y_2+6y_3,$$

$$\frac{dy_3}{dt} = -2y_1-y_2-3y_3,$$

may be written in the matrix differential form

$$\frac{dY}{dt}=AY,$$

where $Y = \begin{pmatrix} y_1 \\ y_2 \\ y_3 \end{pmatrix}$. Evaluate A^2 and A^3. Hence obtain the solution of the given set of equations subject to the conditions $y_1 = 0$, $y_2 = -1$, $y_3 = 1$ at $t = 0$.

28. Verify that the solution of the matrix equation

$$\frac{dY(t)}{dt} = A Y(t) + Y(t)B,$$

where A and B are constant square matrices of the same order, and $Y(t)$ is a square matrix, is

$$Y(t) = e^{At} Y(0) e^{Bt},$$

$Y(0)$ being the matrix $Y(t)$ at $t = 0$.

CHAPTER 22

Series Solution of Ordinary Differential Equations

22.1 The Leibnitz-Maclaurin Method

As mentioned in the last chapter, linear differential equations with variable coefficients can often be solved by assuming a solution in the form of a power series in x. One of the simplest ways of doing this is to use the Leibnitz-Maclaurin method as illustrated by the following examples.

Example 1. To solve the equation

$$(1-x^2)\frac{d^2y}{dx^2} - 5x\frac{dy}{dx} - 3y = 0 \qquad (1)$$

we first differentiate n times using Leibnitz's formula (Chapter 3, 3.6) to obtain

$$(1-x^2)y^{(n+2)} - x(2n+5)y^{(n+1)} - (n+1)(n+3)y^{(n)} = 0, \qquad (2)$$

where, in general, $y^{(r)}$ stands for d^ry/dx^r. If now we assume a solution of (1) in the form of a Maclaurin expansion of y then

$$y = y(0) + xy^{(1)}(0) + \frac{x^2}{2!}y^{(2)}(0) + \ ... \ + \frac{x^r}{r!}y^{(r)}(0) + \ ..., \qquad (3)$$

where $y^{(r)}(0)$ denotes the value of d^ry/dx^r at $x = 0$. The values of these differential coefficients may now be found with the help of the recurrence relation

$$y^{(n+2)}(0) = (n+1)(n+3)y^{(n)}(0), \quad (n \geqq 0), \qquad (4)$$

430

obtained from (2) by putting $x = 0$. Hence we have

$$\left.\begin{aligned}
y^{(2)}(0) &= 1 \cdot 3 \cdot y(0), \\
y^{(3)}(0) &= 2 \cdot 4 \cdot y^{(1)}(0), \\
y^{(4)}(0) &= 3 \cdot 5 \cdot y^{(2)}(0) = 1 \cdot 3^2 \cdot 5 \cdot y(0), \\
y^{(5)}(0) &= 4 \cdot 6 \cdot y^{(3)}(0) = 2 \cdot 4^2 \cdot 6 \cdot y^{(1)}(0), \\
y^{(6)}(0) &= 5 \cdot 7 \cdot y^{(4)}(0) = 1 \cdot 3^2 \cdot 5^2 \cdot 7 \cdot y(0),
\end{aligned}\right\} \quad (5)$$

and so on. In fact the values of all the differential coefficients at $x = 0$ can, in this way, be expressed in terms of $y(0)$ and $y^{(1)}(0)$. Consequently (3) becomes

$$y = y(0)\left(1 + \frac{1 \cdot 3}{2!}x^2 + \frac{1 \cdot 3^2 \cdot 5}{4!}x^4 + \frac{1 \cdot 3^2 \cdot 5^2 \cdot 7}{6!}x^6 + \ldots\right)$$

$$+ y^{(1)}(0)\left(x + \frac{2 \cdot 4}{3!}x^3 + \frac{2 \cdot 4^2 \cdot 6}{5!}x^5 + \frac{2 \cdot 4^2 \cdot 6^2 \cdot 8}{7!}x^7 + \ldots\right). (6)$$

Equation (6) may be taken as the general solution of (1) since it contains two arbitrary constants, $y(0)$ and $y^{(1)}(0)$, which are fixed by specifying boundary conditions on the solution. Suppose, for example, that (1) is to be solved subject to $y(0) = 0$, $y^{(1)}(0) = 1$. Then (6) becomes

$$y = x + \frac{2 \cdot 4}{3!}x^3 + \frac{2 \cdot 4^2 \cdot 6}{5!}x^5 + \frac{2 \cdot 4^2 \cdot 6^2 \cdot 8}{7!}x^7 + \ldots$$

$$+ \frac{2 \cdot 4^2 \cdot 6^2 \ldots (2r)^2(2r+2)}{(2r+1)!}x^{2r+1} + \ldots \quad (7)$$

which converges, by the ratio test, for $x < 1$.

Since these solutions have been obtained by expanding y about the point $x = 0$ (i.e. as a Maclaurin series) they are often referred to as the solutions near $x = 0$. To obtain solutions near any other point, say $x = a$, (3) must be replaced by the Taylor series

$$y = y(a) + (x-a)y^{(1)}(a) + \frac{(x-a)^2}{2!}y^{(2)}(a) + \ldots + \frac{(x-a)^r}{r!}y^{(r)}(a) + \ldots,$$

$$(8)$$

whilst (4) must be replaced by the recurrence relation

$$(1-a^2)y^{(n+2)}(a)-(2n+5)ay^{(n+1)}(a)-(n+1)(n+3)y^{(n)}(a) = 0 \qquad (9)$$

obtained from (2) by putting $x = a$. Throughout this chapter, however, we shall only be concerned with solutions near $x = 0$.

Example 2. To obtain a solution of

$$x\frac{d^2y}{dx^2}+(1+x)\frac{dy}{dx}+2y = 0 \qquad (10)$$

near $x = 0$, we first differentiate the equation n times using Leibnitz's formula. Hence we find

$$xy^{(n+2)}+(1+n+x)y^{(n+1)}+(n+2)y^{(n)} = 0, \qquad (11)$$

which, at $x = 0$, becomes

$$y^{(n+1)}(0) = -\frac{n+2}{n+1}y^{(n)}(0). \qquad (12)$$

This recurrence relation enables the coefficients $y^{(r)}(0)$ in the Maclaurin expansion

$$y = y(0)+xy^{(1)}(0)+\frac{x^2}{2!}y^{(2)}(0)+ \ \dots \ +\frac{x^r}{r!}y^{(r)}(0)+ \ \dots \qquad (13)$$

to be found in terms of $y(0)$. For example,

$$\left.\begin{aligned}
y^{(1)}(0) &= -2y(0),\\
y^{(2)}(0) &= -\tfrac{3}{2}y^{(1)}(0) = 3y(0),\\
y^{(3)}(0) &= -\tfrac{4}{3}y^{(2)}(0) = -4y(0),\\
y^{(4)}(0) &= -\tfrac{5}{4}y^{(3)}(0) = 5y(0),
\end{aligned}\right\} \qquad (14)$$

and so on. Hence (13) becomes

$$y = y(0)\left(1-2x+\frac{3}{2!}x^2-\frac{4}{3!}x^3+\frac{5}{4!}x^4+ \ \dots \ +(-1)^r\left(\frac{r+1}{r!}\right)x^r+ \ \dots \right),$$
$$(15)$$

where $y(0)$ is an arbitrary constant whose value will in general be determined by specifying a boundary condition on y. However, unlike the solution obtained in Example 1, (15) is not the general solution of (10) since it only contains one arbitrary constant ($y(0)$). Nevertheless as we have seen in Chapter 21, 21.9 once one solution is known it is a comparatively easy task to obtain another by a straightforward substitution. Suppose we refer to (15) as y_1. Then another solution, y_2, may be found by putting

$$y_2 = u(x)y_1 \tag{16}$$

into (10) and solving for $u(x)$. In this way we have

$$x(u^{(2)}y_1 + 2u^{(1)}y_1^{(1)} + uy_1^{(2)}) + (1+x)(u^{(1)}y_1 + uy_1^{(1)}) + 2uy_1 = 0, \tag{17}$$

or, since y_1 is already a solution of (10),

$$x(u^{(2)}y_1 + 2u^{(1)}y_1^{(1)}) + (1+x)u^{(1)}y_1 = 0. \tag{18}$$

Writing (18) as

$$\frac{u^{(2)}}{u^{(1)}} + 2\frac{y_1^{(1)}}{y_1} + \frac{1}{x} + 1 = 0, \tag{19}$$

and integrating, we find

$$xe^x y_1^2 \frac{du}{dx} = c, \tag{20}$$

where c is an arbitrary constant of integration.

Consequently

$$u = c \int \frac{e^{-x}}{xy_1^2} \, dx \tag{21}$$

and

$$y_2 = cy_1 \int \frac{e^{-x}}{xy_1^2} \, dx. \tag{22}$$

The general solution of (10) is now an arbitrary linear combination of y_1 and y_2

$$y = Ay_1 + By_1 \int \frac{e^{-x}}{xy_1^2} \, dx, \tag{23}$$

where A and B are arbitrary constants, and where y_1 is defined by (15).

22.2 The Frobenius Method

The type of equation to be solved by this method is assumed to have the form

$$\frac{d^2y}{dx^2}+p(x)\frac{dy}{dx}+q(x)y=0, \qquad (24)$$

where $p(x)$ and $q(x)$ are given functions of x.

We now wish to obtain solutions in the neighbourhood of $x = 0$. In order to do this $p(x)$ and $q(x)$ must be such that either both $p(0)$ and $q(0)$ are finite, in which case $x = 0$ is called an ordinary point of the equation, or both $xp(x)$ and $x^2q(x)$ remain finite at $x = 0$, in which case $x = 0$ is called a regular singular point. If $p(x)$ and $q(x)$ do not satisfy either of these conditions the point $x = 0$ is called an irregular singular point and the Frobenius method of solution about $x = 0$ is not then applicable. For example

$$x = 0 \text{ is an ordinary point of} \qquad \frac{d^2y}{dx^2}+x\frac{dy}{dx}+2y=0, \qquad (25)$$

$$x = 0 \text{ is a regular singular point of} \qquad \frac{d^2y}{dx^2}+\frac{3}{x}\frac{dy}{dx}+\frac{1}{x^2}y=0, \qquad (26)$$

$$x = 0 \text{ is an irregular singular point of} \frac{d^2y}{dx^2}+\frac{1}{x^2}\frac{dy}{dx}+xy=0. \qquad (27)$$

The essence of the Frobenius method is to assume a series solution of the type

$$y = \sum_{r=0}^{\infty} a_r x^{m+r} = x^m(a_0+a_1x+a_2x^2+ \ldots +a_rx^r+ \ldots), \qquad (28)$$

where $a_0, a_1, a_2, \ldots a_r, \ldots$, and m are constants to be determined. This is more general than assuming just a Maclaurin expansion since m may have non-integral values. The method is best illustrated by the following examples.

Example 3. Consider the equation

$$4x\frac{d^2y}{dx^2}+2\frac{dy}{dx}+y=0. \qquad (29)$$

Assuming a solution of the form

$$y = \sum_{r=0}^{\infty} a_r x^{m+r} \tag{30}$$

we obtain directly

$$\frac{dy}{dx} = \sum_{r=0}^{\infty} (m+r) a_r x^{m+r-1} \tag{31}$$

and

$$\frac{d^2 y}{dx^2} = \sum_{r=0}^{\infty} (m+r)(m+r-1) a_r x^{m+r-2}. \tag{32}$$

Hence substituting (30), (31) and (32) into (29) we find

$$4 \sum_{r=0}^{\infty} (m+r)(m+r-1) a_r x^{m+r-1} + 2 \sum_{r=0}^{\infty} (m+r) a_r x^{m+r-1}$$
$$+ \sum_{r=0}^{\infty} a_r x^{m+r} = 0, \tag{33}$$

which, by adding the first two terms, may be written as

$$2 \sum_{r=0}^{\infty} (m+r)(2m+2r-1) a_r x^{m+r-1} + \sum_{r=0}^{\infty} a_r x^{m+r} = 0. \tag{34}$$

Hence expanding (34) as

$$2m(2m-1) a_0 x^{m-1} + 2 \sum_{r=1}^{\infty} (m+r)(2m+2r-1) a_r x^{m+r-1} + \sum_{r=0}^{\infty} a_r x^{m+r}, \tag{35}$$

and writing $r+1$ for r in the second term of (35) we find

$$2m(2m-1) a_0 x^{m-1} + 2 \sum_{r=0}^{\infty} (m+r+1)(2m+2r+1) a_{r+1} x^{m+r}$$
$$+ \sum_{r=0}^{\infty} a_r x^{m+r} = 0. \tag{36}$$

By combining the last two terms of (36) we now find

$$2m(2m-1) a_0 x^{m-1}$$
$$+ \sum_{r=0}^{\infty} \{2(m+r+1)(2m+2r+1) a_{r+1} + a_r\} x^{m+r} = 0. \tag{37}$$

If (37) is to be a solution of (29) for all x, then the coefficients of all powers of x must separately vanish. The lowest power of x in (37) is x^{m-1}; consequently

$$2m(2m-1)a_0 = 0. \tag{38}$$

This equation is called the indicial equation in that it determines the values of the index m; in this case (assuming $a_0 \neq 0$), $m = 0$ or $\frac{1}{2}$. The appearance of an indicial equation is a standard feature of the Frobenius method. The requirement that coefficients of higher powers of x (i.e. x^{m+r}, $r = 0, 1, 2 \ldots$) leads (from the second term of (37)) to the recurrence relation

$$2(m+r+1)(2m+2r+1)a_{r+1}+a_r = 0 \quad (r = 0, 1, 2, \ldots). \tag{39}$$

Hence taking $m = 0$, (39) gives

$$a_1 = -\frac{a_0}{2 \cdot 1 \cdot 1} = -\frac{a_0}{2!}, \tag{40}$$

$$a_2 = -\frac{a_1}{2 \cdot 2 \cdot 3} = -\frac{a_1}{12} = \frac{a_0}{4!}, \tag{41}$$

$$a_3 = -\frac{a_2}{2 \cdot 3 \cdot 5} = -\frac{a_2}{30} = -\frac{a_0}{6!}, \tag{42}$$

and, in general,

$$a_r = \frac{(-1)^r}{(2r)!} a_0. \tag{43}$$

Consequently one solution of (29) is

$$y = x^0 \left(a_0 - \frac{a_0}{2!}x + \frac{a_0}{4!}x^2 - \frac{a_0}{6!}x^3 + \ldots + \frac{(-1)^r a_0}{(2r)!}x^r + \ldots \right), \tag{44}$$

which is easily recognisable as the series form of

$$y = a_0 \cos \sqrt{x}, \tag{45}$$

where a_0 is an arbitrary constant.

A second solution of (29) may now be obtained by considering the case of $m = \frac{1}{2}$. Denoting the coefficients a_r in this solution by b_r, the recurrence relation (39) gives (with $m = \frac{1}{2}$)

$$b_1 = -\frac{b_0}{2 \cdot (\frac{3}{2}) \cdot 2} = -\frac{b_0}{3!}, \tag{46}$$

$$b_2 = -\frac{b_1}{2 \cdot (\frac{5}{2}) \cdot 4} = -\frac{b_1}{20} = \frac{b_0}{5!}, \tag{47}$$

$$b_3 = -\frac{b_2}{2 \cdot (\frac{7}{2}) \cdot 6} = -\frac{b_2}{42} = -\frac{b_0}{7!}, \tag{48}$$

and, in general,

$$b_r = \frac{(-1)^r}{(2r+1)!} b_0. \tag{49}$$

Consequently another solution of (29) is

$$y = x^{\frac{1}{2}}\left(b_0 - \frac{b_0}{3!}x + \frac{b_0}{5!}x^2 - \frac{b_0}{7!}x^3 + \ldots + \frac{(-1)^r}{(2r+1)!}b_0 x^r + \ldots\right), \tag{50}$$

which is equivalent to

$$y = b_0 \sin \sqrt{x}, \tag{51}$$

where b_0 is an arbitrary constant.

The general solution of (29) is therefore given by

$$y = A \cos \sqrt{x} + B \sin \sqrt{x}, \tag{52}$$

where A and B are arbitrary constants.

Example 4. To solve the equation

$$\frac{d^2y}{dx^2} = xy, \tag{53}$$

we again assume a solution of the form

$$y = \sum_{r=0}^{\infty} a_r x^{m+r} \tag{54}$$

Hence differentiating (54) and substituting in (53) we obtain

$$\sum_{r=0}^{\infty} a_r(m+r)(m+r-1)x^{m+r-2} = \sum_{r=0}^{\infty} a_r x^{m+r+1}, \tag{55}$$

which gives

$$a_0 m(m-1)x^{m-2} + a_1(m+1)mx^{m-1} + a_2(m+2)(m+1)x^m$$

$$+ \sum_{r=3}^{\infty} a_r(m+r)(m+r-1)x^{m+r-2} - \sum_{r=0}^{\infty} a_r x^{m+r+1} = 0. \qquad (56)$$

Writing $r+3$ for r in the first summation term, (56) becomes

$$a_0 m(m-1)x^{m-2} + a_1(m+1)mx^{m-1} + a_2(m+2)(m+1)x^m$$

$$+ \sum_{r=0}^{\infty} \{(m+r+3)(m+r+2)a_{r+3} - a_r\}x^{m+r+1} = 0 \qquad (57)$$

from which we have (equating coefficients of all powers of x equal to zero)

$$a_0 m(m-1) = 0, \qquad (58)$$

$$a_1(m+1)m = 0, \qquad (59)$$

$$a_2(m+2)(m+1) = 0, \qquad (60)$$

and

$$a_{r+3} = \frac{a_r}{(m+r+3)(m+r+2)}, \quad (r = 0, 1, 2 \ldots). \qquad (61)$$

The indicial equation (58) gives (assuming $a_0 \neq 0$) $m = 0$ or 1. Taking the case of $m = 0$ first, we see from (59) that a_1 is arbitrary, whilst from (60), $a_2 = 0$. The recurrence relation (61) gives (with $a_2 = 0$)

$$
\left.
\begin{aligned}
a_3 &= \frac{a_0}{3 \cdot 2}, \\[4pt]
a_4 &= \frac{a_1}{4 \cdot 3}, \\[4pt]
a_5 &= \frac{a_2}{5 \cdot 4} = 0, \\[4pt]
a_6 &= \frac{a_3}{6 \cdot 5} = \frac{a_0}{6 \cdot 5 \cdot 3 \cdot 2}, \\[4pt]
a_7 &= \frac{a_4}{7 \cdot 6} = \frac{a_1}{7 \cdot 6 \cdot 4 \cdot 3}, \\[4pt]
a_8 &= \frac{a_5}{8 \cdot 7} = 0,
\end{aligned}
\right\}
\qquad (62)
$$

and so on. Hence the solution of (53) corresponding to $m = 0$ is

$$y = a_0 + a_1 x + \frac{a_0}{3 \cdot 2} x^3 + \frac{a_1}{4 \cdot 3} x^4 + \frac{a_0}{6 \cdot 5 \cdot 3 \cdot 2} x^6 + \frac{a_1}{7 \cdot 6 \cdot 4 \cdot 3} x^7 + \dots \tag{63}$$

$$= a_0\left(1 + \frac{x^3}{3 \cdot 2} + \frac{x^6}{6 \cdot 5 \cdot 3 \cdot 2} + \dots\right) + a_1\left(x + \frac{x^4}{4 \cdot 3} + \frac{x^7}{7 \cdot 6 \cdot 4 \cdot 3} + \dots\right), \tag{64}$$

where a_0 and a_1 are arbitrary independent constants. We must finally investigate the remaining possibility of $m = 1$. In this case, (59) now gives $a_1 = 0$, whilst (60) gives $a_2 = 0$ as before. The recurrence relation (61) gives (with $a_1 = 0$)

$$\begin{rcases} a_3 = \dfrac{a_0}{4 \cdot 3}, \\[2mm] a_4 = \dfrac{a_1}{5 \cdot 4} = 0, \\[2mm] a_5 = \dfrac{a_2}{6 \cdot 5} = 0, \\[2mm] a_6 = \dfrac{a_3}{7 \cdot 6} = \dfrac{a_0}{7 \cdot 6 \cdot 4 \cdot 3}, \\[2mm] a_7 = \dfrac{a_4}{8 \cdot 7} = 0, \\[2mm] a_8 = \dfrac{a_5}{9 \cdot 8} = 0, \\[2mm] a_9 = \dfrac{a_6}{10 \cdot 9} = \dfrac{a_0}{10 \cdot 9 \cdot 7 \cdot 6 \cdot 4 \cdot 3}, \end{rcases} \tag{65}$$

and so on.

Hence the solution corresponding to $m = 1$ is

$$y = x\left(a_0 + \frac{a_0}{4 \cdot 3} x^3 + \frac{a_0}{7 \cdot 6 \cdot 4 \cdot 3} x^6 + \dots\right) \tag{66}$$

which (apart from an arbitrary constant) is just the second series in the solution (64). Nothing new is gained therefore from the $m = 1$

solution, and consequently (64) represents the general solution of (53).

We finally state here that the Frobenius method illustrated by the last two examples does not always give two independent series solutions. It can be proved that if the two roots of the indicial equation are the same, not more than one independent series solution exists; the same is usually true (with certain exceptions) if the two roots differ by an integer.

22.3 Bessel's Equation

The equation

$$x^2 \frac{d^2y}{dx^2} + x \frac{dy}{dx} + (x^2 - v^2)y = 0, \tag{67}$$

where v is a real constant, is called Bessel's equation and its solutions are the Bessel functions of order v. Applying the Frobenius method and assuming a series solution of (67) of the type

$$y = \sum_{r=0}^{\infty} a_r x^{m+r} \tag{68}$$

we obtain the recurrence relation

$$a_{r+2} = -\frac{a_r}{(r+2+m)^2 - v^2}, \quad (r = 0, 1, 2 \ldots) \tag{69}$$

together with

$$a_1 = 0, \tag{70}$$

and

$$m = \pm v \tag{71}$$

from the indicial equation.

Since $a_1 = 0$, (69) gives $a_3 = a_5 = \ldots = 0$, and hence, with $m = +v$, we have the series solution

$$y = a_0 x^v \left\{ 1 - \frac{x^2}{2(2v+2)} + \frac{x^4}{2 \cdot 4(2v+2)(2v+4)} - \ldots \right\}, \tag{72}$$

provided v is not a negative integer.

Similarly with $m = -v$ we obtain the series

$$y = a_0 x^{-v} \left\{ 1 + \frac{x^2}{2(2v-2)} + \frac{x^4}{2 \cdot 4(2v-2)(2v-4)} + \ldots \right\} \qquad (73)$$

provided v is not a positive integer.

In defining Bessel functions it is usual to write a_0 as

$$a_0 = \frac{1}{2^v \Gamma(v+1)}, \qquad (74)$$

where $\Gamma(v+1)$ is the gamma function defined in Chapter 13. With this form of a_0, (72) now defines the Bessel function of the first kind and order v, $J_v(x)$, such that

$$y = J_v(x) = \sum_{r=0}^{\infty} \frac{(-1)^r}{r! \Gamma(v+r+1)} \left(\frac{x}{2} \right)^{v+2r} \qquad (75)$$

Similarly the second solution (73) becomes

$$y = J_{-v}(x) = \sum_{r=0}^{\infty} \frac{(-1)^r}{r! \Gamma(-v+r+1)} \left(\frac{x}{2} \right)^{-v+2r} \qquad (76)$$

The general solution of Bessel's equation for non-integral v is therefore

$$y = A J_v(x) + B J_{-v}(x), \qquad (77)$$

where A and B are arbitrary constants.

If $v = n$, where n is a positive integer, then since $\Gamma(n+r+1) = (n+r)!$, (75) becomes

$$J_n(x) = \sum_{r=0}^{\infty} \frac{(-1)^r}{r!(n+r)!} \left(\frac{x}{2} \right)^{n+2r}. \qquad (78)$$

Likewise

$$J_{-n}(x) = \sum_{r=0}^{\infty} \frac{(-1)^r}{r!(-n+r)!} \left(\frac{x}{2} \right)^{-n+2r}. \qquad (79)$$

Now the first n terms of this series are zero since (by the results of Chapter 13) the Γ-function or factorial function of negative integers is infinite. Hence putting $r = n+p$ in (79) we have

$$J_{-n}(x) = \sum_{r=n}^{\infty} \frac{(-1)^r}{r!(-n+r)!} \left(\frac{x}{2}\right)^{-n+2r} \tag{80}$$

$$= \sum_{p=0}^{\infty} \frac{(-1)^{n+p}}{p!(n+p)!} \left(\frac{x}{2}\right)^{n+2p} \tag{81}$$

$$= (-1)^n J_n(x). \tag{82}$$

We see then that $J_n(x)$ and $J_{-n}(x)$ are linearly dependent. Consequently the Frobenius method gives only one solution in these circumstances, and therefore (77) cannot be taken as the general

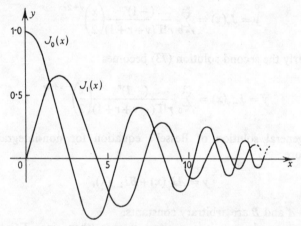

Fig. 22.1

solution of the Bessel equation when $v = n$. A second solution, however, may be obtained by other methods—for example, by following the technique of Example 20, Chapter 21. This solution, known as the Weber function, will not be discussed here.

Finally we note that from (78)

$$J_0(x) = 1 - \frac{x^2}{(1!)^2 2^2} + \frac{x^4}{(2!)^2 2^4} - \frac{x^6}{(3!)^2 2^6} + \cdots \tag{83}$$

and

$$J_1(x) = \frac{x}{2} - \frac{x^3}{2^3 \cdot 1! 2!} + \frac{x^5}{2^5 \cdot 2! 3!} - \frac{x^7}{2^7 \cdot 3! 4!} + \dots \quad (84)$$

from which it follows that

$$-\frac{dJ_0(x)}{dx} = J_1(x). \quad (85)$$

Graphs of $J_0(x)$ and $J_1(x)$ are shown in Fig. 22.1.

22.4 Legendre's Equation

The equation

$$(1-x^2)\frac{d^2 y}{dx^2} - 2x\frac{dy}{dx} + l(l+1)y = 0, \quad (86)$$

where l is a real constant, is Legendre's equation and may be solved by applying the Frobenius method in the usual way. It is left to the reader to show that the two series so obtained are

$$y = a_0\left\{1 - \frac{l(l+1)}{2!}x^2 + \frac{l(l-2)(l+1)(l+3)}{4!}x^4 - \dots\right\} \quad (87)$$

and

$$y = a_1\left\{x - \frac{(l-1)(l+2)}{3!}x^3 + \frac{(l-1)(l-3)(l+2)(l+4)}{5!}x^5 + \dots\right\}, \quad (88)$$

where a_0 and a_1 are arbitrary constants.

When $l = n$, where n is an integer, either one or the other of these two series terminates. For example, if $l = 2$ all terms in (87) beyond x^2 are zero. Similarly if $l = 3$, all terms of (88) beyond x^3 are zero. The resulting polynomials in x (denoted by $P_n(x)$) are called Legendre polynomials (a_0 and a_1 being chosen so that each polynomial has the value unity when $x = 1$). The first few of these polynomials are

$$\left.\begin{array}{l} P_0(x) = 1, \quad P_1(x) = x, \quad P_2(x) = \frac{1}{2}(3x^2 - 1), \\ P_3(x) = \frac{1}{2}(5x^3 - 3x). \end{array}\right\} \quad (89)$$

The non-terminating series is of no interest here.

PROBLEMS 22

1. If $(dy/dx) - y = x^2$ and $y = 0$ at $x = 0$, obtain a series solution (a) by the Leibnitz-Maclaurin method and (b) by the use of a trial series. Compare these results with the solution obtained by direct integration of the equation.

2. If $4(1+x)\dfrac{d^2y}{dx^2} + 2\dfrac{dy}{dx} + y = 0$, show that at $x = 0$

$$4\frac{d^{n+2}y}{dx^{n+2}} = -(4n+2)\frac{d^{n+1}y}{dx^{n+1}} - \frac{d^ny}{dx^n}.$$

Hence, given $y = 0$, $dy/dx = 1$ at $x = 0$, find a series for y up to terms in x^5.

3. Use the Leibnitz-Maclaurin method to obtain a series solution up to the term in x^4 of the equation

$$x^2\frac{d^2y}{dx^2} + x\frac{dy}{dx} - (1-x)y = 0$$

given $dy/dx = 1$ at $x = 0$.

4. Find the general series solutions of the following equations

(a) $2x\dfrac{d^2y}{dx^2} + (3-2x)\dfrac{dy}{dx} + 4y = 0$,

(b) $2(x-x^2)\dfrac{d^2y}{dx^2} + (1-9x)\dfrac{dy}{dx} - 3y = 0$,

(c) $3x\dfrac{d^2y}{dx^2} + 2\dfrac{dy}{dx} + y = 0$.

5. Obtain a Maclaurin series solution of the equation

$$x^2\frac{d^2y}{dx^2} + x\frac{dy}{dx} + (x^2-1)y = 0.$$

444

If this solution is denoted by y_1, show that a second solution is $y_2 = u(x)y_1$, where

$$u = \int \frac{dx}{xy_1^2}.$$

6. Show that the equation

$$(2x+x^2)\frac{d^2y}{dx^2}+\frac{dy}{dx}-2y = 0$$

has a non-zero solution of the form $y = a+bx+cx^2$, and find the constants b and c in terms of a. Show also that there is another solution of the form

$$y = \sum_{n=0}^{\infty} a_n x^{n+\frac{1}{2}},$$

where $4(n+1)a_{n+1} = (3-2n)a_n$.

Determine the radius of convergence of the series. (L.U.)

7. Show that the differential equation

$$(1-x^2)\frac{d^2y}{dx^2}-3x\frac{dy}{dx}+(k^2-1)y = 0$$

(where k is a constant) has solutions of the form

$$\sum_{n=0}^{\infty} a_n x^{n+c};$$

provided that $c = 0$ or 1.

Hence obtain two independent solutions, one of them in the form

$$y = 1+\sum_{n=1}^{\infty} \frac{(1-k^2)(3^2-k^2)\ldots\{(2n-1)^2-k^2\}}{(2n)!}x^{2n},$$

and write down the general solution. (L.U.)

8. Obtain the solution

$$y = J_0(x) = 1-\frac{x^2}{2^2}+\frac{x^4}{2^2.4^2}-\ldots+\frac{(-1)^n x^{2n}}{2^{2n}(n!)^2}+\ldots$$

of the differential equation

$$x\frac{d^2y}{dx^2}+\frac{dy}{dx}+xy=0.$$

Show that $y=uJ_0$ is a second solution if

$$u=\int\frac{dx}{xJ_0^2},$$

and expand u in the form $a+\log_e x+b_2x^2+\dots$, where a is arbitrary and b_2 is to be found. (C.U.)

9. Show that $J_n(x)/x^n$ is a solution of

$$\frac{d^2y}{dx^2}+\left(\frac{1+2n}{x}\right)\frac{dy}{dx}+y=0,$$

and that $\sqrt{(x)}J_n(kx)$ is a solution of

$$\frac{d^2y}{dx^2}+\left(k^2-\frac{4n^2-1}{4x^2}\right)y=0,$$

where, in both cases, n is a positive integer.

10. Assuming the series for $J_0(x)$ given in Problem 8 show that

$$\frac{\sin x}{x}=\int_0^{\pi/2}J_0(x\cos\theta)\cos\theta\,d\theta$$

11. Using the series for $J_\nu(x)$ and $J_{-\nu}(x)$ given in 22.3, show that

$$J_{\frac{1}{2}}(x)=\sqrt{\left(\frac{2}{\pi x}\right)}\sin x,$$

$$J_{-\frac{1}{2}}(x)=\sqrt{\left(\frac{2}{\pi x}\right)}\cos x.$$

12. By putting $y=ve^{-1/4x^2}$, obtain the solution of the equation

$$\frac{d^2y}{dx^2}+(2n+\tfrac{1}{2}-\tfrac{1}{4}x^2)y=0$$

(n integral) in which $y=1$ when $x=0$, and in which v is a polynomial in x. (C.U.)

13. Rodrigues' formula for the Legendre polynomials is

$$P_n(x) = \frac{1}{2^n n!} \frac{d^n}{dx^n} (x^2 - 1)^n.$$

Verify this formula for $n = 0, 1, 2, 3$.

14. If $P_n(z)$ satisfies Legendre's equation

$$(1 - z^2) \frac{d^2 P_n}{dz^2} - 2z \frac{dP_n}{dz} + n(n+1)P_n = 0,$$

show that $V = \dfrac{d^m P_n}{dz^m}$ satisfies the equation

$$(1 - z^2) \frac{d^2 V}{dz^2} - 2(m+1)z \frac{dV}{dz} + \{n(n+1) - m(m+1)\}V = 0,$$

and that $W = (1 - z^2)^{m/2} \dfrac{d^m P_n}{dz^m}$ satisfies

$$(1 - z^2) \frac{d^2 W}{dz^2} - 2z \frac{dW}{dz} + \left\{ n(n+1) - \frac{m^2}{1 - z^2} \right\} W = 0.$$

Hence show that $\sin \theta \dfrac{dP_n(\cos \theta)}{d(\cos \theta)}$ is a solution of

$$\frac{d^2 W}{d\theta^2} + \frac{d}{d\theta}(W \cot \theta) + n(n+1)W = 0. \qquad \text{(C.U.)}$$

15. By means of the substitution

$$P(\theta) = Q(\theta)(\sin \theta)^{-\frac{1}{2}}$$

in Legendre's equation

$$\frac{1}{\sin \theta} \frac{d}{d\theta}\left(\sin \theta \frac{dP}{d\theta} \right) + n(n+1)P = 0,$$

show that, when $\operatorname{cosec} \theta \ll (2n+1)$, the solutions of Legendre's equation are of the form

$$(\sin \theta)^{-\frac{1}{2}} \frac{\sin}{\cos} \left\{ (n + \tfrac{1}{2})\theta \right\}.$$

16. By writing Bessel's equation in the form

$$\frac{d^2y}{dx^2}+\frac{1}{x}\frac{dy}{dx}+\left(1-\frac{v^2}{x^2}\right)y = 0$$

and neglecting, for large x, the term v^2/x^2, obtain the equation

$$\frac{d^2u}{dx^2}+\left(1+\frac{1}{4x^2}\right)u = 0,$$

where $u = \sqrt{(x)}y$. Hence neglecting, for large x, the term $1/4x^2$, show that an asymptotic form $(x \gg 1)$ of the Bessel functions is

$$\frac{1}{\sqrt{(x)}}(A \cos x + B \sin x).$$

CHAPTER 23

The Laplace Transformation

23.1 Definition

Among the many and varied applications of the Laplace transformation, perhaps the most important is to the solution of linear differential equations (both ordinary and partial). However, before discussing this particular application (see 23.5) we shall first define the Laplace transformation and evaluate the transforms of some simple functions.

The Laplace transform $\bar{f}(p)$ of a function $f(x)$, where $x > 0$, is defined by the integral

$$\bar{f}(p) = \int_0^\infty e^{-px} f(x)\, dx, \tag{1}$$

where the variable p is assumed here to be real (although in general it may be complex). It is sometimes convenient to adopt the notation

$$\bar{f}(p) = L\{f(x)\}, \tag{2}$$

where $L\{\ \}$ represents symbolically the operation of taking the Laplace transform of whatever function occurs inside the bracket. Furthermore $L\{\ \}$ is a linear operator since from (1) we have

$$L\{kf(x)\} = kL\{f(x)\}, \tag{3}$$

where k is any constant, and

$$L\{\alpha f(x) + \beta g(x)\} = \int_0^\infty e^{-px}(\alpha f(x) + \beta g(x))\, dx \tag{4}$$

$$= \alpha \int_0^\infty e^{-px} f(x)\, dx + \beta \int_0^\infty e^{-px} g(x)\, dx \tag{5}$$

$$= \alpha L\{f(x)\} + \beta L\{g(x)\}, \tag{6}$$

where α and β are arbitrary constants and $g(x)$ is an arbitrary function defined (like $f(x)$) for $x > 0$.

449

23.2 Simple Transforms

Using the definition (1) we may now obtain the following transforms

(a) If $f(x) = 1$ for $x > 0$, then

$$L\{1\} = \bar{f}(p) = \int_0^\infty e^{-px}\, dx = \frac{1}{p}, \tag{7}$$

where, for the convergence of the integral, $p > 0$.

(b) If $f(x) = e^{ax}$, where a is a real constant, then

$$L\{e^{ax}\} = \bar{f}(p) = \int_0^\infty e^{-px} e^{ax}\, dx = \frac{1}{p-a}, \tag{8}$$

where $p > a$ for convergence.

(c) If $f(x) = \sin ax$, where a is a real constant, then

$$L\{\sin ax\} = \bar{f}(p) = \int_0^\infty e^{-px} \sin ax\, dx = \frac{a}{p^2 + a^2}, \tag{9}$$

provided $p > 0$.

(d) If $f(x) = \cos ax$, then

$$L\{\cos ax\} = \bar{f}(p) = \int_0^\infty e^{-px} \cos ax = \frac{p}{p^2 + a^2}, \tag{10}$$

provided $p > 0$.

(e) If $f(x) = x^n$, where $n = 1, 2, 3, \ldots$, then

$$L\{x^n\} = \bar{f}(p) = \int_0^\infty e^{-px} x^n\, dx = \frac{n!}{p^{n+1}}, \tag{11}$$

provided $p > 0$.

(f) If $f(x) = \sinh ax$, where a is a real constant, then

$$L\{\sinh ax\} = \bar{f}(p) = \int_0^\infty e^{-px} \sinh ax\, dx = \frac{a}{p^2 - a^2}, \tag{12}$$

provided $p > a$.

(g) If $f(x) = \cosh ax$, then

$$L\{\cosh ax\} = \bar{f}(p) = \int_0^\infty e^{-px} \cosh ax = \frac{p}{p^2 - a^2}, \qquad (13)$$

again provided $p > a$.

Transforms of many other functions may be obtained in a similar way, but in some cases it is convenient to use the linearity property of $L\{\ \}$ as expressed by (6). For example, the transform of $\cosh ax$ can be obtained in this way by writing

$$L\{\cosh ax\} = L\left\{\frac{e^{ax} + e^{-ax}}{2}\right\} = \tfrac{1}{2}L\{e^{ax}\} + \tfrac{1}{2}L\{e^{-ax}\}$$

$$= \frac{1}{2}\left(\frac{1}{p-a}\right) + \frac{1}{2}\left(\frac{1}{p+a}\right) = \frac{p}{p^2 - a^2}. \qquad (14)$$

Another useful result (often known as a shift theorem) is that if $\bar{f}(p) = L\{f(x)\}$, then

$$\bar{f}(p+a) = L\{e^{-ax}f(x)\}, \qquad (15)$$

where a is a real constant. This is easily proved since by (1)

$$L\{e^{-ax}f(x)\} = \int_0^\infty e^{-px}e^{-ax}f(x)\,dx = \int_0^\infty e^{-(p+a)x}f(x)\,dx$$

$$= \bar{f}(p+a). \qquad (16)$$

The following examples illustrate the use of this result.

Example 1. Since by (11)

$$L\{x^n\} = \frac{n!}{p^{n+1}}, \quad (p > 0), \qquad (17)$$

it follows from (15) that

$$L\{x^n e^{-ax}\} = \frac{n!}{(p+a)^{n+1}}, \quad (p > -a). \qquad (18)$$

451

Example 2. Since by (9) and (10)

$$L\{\sin ax\} = \frac{a}{p^2 + a^2}, \quad L\{\cos ax\} = \frac{p}{p^2 + a^2}, \tag{19}$$

(15) gives

$$L\{e^{-ax} \sin bx\} = \frac{b}{(p+a)^2 + b^2}, \quad (p > -a) \tag{20}$$

and

$$L\{e^{-ax} \cos bx\} = \frac{p+a}{(p+a)^2 + b^2}, \quad (p > -a) \tag{21}$$

where a and b are arbitrary constants.

A short list of Laplace transforms is given in Table 3.

TABLE 3

$f(x)$	$\bar{f}(p) = L\{f(x)\}$	$f(x)$	$\bar{f}(p) = L\{f(x)\}$
1	$\dfrac{1}{p}, \quad p > 0$	$x^n e^{-ax}$	$\dfrac{n!}{(p+a)^{n+1}}, \quad p > -a$
$x^n,$ $(n = 1, 2, 3 \ldots)$	$\dfrac{n!}{p^{n+1}}, \quad p > 0$	$\dfrac{1}{\sqrt{x}}$	$\sqrt{\dfrac{\pi}{p}}, \quad p > 0$
$e^{ax}(a \neq 0)$	$\dfrac{1}{p-a}, \quad p > a$	$\sqrt{x}$	$\dfrac{1}{2p}\sqrt{\dfrac{\pi}{p}}, \quad p > 0$
$\sin ax$	$\dfrac{a}{p^2 + a^2}, \quad p > 0$	$x \sin ax$	$\dfrac{2ap}{(p^2 + a^2)^2}, \quad p > 0$
$\cos ax$	$\dfrac{p}{p^2 + a^2}, \quad p > 0$	$e^{-ax} \sin bx$	$\dfrac{b}{(p+a)^2 + b^2}, \quad p > -a$
$\cosh ax$	$\dfrac{p}{p^2 - a^2}, \quad p > a$	$e^{-ax} \cos bx$	$\dfrac{p+a}{(p+a)^2 + b^2}, \quad p > -a$

23.3 Inverse Transforms

We now introduce the inverse operator $L^{-1}\{\ \}$ which is such that if

$$L\{f(x)\} = \bar{f}(p) \tag{22}$$

then

$$f(x) = L^{-1}\{\bar{f}(p)\}. \tag{23}$$

In other words given any $\bar{f}(p)$, $L^{-1}\{\bar{f}(p)\}$ is the function from which $\bar{f}(p)$ may be derived by making a Laplace transformation. Clearly from (22) and (23)

$$LL^{-1} = L^{-1}L = 1, \tag{24}$$

whilst in virtue of the linearity of the operator $L\{\ \}$ we have

$$\left.\begin{array}{l} L^{-1}\{\kappa f(x)\} = \kappa L^{-1}\{f(x)\}, \\ L^{-1}\{\alpha f(x) + \beta g(x)\} = \alpha L^{-1}\{f(x)\} + \beta L^{-1}\{g(x)\}, \end{array}\right\} \tag{25}$$

where α, β and κ are arbitrary constants. Hence $L^{-1}\{\ \}$ is also a linear operator.

Inverse transforms may be most conveniently found from a table of standard transforms. For example, since (from (7))

$$L\{1\} = \frac{1}{p} \tag{26}$$

we have

$$L^{-1}\left\{\frac{1}{p}\right\} = 1. \tag{27}$$

Similarly from (10)

$$L^{-1}\left\{\frac{p}{p^2 + a^2}\right\} = \cos ax. \tag{28}$$

When $\bar{f}(p)$ is a rational function of p but is not immediately recognisable as a standard type it may very often be expressed, using partial fractions, as the sum of a number of terms which may be inverted at sight. The following examples illustrate this point.

Example 3. To find

$$L^{-1}\left\{\frac{1}{(p+a)(p+b)}\right\}, \tag{29}$$

where a and b are constants, we write

$$\frac{1}{(p+a)(p+b)} = \frac{A}{p+a} + \frac{B}{p+b}. \tag{30}$$

Comparing coefficients of powers of p on each side of (30) we find

$$B = -A = \frac{1}{a-b} \tag{31}$$

provided $a \neq b$.

Consequently using (26)

$$L^{-1}\left\{\frac{1}{(p+a)(p+b)}\right\} = \frac{1}{(b-a)}L^{-1}\left\{\frac{1}{p+a}\right\} - \frac{1}{(b-a)}L^{-1}\left\{\frac{1}{p+b}\right\} \tag{32}$$

$$= \frac{1}{(b-a)}(e^{-ax} - e^{-bx}). \tag{33}$$

Example 4. To find

$$L^{-1}\left\{\frac{p}{(p^2+a^2)(p^2+b^2)}\right\}, \tag{34}$$

where a and b are constants, we write

$$\frac{p}{(p^2+a^2)(p^2+b^2)} = \frac{1}{(b^2-a^2)}\left(\frac{p}{p^2+a^2} - \frac{p}{p^2+b^2}\right), \quad (b \neq a). \tag{35}$$

Consequently using (25)

$$L^{-1}\left\{\frac{p}{(p^2+a^2)(p^2+b^2)}\right\} = \frac{1}{b^2-a^2}(\cos ax - \cos bx). \tag{36}$$

Example 5. To find

$$L^{-1}\left\{\frac{1}{p^2(p^2+a^2)}\right\}, \tag{37}$$

where a is a non-zero constant, we write

$$\frac{1}{p^2(p^2+a^2)} = \frac{A}{p^2} + \frac{B}{p^2+a^2}, \tag{38}$$

which gives $A = -B = 1/a^2$. Hence

$$L^{-1}\left\{\frac{1}{p^2(p^2+a^2)}\right\} = \frac{1}{a^2}L^{-1}\left\{\frac{1}{p^2}\right\} - \frac{1}{a^2}L^{-1}\left\{\frac{1}{p^2+a^2}\right\}, \tag{39}$$

$$= \frac{x}{a^2} - \frac{1}{a^3}L^{-1}\left\{\frac{a}{p^2+a^2}\right\}, \tag{40}$$

$$= \frac{x}{a^2} - \frac{1}{a^3}\sin ax. \tag{41}$$

Inverse transforms may in general be obtained using various techniques of complex variable theory, in particular, contour integration (see the remarks of Chapter 4, 4.5). However, we shall not discuss these methods here.

23.4 Transforms of Differential Coefficients

As a preliminary to the solution of ordinary differential equations by the Laplace transform method, we now evaluate the transforms of the differential coefficients of a function $y(x)$. Consider first the transform of dy/dx. Then

$$L\left\{\frac{dy}{dx}\right\} = \int_0^\infty e^{-px}\frac{dy}{dx}\,dx = \left[ye^{-px}\right]_0^\infty + p\int_0^\infty e^{-px}y\,dx \tag{42}$$

$$= -y(0) + pL\{y\}, \quad \text{assuming } e^{-px}y \to 0 \text{ as } x \to \infty, \tag{43}$$

$$= -y(0) + p\bar{y}(p), \tag{44}$$

where $y(0)$ is the value of $y(x)$ at $x = 0$. Similarly

$$L\left\{\frac{d^2y}{dx^2}\right\} = \int_0^\infty e^{-px}\frac{d^2y}{dx^2}\,dx = \left[e^{-px}\frac{dy}{dx}\right]_0^\infty + p\int_0^\infty e^{-px}\frac{dy}{dx}\,dx. \tag{45}$$

Using (44) this becomes, (assuming $e^{-px}\dfrac{dy}{dx}\to 0$ as $x\to\infty$),

$$L\left\{\frac{d^2y}{dx^2}\right\} = p^2\bar{y}(p) - py(0) - y^{(1)}(0), \tag{46}$$

where $y^{(1)}(0)$ is the value of dy/dx at $x = 0$.

Transforms of higher derivatives may be evaluated in the same way, and in general we find

$$L\left\{\frac{d^ny}{dx^n}\right\} = p^n\bar{y}(p) - p^{n-1}y(0) - p^{n-2}y^{(1)}(0)\ldots - py^{(n-2)}(0) - y^{(n-1)}(0),$$
$$\tag{47}$$

where $y^{(r)}(0)$ is the value of d^ry/dx^r at $x = 0$.

23.5. Solution of Ordinary Differential Equations

The essential process involved in solving ordinary linear differential equations by the Laplace transform method is to first convert the equation into an algebraic equation in $\bar{y}(p)$ using the results of the last section and to then solve for $\bar{y}(p)$. If this solution can be inverted, the function $y(x)$ so obtained will be a solution of the differential equation. The following examples illustrate this method.

Example 6. To solve

$$(D+2)y = \cos x \tag{48}$$

given that $y = 1$ at $x = 0$ (i.e. $y(0) = 1$), we take the Laplace transform of (48) and obtain (using (44) and (10))

$$p\bar{y}(p) - y(0) + 2\bar{y}(p) = \frac{p}{p^2+1}. \tag{49}$$

Hence rearranging we find

$$\bar{y}(p) = \frac{p}{(p+2)(p^2+1)} + \frac{y(0)}{p+2} \tag{50}$$

$$= \frac{p}{(p+2)(p^2+1)} + \frac{1}{p+2}, \quad \text{since } y(0) = 1. \tag{51}$$

We now invert this equation by first writing

$$\frac{p}{(p+2)(p^2+1)} = \frac{A}{p+2} + \frac{Bp+C}{p^2+1}, \tag{52}$$

where A, B and C are constants. Proceeding in the usual way we find $A = -B = -\frac{2}{5}$ and $C = \frac{1}{5}$, which gives

$$\bar{y}(p) = -\frac{2}{5}\left(\frac{1}{p+2}\right) + \frac{1}{5}\left(\frac{1+2p}{p^2+1}\right) + \frac{1}{p+2} \tag{53}$$

$$= \frac{1}{5}\left(\frac{1}{p^2+1}\right) + \frac{2}{5}\left(\frac{p}{p^2+1}\right) + \frac{3}{5}\left(\frac{1}{p+2}\right). \tag{54}$$

Consequently

$$y(x) = L^{-1}\{\bar{y}(p)\} = \frac{1}{5}\sin x + \frac{2}{5}\cos x + \frac{3}{5}e^{-2x}. \tag{55}$$

Example 7. To solve

$$(D^2+a^2)y = \sin bx, \tag{56}$$

where a and b are constants ($a \neq b$, $a \neq 0$), we take the Laplace transform of (56) using (46) and (9). Then

$$p^2\bar{y}(p) - py(0) - y^{(1)}(0) + a^2\bar{y}(p) = \frac{b}{p^2+b^2}, \tag{57}$$

which gives

$$\bar{y}(p) = \frac{b}{(p^2+a^2)(p^2+b^2)} + \frac{py(0)}{p^2+a^2} + \frac{y^{(1)}(0)}{p^2+a^2}, \tag{58}$$

where $y(0)$ and $y^{(1)}(0)$ are respectively the values of y and dy/dx at $x = 0$. Using partial fractions (58) may now be written as

$$\bar{y}(p) = \frac{b}{b^2-a^2}\left(\frac{1}{p^2+a^2} - \frac{1}{p^2+b^2}\right) + y(0)\left(\frac{p}{p^2+a^2}\right) + \frac{y^{(1)}(0)}{a}\left(\frac{a}{p^2+a^2}\right), \tag{59}$$

which gives, on inversion, the solution

$$y(x) = L^{-1}\{\bar{y}(p)\}$$
$$= \frac{b\sin ax}{a(b^2-a^2)} - \frac{\sin bx}{b^2-a^2} + y(0)\cos ax + \frac{y^{(1)}(0)\sin ax}{a}, \tag{60}$$

or

$$y(x) = y(0)\cos ax + A\sin ax + \frac{\sin bx}{a^2-b^2}, \tag{61}$$

where

$$A = \frac{1}{a}\left(y^{(1)}(0) + \frac{b}{b^2 - a^2}\right). \tag{62}$$

Example 8. As mentioned in Chapter 21, 21.11, the Laplace transform method may be applied usefully to linear simultaneous differential equations with constant coefficients. This is illustrated by the following:

Consider the equations

$$\left.\begin{array}{l}(D^2 + 2)y - x = 0,\\ (D^2 + 2)x - y = 0,\end{array}\right\} \tag{63}$$

where x and y are the dependent variables and $D \equiv \dfrac{d}{dt}$, t being the independent variable. We now take the Laplace transform of each equation to get

$$\left.\begin{array}{l}p^2\bar{y}(p) - py(0) - y^{(1)}(0) + 2\bar{y}(p) - \bar{x}(p) = 0,\\ p^2\bar{x}(p) - px(0) - x^{(1)}(0) + 2\bar{x}(p) - \bar{y}(p) = 0,\end{array}\right\} \tag{64}$$

where, as usual, $x(0)$, $y(0)$ are the values of x and y at $t = 0$, and $x^{(1)}(0)$, $y^{(1)}(0)$ are the values of the first derivatives of x and y at $t = 0$.

We assume for simplicity the boundary conditions $x(0) = 2$, $y(0) = 0$, $x^{(1)}(0) = y^{(1)}(0) = 0$. Then (64) become

$$\left.\begin{array}{l}(p^2 + 2)\bar{y}(p) - \bar{x}(p) = 0,\\ (p^2 + 2)\bar{x}(p) - \bar{y}(p) = 2p.\end{array}\right\} \tag{65}$$

Eliminating $\bar{y}(p)$ from these two algebraic equations we obtain

$$\bar{x}(p) = \frac{2p(p^2 + 2)}{(p^2 + 1)(p^2 + 3)}. \tag{66}$$

Inverting this by the method of partial fractions we find

$$x(t) = \cos t + \cos\sqrt{3}t. \tag{67}$$

Likewise by eliminating $\bar{x}(p)$ from (65) we finally obtain

$$y(t) = \cos t - \cos\sqrt{3}t. \tag{68}$$

PROBLEMS 23

1. Verify the following results

   (a) $L\{a+bx\} = \dfrac{ap+b}{p^2}$,

   (b) $L\{x^{-\frac{1}{2}}\} = \sqrt{\dfrac{\pi}{p}}$,

   (c) $L\{x \cos ax\} = \dfrac{p^2 - a^2}{(p^2 + a^2)^2}$,

   (d) $L\{e^{ax}x^{-\frac{1}{2}}(1+2ax)\} = \dfrac{p\sqrt{\pi}}{(p-a)^{3/2}}$.

2. Evaluate $L^{-1}\{\bar{f}(p)\}$ when $\bar{f}(p)$ has the following forms

   (a) $\dfrac{p^2}{(p^2+1)(p^2+2)}$,   (b) $\dfrac{1}{p(p-1)^3}$.

3. Solve the following equations by the Laplace transform method

   (a) $(D^2 + 4D + 8)y = \cos 2x$, given that $y = 2$ and $dy/dx = 1$ at $x = 0$;

   (b) $(D+1)y + Dz = 0$,
       $(D-1)y + 2Dz = e^{-x}$,

   given that $y = \frac{1}{2}$ and $z = 0$ at $x = 0$.

   (In both cases $D \equiv d/dx$.)

4. On the assumption that

$$L\left\{\int_0^\infty f(x, u)\, du\right\} = \int_0^\infty L\{f(x, u)\}\, du,$$

where x is a parameter, show that

$$\int_0^\infty \frac{\sin xu}{u}\, du = \frac{\pi}{2},$$

459

and that

$$\int_0^\infty \cos(xu^2)\,du = \frac{\Gamma(\frac{3}{4})\Gamma(\frac{1}{4})}{4\sqrt{(\pi x)}} = \sqrt{\left(\frac{\pi}{8x}\right)},$$

using the result of Chapter 13, Problem 9.

5. The convolution theorem states that under certain conditions

$$L\left\{\int_0^x f(x-y)g(y)\,dy\right\} = \bar{f}(p)\bar{g}(p),$$

where $\bar{f}(p) = L\{f(x)\}$ and $\bar{g}(p) = L\{g(x)\}$.

Show, using the Laplace transform method, that the equation

$$(D^2+4)y = f(x),$$

where $f(x)$ is an arbitrary function and where $y(0) = 1, y^{(1)}(0) = 1$, becomes

$$\bar{y}(p) = \frac{p+1}{p^2+4} + \frac{\bar{f}(p)}{p^2+4}.$$

Hence, using the convolution theorem, show that

$$y(x) = \cos 2x + \tfrac{1}{2}\sin 2x + \tfrac{1}{2}\int_0^x f(x')\sin 2(x-x')\,dx',$$

where x' is an arbitrary integration parameter.

CHAPTER 24

Partial Differential Equations

24.1 Introduction

Any equation containing partial differential coefficients is called a partial differential equation, the order of the equation being equal (by analogy with ordinary differential equations) to the order of the highest partial differential coefficient occurring in it. For example, the equations

$$3y^2 \frac{\partial u}{\partial x} + \frac{\partial u}{\partial y} = 2u, \tag{1}$$

$$\frac{\partial^2 u}{\partial x^2} + f(x, y) \frac{\partial^2 u}{\partial y^2} = 0, \tag{2}$$

(where $f(x, y)$ is an arbitrary function) are typical partial differential equations of the first and second orders respectively, x and y being independent variables and u the function to be found. Both these equations are linear in the sense that both u and its derivatives occur only to the first power, and that products of u and its derivatives are absent. A typical non-linear equation in two independent variables is

$$u \frac{\partial^2 u}{\partial x^2} + \left(\frac{\partial u}{\partial y}\right)^2 = u^2. \tag{3}$$

However, we shall not consider equations of this type here. In general, the solution of partial differential equations presents a much more difficult problem than the solution of ordinary differential equations, and except for certain special types of linear partial differential equations no general method of solution is available. In this chapter we shall therefore only indicate how a few particularly simple linear partial differential equations (which nevertheless are of physical interest) can be solved.

We have already seen in Chapter 21 that the general solution of an ordinary differential equation contains arbitrary constants. The general solution of a partial differential equation, however, usually contains arbitrary functions. To illustrate this point we consider the converse problem of the formation of partial differential equations from given functions.

For example, if

$$u = yf(x), \tag{4}$$

where $f(x)$ is an arbitrary function of x, then

$$\frac{\partial u}{\partial y} = f(x), \tag{5}$$

and consequently eliminating $f(x)$ from (4) and (5) we obtain the first order partial differentiation equation

$$y\frac{\partial u}{\partial y} = u. \tag{6}$$

Any function u of the type given by (4) is therefore a solution of (6) irrespective of the form of $f(x)$. Similarly, if

$$u = f(x+y) + g(x-y), \tag{7}$$

where $f(x+y)$, $g(x-y)$ are arbitrary functions of $x+y$ and $x-y$ respectively, then

$$\frac{\partial u}{\partial x} = f'(x+y) + g'(x-y), \tag{8}$$

$$\frac{\partial^2 u}{\partial x^2} = f''(x+y) + g''(x-y), \tag{9}$$

$$\frac{\partial u}{\partial y} = f'(x+y) - g'(x-y), \tag{10}$$

$$\frac{\partial^2 u}{\partial y^2} = f''(x+y) + g''(x-y), \tag{11}$$

where dashes denote differentiation with respect to the appropriate argument (i.e. $x+y$ or $x-y$). Hence eliminating the arbitrary functions by equating (9) and (11) we obtain the second order partial differential equation

$$\frac{\partial^2 u}{\partial x^2} = \frac{\partial^2 u}{\partial y^2}. \tag{12}$$

The function u defined by (7) therefore satisfies (12) irrespective of the functional forms of $f(x+y)$ and $g(x-y)$.

In both these examples the order of the resulting partial differential equation is equal to the number of arbitrary functions contained in the initial expression for u. This would seem to suggest (by analogy with ordinary differential equations) that the general solution of an nth order partial differential equation should contain just n arbitrary functions. Although this is not always the case, it is a legitimate definition of the general solution of a partial differential equation for the class of equations considered in this chapter (linear, with constant coefficients). Finally we remark that in those physical problems which lead to a partial differential equation the arbitrary functions in the solution must usually be chosen to satisfy certain conditions called boundary conditions which are dictated by the particular problem under consideration.

24.2 Second Order Constant Coefficient Equations

Any equation of the type

$$a\frac{\partial^2 u}{\partial x^2} + 2h\frac{\partial^2 u}{\partial x\,\partial y} + b\frac{\partial^2 u}{\partial y^2} + 2f\frac{\partial u}{\partial x} + 2g\frac{\partial u}{\partial y} + cu = 0, \qquad (13)$$

where a, h, b, f, g and c are constants, is a linear second order constant coefficient partial differential equation in two variables (x and y). By comparison with the equation of the general conic

$$ax^2 + 2hxy + by^2 + 2fx + 2gy + c = 0 \qquad (14)$$

we say that (13) is of

$$\left.\begin{array}{l} \text{elliptic} \\ \text{parabolic} \\ \text{hyperbolic} \end{array}\right\} \text{ type when } \left\{\begin{array}{l} ab - h^2 > 0, \\ ab - h^2 = 0, \\ ab - h^2 < 0. \end{array}\right. \qquad (15)$$

For example, Laplace's equation in two variables (see Chapter 9, 9.4).

$$\frac{\partial^2 u}{\partial x^2} + \frac{\partial^2 u}{\partial y^2} = 0 \qquad (16)$$

may be obtained from (13) by putting $a = 1$, $h = 0$, $b = 1$, $f = g = c$ $= 0$, and hence, since $ab - h^2 > 0$, is of elliptic type. Similarly the equation

$$\frac{\partial^2 u}{\partial x^2} - k^2 \frac{\partial^2 u}{\partial y^2} = 0, \tag{17}$$

(where k is a real constant) may be obtained from (13) by putting $a = 1$, $h = 0$, $b = -k^2$, $f = g = c = 0$. Hence, since $ab - h^2 = -k^2$ < 0, the equation is of hyperbolic type. However the equation

$$\frac{\partial^2 u}{\partial x^2} - k \frac{\partial u}{\partial y} = 0 \tag{18}$$

is of parabolic type, since $a = 1$, $h = 0$, $b = 0$, $g = -\frac{1}{2}k$, $f = c = 0$, and $ab - h^2 = 0$.

As we shall see in the next section the form of the general solution of a given second order constant coefficient partial differential equation depends very much on whether the equation is of elliptic, parabolic, or hyperbolic type.

24.3 Euler's Equation

The equation

$$a \frac{\partial^2 u}{\partial x^2} + 2h \frac{\partial^2 u}{\partial x \, \partial y} + b \frac{\partial^2 u}{\partial y^2} = 0, \tag{19}$$

where a, h and b are constants, is a special case of (13) (obtained by putting $f = g = c = 0$), and is usually known as Euler's equation. The general solution of this equation may be obtained in the following way.

We first define two new independent variables ζ and η by the linear relations

$$\left. \begin{array}{l} \xi = px + qy, \\ \eta = rx + sy, \end{array} \right\} \tag{20}$$

where p, q, r and s are arbitrary constants. Then

$$\frac{\partial u}{\partial x} = \frac{\partial u}{\partial \xi} \frac{\partial \xi}{\partial x} + \frac{\partial u}{\partial \eta} \frac{\partial \eta}{\partial x} = p \frac{\partial u}{\partial \xi} + r \frac{\partial u}{\partial \eta}, \tag{21}$$

$$\frac{\partial u}{\partial y} = \frac{\partial u}{\partial \xi}\frac{\partial \xi}{\partial y} + \frac{\partial u}{\partial \eta}\frac{\partial \eta}{\partial y} = q\frac{\partial u}{\partial \xi} + s\frac{\partial u}{\partial \eta}, \tag{22}$$

$$\frac{\partial^2 u}{\partial x^2} = \frac{\partial}{\partial x}\left(\frac{\partial u}{\partial x}\right) = \left(p\frac{\partial}{\partial \xi} + r\frac{\partial}{\partial \eta}\right)\left(p\frac{\partial u}{\partial \xi} + r\frac{\partial u}{\partial \eta}\right) \tag{23}$$

$$= p^2\frac{\partial^2 u}{\partial \xi^2} + 2pr\frac{\partial^2 u}{\partial \xi\,\partial \eta} + r^2\frac{\partial^2 u}{\partial \eta^2}, \tag{24}$$

$$\frac{\partial^2 u}{\partial y^2} = \frac{\partial}{\partial y}\left(\frac{\partial u}{\partial y}\right) = \left(q\frac{\partial}{\partial \xi} + s\frac{\partial}{\partial \eta}\right)\left(q\frac{\partial u}{\partial \xi} + s\frac{\partial u}{\partial \eta}\right) \tag{25}$$

$$= q^2\frac{\partial^2 u}{\partial \xi^2} + 2sq\frac{\partial^2 u}{\partial \xi\,\partial \eta} + s^2\frac{\partial^2 u}{\partial \eta^2}, \tag{26}$$

and

$$\frac{\partial^2 u}{\partial x\,\partial y} = \frac{\partial}{\partial x}\left(\frac{\partial u}{\partial y}\right) = \left(p\frac{\partial}{\partial \xi} + r\frac{\partial}{\partial \eta}\right)\left(q\frac{\partial u}{\partial \xi} + s\frac{\partial u}{\partial \eta}\right) \tag{27}$$

$$= pq\frac{\partial^2 u}{\partial \xi^2} + (rq + sp)\frac{\partial^2 u}{\partial \xi\,\partial \eta} + rs\frac{\partial^2 u}{\partial \eta^2}. \tag{28}$$

Substituting these expressions for the second partial derivatives into (19) we find

$$(ap^2 + bq^2 + 2hpq)\frac{\partial^2 u}{\partial \xi^2} + 2\{apr + bsq + h(rq + sp)\}\frac{\partial^2 u}{\partial \xi\,\partial \eta}$$
$$+ (ar^2 + bs^2 + 2hrs)\frac{\partial^2 u}{\partial \eta^2} = 0. \tag{29}$$

We now choose the arbitrary constants p, q, r, s such that $p = r = 1$, and such that q and s are the two roots X_1 and X_2 of the equation

$$a + 2hX + bX^2 = 0. \tag{30}$$

Consequently (29) becomes

$$\{a + h(X_1 + X_2) + bX_1X_2\}\frac{\partial^2 u}{\partial \xi\,\partial \eta} = 0. \tag{31}$$

However, since by Chapter 20, 20.2,

$$X_1 + X_2 = -\frac{2h}{b},$$

$$X_1 X_2 = \frac{a}{b},$$

(32)

(31) may be written as

$$\frac{2}{b}(ab - h^2)\frac{\partial^2 u}{\partial \xi \, \partial \eta} = 0.$$

(33)

Hence, provided (19) is not parabolic in the sense that $ab - h^2 \neq 0$, (33) gives

$$\frac{\partial^2 u}{\partial \xi \, \partial \eta} = 0,$$

(34)

which by straightforward integration has the general solution

$$u = F(\xi) + G(\eta),$$

(35)

where F and G are arbitrary functions. Consequently since $\xi = x + X_1 y$, $\eta = x + X_2 y$ (X_1 and X_2 being the two roots of (30)), the general solution of (19) (provided the equation is not of parabolic type) is

$$u = F(x + X_1 y) + G(x + X_2 y).$$

(36)

Finally we see from (30) that the nature of the roots X_1 and X_2 which appear in this solution depends on whether the equation is of hyperbolic or elliptic type; for when $ab - h^2 < 0$ (hyperbolic) X_1 and X_2 are real, whereas when $ab - h^2 > 0$ (elliptic) X_1 and X_2 are necessarily complex.

When (19) is of parabolic type ($ab - h^2 = 0$) the general solution may be obtained from (29) by putting $p = 1$, and (for the moment) leaving q, r and s arbitrary. Then

$$(a + bq^2 + 2hq)\frac{\partial^2 u}{\partial \xi^2} + 2\{ar + bsq + h(rq + s)\}\frac{\partial^2 u}{\partial \xi \, \partial \eta}$$

$$+ (ar^2 + bs^2 + 2hrs)\frac{\partial^2 u}{\partial^2 \eta} = 0. \quad (37)$$

If q is now chosen to be the root of

$$a + bq^2 + 2hq = 0 \qquad (38)$$

then, since $ab - h^2 = 0$ by assumption,

$$q = -\frac{h}{b} \text{ (twice).} \qquad (39)$$

Hence in virtue of (38) the first term of (37) is zero. Likewise the second term of (37) is also zero since, by (39)),

$$ar + bsq + h(rq + s) = (ab - h^2)\frac{r}{b} = 0. \qquad (40)$$

Consequently (37) becomes (provided r and s are not both zero)

$$\frac{\partial^2 u}{\partial \eta^2} = 0. \qquad (41)$$

By direct integration the general solution of (41) is

$$u = F(\xi) + \eta G(\xi), \qquad (42)$$

where again F and G are arbitrary functions. Hence, since $p = 1$, $q = -h/b \ (= X$, say), we have from (20)

$$\begin{aligned} \xi &= x + Xy, \\ \eta &= rx + sy, \end{aligned} \qquad (43)$$

where r and s are arbitrary (but not both zero). The solution (42) is therefore

$$u = F(x + Xy) + (rx + sy)G(x + Xy), \qquad (44)$$

which is accordingly the general solution of (19) when $ab - h^2 = 0$.

We now illustrate these results by the following examples.

Example 1. Laplace's equation in two variables

$$\frac{\partial^2 u}{\partial x^2} + \frac{\partial^2 u}{\partial y^2} = 0 \qquad (45)$$

is a special case of Euler's equation obtained by putting $a = b = 1$ and $h = 0$. Hence (30) becomes

$$1 + X^2 = 0 \qquad (46)$$

467

giving $X_1 = i$, $X_2 = -i$. By (36) the general solution of (45) is therefore

$$u = F(x+iy)+G(x-iy). \tag{47}$$

The appearance of imaginary quantities in (47) is in accordance with the elliptic character of (45).

Since all solutions of (45) must have the form (47) it is instructive to see how the solution

$$u = \tan^{-1}\frac{y}{x} \tag{48}$$

deduced in Chapter 9, Example 6, can be written in this way. Using complex numbers and putting $z = re^{i\theta} = x+iy$, we have $\tan^{-1}(y/x) = \theta$ and

$$\frac{z}{\bar{z}} = \frac{re^{i\theta}}{re^{-i\theta}} = e^{2i\theta}. \tag{49}$$

Consequently

$$\theta = \tan^{-1}\frac{y}{x} = \frac{1}{2i}\log_e z - \frac{1}{2i}\log_e \bar{z} \tag{50}$$

$$= \frac{1}{2i}\log_e (x+iy) - \frac{1}{2i}\log_e (x-iy), \tag{51}$$

which is of the form (47) as required.

Example 2. The equation

$$2\frac{\partial^2 u}{\partial x^2} + 3\frac{\partial^2 u}{\partial x\,\partial y} + \frac{\partial^2 u}{\partial y^2} = 0 \tag{52}$$

is of hyperbolic type since $a = 2$, $h = \frac{3}{2}$, $b = 1$ and hence $ab - h^2 < 0$. By (30) we have

$$2 + 3X + X^2 = 0, \tag{53}$$

which gives $X_1 = -1$, $X_2 = -2$.

Hence the general solution of (52) is

$$u = F(x-y)+G(x-2y). \tag{54}$$

Example 3. The equation

$$\frac{\partial^2 u}{\partial x^2} + 4 \frac{\partial^2 u}{\partial x \, \partial y} + 4 \frac{\partial^2 u}{\partial y^2} = 0 \tag{55}$$

is of parabolic type since $a = 1$, $h = 2$, $b = 4$ and $ab - h^2 = 0$. By (38) therefore we find

$$1 + 4X + 4X^2 = 0 \tag{56}$$

or

$$X = -\tfrac{1}{2} \quad \text{(twice)}.$$

Hence by (42) the general solution of (55) is

$$u = F(x - \tfrac{1}{2}y) + (rx + sy)G(x - \tfrac{1}{2}y), \tag{57}$$

where r and s are arbitrary constants.

We finally remark that the general solution of the inhomogeneous Euler equation

$$a \frac{\partial^2 u}{\partial x^2} + 2h \frac{\partial^2 u}{\partial x \, \partial y} + b \frac{\partial^2 u}{\partial y^2} = f(x, y), \tag{58}$$

where $f(x, y)$ is a given function, may be obtained by a D-operator technique similar to that used in Chapter 21 for ordinary differential equations. However, we shall not discuss this method here.

24.4 Separation of Variables

Although it is always satisfying from the point of view of a pure mathematician to be able to derive general solutions of partial differential equations, such solutions are usually of little value when given boundary conditions are to be imposed on the solution. For example, it is usually a somewhat difficult matter to choose the functions F and G of the solution (see (47)) of

$$\frac{\partial^2 u}{\partial x^2} + \frac{\partial^2 u}{\partial y^2} = 0 \tag{59}$$

such that the equation is satisfied inside a square region defined by the lines $x = 0$, $x = a$, $y = 0$, $y = b$ and such that u takes prescribed values on the boundary of this region.

To overcome this difficulty it is best to obtain a less general type of solution which is governed by the type of boundary conditions to be imposed. One method of doing this depends on assuming the solution to be a product of functions each of which contains only one of the independent variables. This is the method of separation of variables.

We shall now illustrate the use of this method by considering the solutions of three equations of physical importance satisfying specified boundary conditions. These three equations are

(a) the one-dimensional wave equation

$$\frac{\partial^2 u}{\partial x^2} - \frac{1}{c^2}\frac{\partial^2 u}{\partial t^2} = 0, \tag{60}$$

(b) the one-dimensional heat conduction (and diffusion) equation

$$\frac{\partial^2 u}{\partial x^2} = \frac{1}{k}\frac{\partial u}{\partial t}, \tag{61}$$

and

(c) Laplace's equation in two-dimensions

$$\frac{\partial^2 u}{\partial x^2} + \frac{\partial^2 u}{\partial y^2} = 0. \tag{62}$$

In these equations c and k are physical constants, x and y are the space variables of a Cartesian coordinate system, and t is the time variable. The physical interpretation of the dependent variable u is, of course, different in each equation.

(a) The wave equation

Example 4. To find the solution of the equation

$$\frac{\partial^2 u}{\partial x^2} = \frac{1}{c^2}\frac{\partial^2 u}{\partial t^2}, \tag{63}$$

which is periodic in x and t, and which satisfies the boundary conditions

$$u(0, t) = u(l, t) = 0, \quad t \geq 0, \tag{64}$$

$$u(x, 0) = f(x), \quad 0 \leq x \leq l, \tag{65}$$

$$\left[\frac{\partial u(x, t)}{\partial t}\right]_{t=0} = g(x), \quad 0 \leqq x \leqq l, \tag{66}$$

where f and g are given functions and l is a given constant.

We now assume a solution of (63) of the form

$$u(x, t) = X(x)T(t), \tag{67}$$

where X is a function of x only and T is a function of t only. In this way (63) becomes

$$\frac{1}{X}\frac{d^2X}{dx^2} = \frac{1}{c^2 T}\frac{d^2T}{dt^2}. \tag{68}$$

This equation is satisfied if we now write

$$\frac{1}{X}\frac{d^2X}{dx^2} = -\frac{w^2}{c^2} \tag{69}$$

and

$$\frac{1}{T}\frac{d^2T}{dt^2} = -w^2, \tag{70}$$

where w is any real number. The solutions of these equations are periodic, as required, in virtue of the negative signs introduced in (69) and (70), and have the well-known forms

$$X(x) = A\cos\frac{wx}{c} + B\sin\frac{wx}{c}, \tag{71}$$

$$T(t) = C\cos wt + D\sin wt, \tag{72}$$

where A, B, C and D are arbitrary constants. Hence (67) becomes

$$u(x, t) = \left(A\cos\frac{wx}{c} + B\sin\frac{wx}{c}\right)(C\cos wt + D\sin wt), \tag{73}$$

which clearly satisfies (63) for all values of the constants A, B, C, D and w.

We must now satisfy the boundary conditions (64)–(66). To do this we first put $x = 0$ in (73). The boundary condition $u(0, t) = 0 (t > 0)$ then gives

$$0 = A(C\cos wt + D\sin wt) \tag{74}$$

471

for all t, which implies

$$A = 0. \tag{75}$$

Secondly, putting $x = l$ in (73), the condition $u(l, t) = 0 (t \geqq 0)$ gives (using (75))

$$0 = B \sin\frac{wl}{c}(C \cos wt + D \sin wt). \tag{76}$$

However, since B cannot be equal to zero without making $u(x, t)$ identically zero, (76) can only be satisfied (for all t) provided

$$\sin \frac{wl}{c} = 0. \tag{77}$$

Hence

$$w = \frac{r\pi c}{l} \quad \text{where} \quad r = 1, 2, 3 \ldots \tag{78}$$

(the case $r = 0$, which gives $w = 0$, being excluded since this again makes $u(x, t)$ identically zero). We now see that there exists an infinite set of discrete values of w (the eigenvalues) and that to each value of w there corresponds a particular solution (the eigenfunction). For putting (75) and (78) into (73) we have

$$\left.\begin{aligned}
&w_1 = \frac{\pi c}{l}, u_1(x, t) = \sin \frac{\pi x}{l}\left(C_1 \sin \frac{\pi ct}{l} + D_1 \cos \frac{\pi ct}{l}\right), \\
&w_2 = \frac{2\pi c}{l}, u_2(x, t) = \sin \frac{2\pi x}{l}\left(C_2 \sin\frac{2\pi ct}{l} + D_2 \cos \frac{2\pi ct}{l}\right), \\
&\quad\vdots \qquad\qquad \vdots \qquad \vdots \qquad \vdots \qquad\qquad \vdots \\
&w_r = \frac{r\pi c}{l}, u_r(x, t) = \sin \frac{r\pi x}{l}\left(C_r \sin \frac{r\pi ct}{l} + D_r \cos \frac{r\pi ct}{l}\right), \\
&\quad\vdots \qquad\qquad \vdots \qquad \vdots \qquad \vdots \qquad\qquad \vdots
\end{aligned}\right\} \tag{79}$$

and so on, $C_1, C_2, \ldots C_r, D_1, D_2, \ldots D_r, \ldots$ being arbitrary constants. Each of these expressions for $u(x, t)$ is a particular solution of the wave equation (63) satisfying the boundary conditions (64). Now since (63) is a linear equation, any linear combination of particular solutions is also a solution. Accordingly we take the linear combination

$$u(x, t) = \sum_{r=1}^{\infty} \left(C_r \sin\frac{r\pi ct}{l} + D_r \cos\frac{r\pi ct}{l} \right) \sin\frac{r\pi x}{l} \qquad (80)$$

as the general solution of (63) satisfying the boundary conditions (64). The arbitrary constants C_r and D_r in this solution must now be chosen so that the boundary conditions at $t = 0$ (i.e. (65) and (66)) are satisfied.

Consider first (65) which requires

$$u(x, 0) = f(x), \quad 0 \leqq x \leqq l. \qquad (81)$$

Then putting $t = 0$ in (80) we have

$$f(x) = \sum_{r=1}^{\infty} D_r \sin\frac{r\pi x}{l}. \qquad (82)$$

Similarly (66), which requires

$$\left[\frac{\partial u(x, t)}{\partial t} \right]_{t=0} = g(x), \quad 0 \leqq x \leqq l, \qquad (83)$$

is satisfied by differentiating (80) with respect to t and then putting $t = 0$. In this way

$$g(x) = \frac{\pi c}{l} \sum_{r=1}^{\infty} rC_r \sin\frac{r\pi x}{l} \qquad (84)$$

The coefficients C_r and D_r may now be determined from (82) and (84) by a Fourier series technique (see Chapter 15). Consequently

$$D_r = \frac{2}{l} \int_0^l f(x) \sin\frac{r\pi x}{l} \, dx \qquad (85)$$

and

$$C_r = \frac{2}{r\pi c} \int_0^l g(x) \sin\frac{r\pi x}{l} \, dx, \qquad (86)$$

where $r = 1, 2, 3 \ldots$.

Hence finally, substituting (85) and (86) into (80), we have the solution

$$u(x, t) = \sum_{r=1}^{\infty} \left\{ \left(\frac{2}{r\pi c} \int_0^l g(x') \sin \frac{r\pi x'}{l} \, dx' \right) \sin \frac{r\pi ct}{l} \sin \frac{r\pi x}{l} \right\}$$

$$+ \sum_{r=1}^{\infty} \left\{ \left(\frac{2}{l} \int_0^l f(x') \sin \frac{r\pi x'}{l} \, dx' \right) \cos \frac{r\pi ct}{l} \sin \frac{r\pi x}{l} \right\}, \quad (87)$$

where we have written x' for the variable of integration to distinguish it from the independent variable x.

The method of separation of variables may be readily extended to the solution of the two- and three-dimensional wave equations

$$\frac{\partial^2 u}{\partial x^2} + \frac{\partial^2 u}{\partial y^2} = \frac{1}{c^2} \frac{\partial^2 u}{\partial t^2} \quad \text{and} \quad \frac{\partial^2 u}{\partial x^2} + \frac{\partial^2 u}{\partial y^2} + \frac{\partial^2 u}{\partial z^2} = \frac{1}{c^2} \frac{\partial^2 u}{\partial t^2}. \quad (88)$$

However, it is not always convenient or desirable to solve these equations using a Cartesian coordinate system. For example, if the solution of the wave equation is required inside a three-dimensional region with spherical symmetry it is usually better to adopt spherical polar coordinates (r, θ, ϕ). Similarly the solution inside a region with cylindrical symmetry is more easily obtained using cylindrical polar coordinates (r, ϕ, z). Whatever coordinate system is eventually chosen the Laplacian operator ∇^2 must always be expressed in terms of the corresponding coordinates (see Chapter 9, 9.11 and Chapter 19, 19.8).

(b) *The heat conduction equation*

Example 5. To obtain the solution of the equation

$$\frac{\partial^2 u}{\partial x^2} = \frac{1}{k} \frac{\partial u}{\partial t} \quad (89)$$

which decreases exponentially with t, and which satisfies the boundary conditions

$$u(0, t) = u(l, t) = 0, \quad t \geq 0, \quad (90)$$

$$u(x, 0) = f(x), \quad 0 \leq x \leq l, \quad (91)$$

where f is a given function and l is a constant. Assuming (as in Example 4) a solution of the form

$$u(x, t) = X(x)T(t) \tag{92}$$

and substituting in (89) we find

$$\frac{1}{X} \frac{d^2X}{dx^2} = \frac{1}{kT} \frac{dT}{dt}. \tag{93}$$

Hence putting

$$\frac{1}{X} \frac{d^2X}{dx^2} = -w^2, \tag{94}$$

$$\frac{1}{kT} \frac{dT}{dt} = -w^2, \tag{95}$$

where w is any real number (the negative sign in (95) ensuring an exponentially decreasing solution in t), we have

$$X = A \cos wx + B \sin wx, \tag{96}$$

and

$$T = Ce^{-w^2kt}, \tag{97}$$

where A, B and C are arbitrary constants. With (96) and (97), (92) now becomes

$$u(x, t) = (A' \cos wx + B' \sin wx)e^{-w^2kt}, \tag{98}$$

where A' and B' are new arbitrary constants. In order to satisfy the boundary conditions (90) we first put $x = 0$ in (98). Consequently

$$0 = A'e^{-w^2kt} \tag{99}$$

for all t, which implies

$$A' = 0. \tag{100}$$

Secondly putting $x = l$ in (98) we find (using (100))

$$0 = (B' \sin wl)e^{-w^2kt}, \tag{101}$$

which leads to non-trivial solutions provided

$$\sin wl = 0. \tag{102}$$

475

Hence

$$w = \frac{r\pi}{l} \quad \text{where} \quad r = 1, 2, 3 \ldots \tag{103}$$

(the case $r = 0$ again being excluded to avoid making $u(x, t)$ identically zero). Putting (100) and (103) into (98) we now obtain the infinity of eigenvalues and corresponding eigenfunctions

$$\left.\begin{array}{llll} w_1 = \dfrac{\pi}{l}, & u_1(x, t) = B_1' e^{-\pi^2 kt/l^2} \; \sin \dfrac{\pi x}{l}, \\[2mm] w_2 = \dfrac{2\pi}{l}, & u_2(x, t) = B_2' e^{-4\pi^2 kt/l^2} \sin \dfrac{2\pi x}{l}, \\ \vdots & \vdots \qquad \vdots \qquad \vdots \\ w_r = \dfrac{r\pi}{l}, & u_r(x, t) = B_r' e^{-r^2\pi^2 kt/l^2} \sin \dfrac{r\pi x}{l}, \\ \vdots & \vdots \qquad \vdots \qquad \vdots \end{array}\right\} \tag{104}$$

and so on, $B_1', B_2', \ldots B_r' \ldots$ being arbitrary constants. We now take the linear combination of solutions (104) given by

$$u(x, t) = \sum_{r=1}^{\infty} B_r' e^{-r^2\pi^2 kt/l^2} \sin \frac{r\pi x}{l} \tag{105}$$

as the general solution of (89) satisfying the boundary conditions (90), where the constants B_r' must now be chosen to satisfy the remaining boundary condition (91).

Hence putting $t = 0$ in (105) we have

$$f(x) = \sum_{r=1}^{\infty} B_r' \sin \frac{r\pi x}{l}, \tag{106}$$

from which it follows, using the Fourier series technique, that

$$B_r' = \frac{2}{l} \int_0^l f(x) \sin \frac{r\pi x}{l} \, dx. \tag{107}$$

476

The solution of (89) subject to the boundary conditions (90) and (91) is therefore

$$u(x, t) = \sum_{r=1}^{\infty} \left\{ \left(\frac{2}{l} \int_0^l f(x') \sin\frac{r\pi x'}{l} \, dx' \right) e^{-r^2\pi^2 kt/l^2} \sin\frac{r\pi x}{l} \right\}, \quad (108)$$

where x' has been written as the variable of integration to avoid confusion with x.

Similar remarks to those made at the end of Example 4 on the solution of the two- and three-dimensional wave equations apply equally to the solution of the two- and three-dimensional heat conduction equations

$$\frac{\partial^2 u}{\partial x^2} + \frac{\partial^2 u}{\partial y^2} = \frac{1}{k}\frac{\partial u}{\partial t} \quad \text{and} \quad \frac{\partial^2 u}{\partial x^2} + \frac{\partial^2 u}{\partial y^2} + \frac{\partial^2 u}{\partial z^2} = \frac{1}{k}\frac{\partial u}{\partial t}. \quad (109)$$

(c) Laplace's equation in two dimensions

Example 6. To obtain the solution of Laplace's equation

$$\frac{\partial^2 u}{\partial x^2} + \frac{\partial^2 u}{\partial y^2} = 0 \quad (110)$$

which is periodic in x inside the rectangular region defined by $0 \leq x \leq a$, $0 \leq y \leq b$, and which satisfies the boundary conditions

$$u(x, y) = \begin{cases} 0 & \text{when} \quad x = 0, \quad 0 \leq y \leq b, \\ 0 & \text{when} \quad x = a, \quad 0 \leq y \leq b, \\ 0 & \text{when} \quad y = b, \quad 0 \leq x \leq a, \\ f(x) & \text{when} \quad y = 0, \quad 0 < x < a. \end{cases} \quad (111)$$

Writing

$$u(x, y) = X(x)Y(y) \quad (112)$$

(110) becomes

$$\frac{1}{X}\frac{d^2 X}{dx^2} + \frac{1}{Y}\frac{d^2 Y}{dy^2} = 0. \quad (113)$$

Hence putting

$$\frac{1}{X}\frac{d^2 X}{dx^2} = -w^2 \quad \text{(for periodic solutions)} \quad (114)$$

477

and

$$\frac{1}{Y}\frac{d^2Y}{dy^2} = +w^2, \tag{115}$$

where w is any real number, we have

$$X = A\cos wx + B\sin wx, \tag{116}$$

$$Y = C\cosh wy + D\sinh wy, \tag{117}$$

A, B, C and D being arbitrary constants.

Consequently (112) becomes

$$u(x, y) = (A\cos wx + B\sin wx)(C\cosh wy + D\sinh wy). \tag{118}$$

In order that this solution shall satisfy the first boundary condition of (111) we must have (putting $x = 0$)

$$0 = A(C\cosh wy + D\sinh wy), \quad (0 \leqq y \leqq b) \tag{119}$$

or

$$A = 0. \tag{120}$$

Similarly the second boundary condition requires (for non-trivial solutions)

$$\sin wa = 0, \tag{121}$$

or

$$w = \frac{r\pi}{a} \quad \text{where} \quad r = 1, 2, 3 \dots. \tag{122}$$

Likewise the third boundary condition of (111) gives (putting $y = b$ in (118))

$$0 = (A\cos wx + B\sin wx)(C\cosh wb + D\sinh wb) \tag{123}$$

in $0 \leqq x \leqq a$, or

$$\frac{C}{D} = -\tanh wb. \tag{124}$$

Hence substituting (120), (122) and (124) into (118) we find the

478

following infinite set of eigenvalues and corresponding eigenfunctions

$$
\left.
\begin{aligned}
&w_1 = \frac{\pi}{a}, \ u_1(x, y) = E_1 \sin \frac{\pi x}{a} \ \sinh \frac{\pi(b-y)}{a}, \\
&w_2 = \frac{2\pi}{a}, \ u_2(x, y) = E_2 \sin \frac{2\pi x}{a} \ \sinh \frac{2\pi(b-y)}{a}, \\
&\quad\vdots \qquad\qquad \vdots \qquad\qquad \vdots \\
&w_r = \frac{r\pi}{a}, \ u_r(x, y) = E_r \sin \frac{r\pi x}{a} \ \sinh \frac{r\pi(b-y)}{a}, \\
&\quad\vdots \qquad\qquad \vdots \qquad\qquad \vdots
\end{aligned}
\right\}
\tag{125}
$$

where $E_1, E_2, \ldots E_r \ldots$ are arbitrary constants.

As before we now take a linear combination of these particular solutions

$$
u(x, y) = \sum_{r=1}^{\infty} E_r \sin \frac{r\pi x}{a} \sinh \frac{r\pi(b-y)}{a} \tag{126}
$$

as the general solution of (110) satisfying the first three boundary conditions of (111). The constants E_r in this solution must now be chosen to satisfy the fourth and last boundary condition of (111). Hence putting $y = 0$ in (126) we have

$$
f(x) = \sum_{r=1}^{\infty} E_r \sinh \frac{r\pi b}{a} \sin \frac{r\pi x}{a}, \tag{127}
$$

from which we find

$$
E_r \sinh \frac{r\pi b}{a} = \frac{2}{a} \int_0^a f(x) \sin \frac{r\pi x}{a} \, dx. \tag{128}
$$

Consequently substituting (128) into (126) the final solution of (110) subject to the boundary conditions (111) is

$$
u(x, y) = \sum_{r=1}^{\infty} \left\{ \left(\frac{2}{a} \int_0^a f(x') \sin \frac{r\pi x'}{a} \, dx' \right) \frac{\sin \frac{r\pi x}{a} \sinh \frac{r\pi(b-y)}{a}}{\sinh \frac{r\pi b}{a}} \right\}, \tag{129}
$$

where, as before, x' is the variable of integration.

Example 7. In Examples 4 and 5 the boundary values at $x = 0$ and $x = l$ have been taken as zero. We now discuss the solution of the heat conduction equation

$$\frac{\partial^2 u}{\partial x^2} = \frac{1}{k}\frac{\partial u}{\partial t} \tag{130}$$

subject to the boundary conditions

$$u(0,t) = U_0, \quad t \geqq 0, \tag{131}$$

$$u(l,t) = U_1, \quad t \geqq 0, \tag{132}$$

and

$$u(x,0) = f(x), \quad 0 \leqq x \leqq l, \tag{133}$$

where U_0, U_1, k and l are given constants and $f(x)$ is a given function.

Writing

$$u(x,t) = v(x) + w(x,t) \tag{134}$$

and inserting into (130) we find

$$\frac{d^2 v}{dx^2} = 0 \tag{135}$$

and

$$\frac{\partial^2 w}{\partial x^2} = \frac{1}{k}\frac{\partial w}{\partial t} \tag{136}$$

together with the boundary conditions (using (131), (132), (133) and (134))

$$v(0) = U_0, \quad v(l) = U_1 \tag{137}$$

and

$$w(0,t) = 0, \quad w(l,t) = 0, \quad t \geqq 0, \tag{138}$$

$$w(x,0) = f(x) - v(x), \quad 0 \leqq x \leqq l. \tag{139}$$

Solving (135) and imposing the boundary conditions (137) we have

$$v(x) = U_0 + \frac{x}{l}(U_1 - U_0). \tag{140}$$

The function $w(x,t)$ is now to be obtained from (136) subject to the boundary conditions (138) and (139). We note that these boundary conditions have zeros on the right-hand sides and accordingly the solution for $w(x,t)$ follows the method of Example 5. Hence (by comparison with (105))

$$w(x,t) = \sum_{r=1}^{\infty} B_r e^{-r^2\pi^2kt/l^2} \sin\frac{r\pi x}{l}, \tag{141}$$

where now

$$B_r = \frac{2}{l}\int_0^l [f(x) - v(x)] \sin\frac{r\pi x}{l} dx \tag{142}$$

$$= \frac{2}{l}\int_0^l \left[f(x) - U_0 - \frac{x}{l}(U_1 - U_0) \right] \sin\frac{r\pi x}{l} dx \tag{143}$$

$$= \frac{2}{l}\int_0^l f(x) \sin\frac{r\pi x}{l} dx + \frac{2}{r\pi}[(-1)^r U_1 - U_0], \tag{144}$$

where

$$r = 1, 2, 3, \ldots$$

The solution of (130) subject to the boundary conditions (131)–(133) is therefore obtained by adding (140) and (141) to get $u(x,t)$, B_r being given by (144). When U_1 and U_0 are zero this solution reduces to the solution (108) of Example 5.

24.5 Other Methods of Solution

Besides the method of separation of variables which we have just discussed, other methods for the solution of linear partial differential equations are known. In particular the method of the Laplace transformation (which was used in Chapter 23 for the solution of ordinary linear differential equations with constant coefficients) may be used for the solution of linear partial differential equations with constant coefficients. However, we shall not discuss this method here. The reader is advised to remember that, in general, it is a difficult if not impossible task to solve partial differential equations analytically, and that in all but simple cases some numerical method is the best, if not the only, way of obtaining a solution satisfying given boundary conditions.

PROBLEMS 24

1. Eliminate the arbitrary functions from the following so obtaining partial differential equations of which the general solutions are

   (a) $u = f(x+y)$, (b) $u = f(xy)$,
   (c) $u = f(x+y) + g(x-y)$, (d) $u = x^n f(y/x)$.

2. State the nature of each of the following equations (i.e. whether elliptic, parabolic or hyperbolic), and obtain their general solutions

   (a) $3\dfrac{\partial^2 u}{\partial x^2} + 4\dfrac{\partial^2 u}{\partial x\,\partial y} - \dfrac{\partial^2 u}{\partial y^2} = 0$,

   (b) $\dfrac{\partial^2 u}{\partial x^2} - 2\dfrac{\partial^2 u}{\partial x\,\partial y} + \dfrac{\partial^2 u}{\partial y^2} = 0$,

   (c) $4\dfrac{\partial^2 u}{\partial x^2} + \dfrac{\partial^2 u}{\partial y^2} = 0$.

3. A function $Z(r, t)$ satisfies

   $$\frac{1}{r^2}\frac{\partial}{\partial r}\left(r^2 \frac{\partial Z}{\partial r}\right) = \frac{1}{c^2}\frac{\partial^2 Z}{\partial t^2},$$

   where c is a constant. By introducing the new dependent variable $W = rZ$, and the independent variables $u = r + ct$ and $v = r - ct$, reduce the differential equation to $\partial^2 W/\partial u\,\partial v = 0$.

   Hence show that the general solution Z is of the form

   $$Z = \frac{U+V}{r},$$

   where U is an arbitrary function of $r + ct$ only, and V is an arbitrary function of $r - ct$ only. (C.U.)

4. If $V = [Ar + (B/r)]f(\theta)$, where A and B are constants, satisfies the equation

   $$r^2\frac{\partial^2 V}{\partial r^2} + r\frac{\partial V}{\partial r} + \frac{\partial^2 V}{\partial \theta^2} = 0,$$

   find the form of the function $f(\theta)$.

5. Find the connection between the constants w and α so that

$$V = Ae^{\alpha x} \sin(wt + \alpha x) + Be^{-\alpha x} \sin(wt - \alpha x)$$

may be a solution of

$$\frac{\partial^2 V}{\partial x^2} = 2\frac{\partial V}{\partial t},$$

where A and B are constants.

Find A and B when $V = 2 \sin t$ for $x = 0$, given that $\alpha > 0$, and $V \to 0$ as $x \to \infty$. (L.U.)

6. The function $V(x, y)$ satisfies the partial differential equation

$$\frac{\partial^2 V}{\partial x^2} = \frac{\partial V}{\partial y}.$$

If $V = XY$, where X depends only on x, and Y depends only on y, and $V = \cos 2x$ when $y = 0$, find V for all x, y. (L.U.)

7. Assuming a solution of the type $V = X(x)T(t)$ of the equation

$$\frac{\partial^2 V}{\partial x^2} = \frac{2}{a}\frac{\partial V}{\partial t} + V,$$

find V subject to the conditions

(i) $V = 0$ when $x = l$ for all t,

(ii) $\partial V/\partial x = -Ce^{-at}$ when $x = 0$, for all t;

a, C are given constants.

Hence show that $\partial V/\partial x = -Ce^{-at} \cos(l-x) \sec l$ for all x, t. (L.U.)

8. Find the ordinary differential equation satisfied by $f(r)$ if $(1/r)f(r) \cos wt$ is a solution of the equation

$$\frac{\partial^2 u}{\partial r^2} + \frac{2}{r}\frac{\partial u}{\partial r} = \frac{1}{c^2}\frac{\partial^2 u}{\partial t^2},$$

where w and c are constants.

Hence show that a solution of this partial differential equation is

$$u = \frac{1}{r}(A \cos nr + B \sin nr) \cos wt,$$

where $n = w/c$ and A and B are arbitrary constants.

Obtain the equation which must be satisfied by w, given (i) u is finite at $r = 0$ for all t, (ii) $\partial u/\partial r = 0$ at $r = a$ for all t, (iii) u is not identically zero. (L.U.)

9. Show that the solution of

$$\frac{\partial^2 V}{\partial x^2} + \frac{\partial^2 V}{\partial y^2} = 0$$

over the region $0 \leq x \leq a$, $0 \leq y \leq a$ with $V = 0$ along the lines $x = 0$, $x = a$, $y = 0$ and $V = V_0$ (= constant) along $y = a$ is

$$V(x, y) = \frac{4V_0}{\pi} \sum_{r=0}^{\infty} \left\{ \frac{\text{cosech}\,(2r+1)\pi}{2r+1} \sinh (2r+1)\frac{\pi y}{a} \sin (2r+1)\frac{\pi x}{a} \right\}.$$

10. Obtain the general form of solutions periodic in t of the equation

$$\frac{\partial^2 V}{\partial t^2} + V - 4\frac{\partial V}{\partial x} = 0.$$

If, when $x = 0$, $V = V_0 \sin 3t$, show that, when $x = 1$, V oscillates between (approximately) $\pm 0{\cdot}135V_0$. (L.U.)

11. Obtain all solutions of the equation

$$\frac{\partial^2 z}{\partial x^2} - \frac{\partial z}{\partial y} = z$$

of the form $z = (A \cos kx + B \sin kx)f(y)$, where A, B and k are constants. Find a solution of the equation for which $z = 0$ when $x = 0$; $z = 0$ when $x = \pi$; $z = x$ when $y = 1$. (L.U.)

12. Show that the equation

$$\frac{\partial V}{\partial t} = k\frac{\partial^2 V}{\partial x^2}$$

484

(where k is a constant) has solutions of the type

$$V = \sum_{r=1}^{\infty} e^{-kr^2t}(A_r \cos rx + B_r \sin rx),$$

where A_r, B_r are constants.

The initial conditions are

$$V(x, 0) = \frac{2V_0 x}{a}, \quad (0 \leq x \leq a/2)$$

$$V(x, 0) = 0, \quad (a/2 < x \leq a)$$

and $V(0, t) = V(a, t) = 0$, where V_0 and a are constants.

Find the form of the Fourier series giving V. (L.U.)

13. Show that

$$V = J_0(kr) \cos (kct + \alpha)$$

is a solution of the equation

$$\frac{\partial^2 V}{\partial t^2} = c^2 \left\{ \frac{\partial^2 V}{\partial r^2} + \frac{1}{r} \frac{\partial V}{\partial r} \right\},$$

where k, c and α are constants. (J_0 is Bessel's function of zero order; see Chapter 22, 22.3.)

14. Show that

$$u(x, t) = A \left(1 - \text{erf} \frac{x}{2\sqrt{(kt)}} \right)$$

is a solution of the equation

$$\frac{\partial^2 u}{\partial x^2} = \frac{1}{k} \frac{\partial u}{\partial t},$$

where A and k are constants. (For a definition of the error function erf x, see Chapter 13, 13.5.)

15. The temperature distribution $T(x, t)$ along a thin bar of length a satifies the equation

$$\frac{\partial^2 T}{\partial x^2} = \frac{1}{k}\frac{\partial T}{\partial t}, \quad (t \geqq 0, 0 \leqq x \leqq a),$$

where k is a constant, t denotes time and x is the distance from one of the ends. Find $T(x, t)$ if the bar is insulated at each end and the initial distribution is given by

$$T(x, 0) = 2T_0 \cos^2(\pi x/a),$$

where T_0 is a constant. Show that the bar eventually attains the uniform temperature T_0. (L.U.)

16. Solve the equation

$$\frac{\partial^2 \theta}{\partial x^2} = \frac{1}{k}\frac{\partial \theta}{\partial t},$$

where k is a constant, subject to the following conditions

(a) θ is finite as $t \to \infty$ for $0 < x < l$,

(b) $\theta = 100$ for $x = 0, t > 0$,

(c) $\dfrac{\partial \theta}{\partial x} = 0$ for $x = l, t > 0$,

(d) $\theta = 100 + \sin\dfrac{\pi x}{2l}$ for $0 < x < l, t = 0$.

CHAPTER 25

Calculus of Variations

25.1 Introduction

One of the simple applications of the differential calculus is to the location of the maxima and minima. The calculus of variations deals with the similar but more complicated problem of maximising or minimising an integral. To see more precisely what is meant by this we consider the following simple geometrical problem which will be solved later on in this chapter. Suppose $P(x_1, y_1)$, $Q(x_2, y_2)$ are two given fixed points in a Cartesian coordinate system. We now wish to find the equation of the curve joining these points such that when this curve is rotated about the x-axis to form a surface of revolution, the surface area is a minimum. If $y(x)$ is any curve joining the given points and if ds is an infinitesimal element of arclength then the surface area S generated by rotating $y(x)$ through 2π radians about the x-axis is given by

$$S = 2\pi \int_P^Q y(x)\, ds = 2\pi \int_{x_1}^{x_2} y \sqrt{\left\{1 + \left(\frac{dy}{dx}\right)^2\right\}}\, dx. \qquad (1)$$

Since S is a function of the function y it is usual to call it a functional and to write it as $S[y]$; the problem is therefore one of choosing the function y such that the functional $S[y]$ has a minimum value. This is, in fact, a typical problem in the calculus of variations (or theory of functionals) where the general aim is to find the stationary values of an integral with respect to a function. However, although it is comparatively easy to find the necessary conditions for an integral to be stationary with respect to a function it is a vastly more complicated problem to formulate sufficient conditions for the stationary value to be a maximum (or minimum). In this chapter we shall only be concerned with formulating a necessary condition for an integral to have a maximum or minimum.

25.2 Euler's Equation

The general problem to be discussed here is that of finding the

487

function y such that the integral

$$I[y] = \int_{x_1}^{x_2} f(x, y, y') \, dx \tag{2}$$

is stationary, where $y' = dy/dx$. It is assumed here that the function f is a given differentiable function of the three variables x, y and y', and that $y = y_1$ at $x = x_1$, $y = y_2$ at $x = x_2$, where x_1, x_2 are given (constant) limits of integration. Suppose $y(x)$ is any curve passing through the two points $(x_1 \, y_1)$, (x_2, y_2) (see Fig. 25.1). Consider

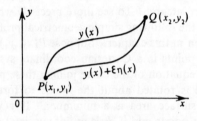

Fig. 25.1

now a small variation in the function y such that $y(x)$ becomes

$$y(x) + \varepsilon \eta(x), \tag{3}$$

where ε is a small parameter and $\eta(x)$ is an arbitrary function such that $\eta(x_1) = \eta(x_2) = 0$. Clearly (3) represents an infinity of curves passing through the end-points (x_1, y_1), (x_2, y_2) each curve being characterised by a particular value of ε (see Fig. 25.1). If we denote by I^* the value of (2) along an arbitrary varied path then

$$\delta I = I^* - I = \int_{x_1}^{x_2} \{f(x, y + \varepsilon \eta, y' + \varepsilon \eta') - f(x, y, y')\} \, dx, \tag{4}$$

where dashes denote differentiation with respect to x. Assuming now that the function f may be expanded in powers of ε we have (using Taylor's expansion of a function of two variables)

$$\delta I = \int_{x_1}^{x_2} \{f(x, y + \varepsilon \eta, y' + \varepsilon \eta') - f(x, y, y')\} \, dx$$

$$= \varepsilon I_1 + \frac{\varepsilon^2}{2!} I_2 + \text{terms in } \varepsilon^3 \text{ and higher}, \tag{5}$$

488

where

$$I_1 = \int_{x_1}^{x_2} \left\{ \eta \frac{\partial f}{\partial y} + \eta' \frac{\partial f}{\partial y'} \right\} dx \qquad (6)$$

and

$$I_2 = \int_{x_1}^{x_2} \left\{ \eta^2 \frac{\partial^2 f}{\partial y^2} + 2\eta\eta' \frac{\partial^2 f}{\partial y \, \partial y'} + \eta'^2 \frac{\partial^2 f}{\partial y'^2} \right\} dx. \qquad (7)$$

Consequently to the first order in ε the variation δI is given by

$$\delta I = \varepsilon I_1. \qquad (8)$$

The sign of δI therefore depends on the sign of ε; in other words the value of the integral (2) may increase for some variations and decrease for others. However, for the integral to have a true maximum or minimum all variations should respectively decrease or increase its value. This can be brought about by requiring

$$I_1 = \int_{x_1}^{x_2} \left(\eta \frac{\partial f}{\partial y} + \eta' \frac{\partial f}{\partial y'} \right) dx = 0, \qquad (9)$$

which, on integrating the last term by parts and remembering that $\eta(x_1) = \eta(x_2) = 0$, gives

$$\int_{x_1}^{x_2} \left\{ \frac{\partial f}{\partial y} - \frac{d}{dx} \left(\frac{\partial f}{\partial y'} \right) \right\} \eta \, dx = 0. \qquad (10)$$

Since $\eta(x)$ is an arbitrary function, (10) can only be satisfied if

$$\frac{\partial f}{\partial y} - \frac{d}{dx} \left(\frac{\partial f}{\partial y'} \right) = 0. \qquad (11)$$

This is Euler's equation for the function y which must be satisfied if (2) is to have a stationary value. There are several cases in which Euler's equation may be simplified, as we shall now see.

Case 1. If f is explicitly independent of y then $\partial f / \partial y = 0$ and (11) becomes

$$\frac{d}{dx} \left(\frac{\partial f}{\partial y'} \right) = 0 \qquad (12)$$

or

$$\frac{\partial f}{\partial y'} = c, \tag{13}$$

where c is an arbitrary constant.

Case 2. If f is explicitly independent of y', then $\partial f/\partial y' = 0$ and (11) becomes

$$\frac{\partial f}{\partial y} = 0. \tag{14}$$

Case 3. If f is explicitly independent of x, then using the identity

$$\frac{df}{dx} = \frac{\partial f}{\partial x} + y'\frac{\partial f}{\partial y} + y''\frac{\partial f}{\partial y'} \tag{15}$$

(11) may be written as

$$\frac{d}{dx}\left(f - y'\frac{\partial f}{\partial y'}\right) = 0, \tag{16}$$

or

$$f - y'\frac{\partial f}{\partial y'} = c, \tag{17}$$

where c is an arbitrary constant.

The following examples illustrate the use of Euler's equation.

Example 1. To find the stationary value of

$$I = \int_A^B \left\{\left(\frac{dy}{dx}\right)^2 + 2yx - y^2\right\} dx, \tag{18}$$

where $A = (0, 0)$, $B = (\pi/2, \pi/2)$ in the xy-plane, we use Euler's equation (11) to obtain

$$2x - 2y - \frac{d}{dx}\left(2\frac{dy}{dx}\right) = 0 \tag{19}$$

or

$$\frac{d^2y}{dx^2} + y = x. \tag{20}$$

Solving (20) we find

$$y = x + C \cos x + D \sin x, \qquad (21)$$

where C and D are arbitrary constants. However, in order that this curve should pass through the two end-points $A(0, 0)$, $B(\pi/2, \pi/2)$ we find that

$$C = D = 0. \qquad (22)$$

Consequently the function which makes (18) stationary is

$$y = x. \qquad (23)$$

The stationary value of I, say I_s, is now obtained by inserting (23) into (18) and integrating. This gives

$$I_s = \int_{x=0}^{x=\pi/2} (1 + 2x^2 - x^2)\, d \qquad (24)$$

$$= \left[x + \frac{x^3}{3} \right]_0^{\pi/2} = \frac{\pi}{2}\left(1 + \frac{\pi^2}{12} \right). \qquad (25)$$

Example 2. We now return to the problem of the minimum surface area discussed in 25.1. To find the function y which produces a stationary value of

$$S = 2\pi \int_{x_1}^{x_2} y \sqrt{\left(1 + \left(\frac{dy}{dx} \right)^2 \right)}\, dx, \qquad (26)$$

we first notice that the integrand is explicitly independent of x. Consequently Euler's equation may be used in the form (see (17)).

$$f - y' \frac{\partial f}{\partial y'} = c, \quad \left(y' = \frac{dy}{dx} \right). \qquad (27)$$

Since $f(y, y') = y\sqrt{(1 + y'^2)}$, (27) becomes

$$y\sqrt{(1 + y'^2)} - \frac{yy'^2}{\sqrt{(1 + y'^2)}} = c, \qquad (28)$$

which simplifies to

$$c \frac{dy}{dx} = \pm\sqrt{(y^2 - c^2)}. \qquad (29)$$

The solution of this equation is

$$y = c \cosh\left(\frac{x}{c} + d\right),\qquad(30)$$

where c and d are constants to be found by requiring (30) to pass through the end-points (x_1, y_1), (x_2, y_2). This curve is called a catenary.

PROBLEMS 25

1. Find the stationary value of the integral

$$\int_A^B \left\{ 2y \sin x + \left(\frac{dy}{dx}\right)^2 \right\} dx,$$

where A and B are the points $(0, \pi)$ and $(\pi, 0)$, respectively in the xy-plane. (L.U.)

2. Find the solution of the Euler equation for the integral

$$I = \int_A^B \left\{ \left(\frac{dy}{dx}\right)^2 \left(1 + \frac{dy}{dx}\right)^2 \right\} dx,$$

where $A = (0, 0)$ and $B = (1, 2)$ in the xy-plane.

3. Solve the Euler equation for the integral

$$\int_A^B \left\{ 12xy + \left(\frac{dy}{dx}\right)^2 \right\} dx,$$

where $A = (0, 0)$ and $B = (1, 1)$.

4. Solve the Euler equation for the integral

$$\int_0^{\pi/2} \left\{ \left(\frac{dy}{dx}\right)^2 + 2xy\left(\frac{dy}{dx}\right) \right\} dx$$

given $y = 0$ when $x = 0$, and $y = 1$ when $x = \pi/2$.

CHAPTER 26

Difference Equations

26.1 Definitions

Suppose $u_1, u_2, u_3, ..., u_n$ are the terms of a sequence. We now introduce a difference operator Δ such that the result of operating with Δ on u_r is defined by

$$\Delta u_r = u_{r+1} - u_r, \tag{1}$$

where $r = 1, 2, 3, ..., (n-1)$. Hence

$$\Delta u_1 = u_2 - u_1, \tag{2}$$

$$\Delta u_2 = u_3 - u_2, \tag{3}$$

and so on. These expressions are usually called the first finite differences of $u_1, u_2, u_3, ..., u_n$. In the same way the second finite differences $\Delta^2 u_r$ are defined by again operating with Δ such that

$$\Delta^2 u_r = \Delta(\Delta u_r) = \Delta(u_{r+1} - u_r) = (u_{r+2} - u_{r+1}) - (u_{r+1} - u_r)$$

$$= u_{r+2} - 2u_{r+1} + u_r. \tag{4}$$

Consequently

$$\Delta^2 u_1 = u_3 - 2u_2 + u_1, \tag{5}$$

$$\Delta^2 u_2 = u_4 - 2u_3 + u_2, \tag{6}$$

$$............................$$

$$............................$$

Higher order finite differences $\Delta^3 u_r, \Delta^4 u_r ..., \Delta^m u_r$ may be defined in a similar way.

Any equation which expresses a relation between finite differences $\Delta u_r, \Delta^2 u_r, ..., \Delta^m u_r$ is called a difference equation, the order of the equation being equal to the highest order finite difference term contained in it. For example,

$$\Delta u_r - 5u_r = 3, \tag{7}$$

$$\Delta^2 u_r - 3\Delta u_r + 2u_r = r^2, \tag{8}$$

$$\Delta^3 u_r + r\Delta u_r + 2^r u_r = 0, \tag{9}$$

493

are examples of difference equations which are respectively of the first, second and third orders. The general mth order linear difference equation has the form

$$a_0 \Delta^m u_r + a_1 \Delta^{m-1} u_r + a_2 \Delta^{m-2} u_r + \ldots + a_{m-1} \Delta u_r + a_m u_r = f(r), \quad (10)$$

where $f(r)$ and $a_0, a_1, a_2, \ldots, a_m$ are given functions of r. By analogy with the terminology of ordinary differential equations (10) is said to be homogeneous if $f(r) = 0$ and to be of the constant coefficient type if $a_0, a_1, a_2, \ldots, a_m$ are numbers independent of r. We see immediately that (7) and (8) are constant coefficient equations, whilst (9) is a homogeneous equation with variable coefficients.

Difference equations may always be written in a form involving successive values of u_r by substituting the definitions of $\Delta u_r, \Delta^2 u_r \ldots$.

For example, using (1), equation (7) becomes

$$u_{r+1} - 6u_r = 3, \quad (11)$$

whilst, using (1) and (4), (8) becomes

$$(u_{r+2} - 2u_{r+1} + u_r) - 3(u_{r+1} - u_r) + 2u_r = r^2 \quad (12)$$

or

$$u_{r+2} - 5u_{r+1} + 6u_r = r^2. \quad (13)$$

When written in this form difference equations are often called recurrence relations. A first order difference equation is therefore a recurrence relation between two successive values of the sequence, namely u_r and u_{r+1}, whilst a second order difference equation is a relation between three successive values u_r, u_{r+1} and u_{r+2}.

26.2 Formation of Difference Equations

The following examples illustrate the formation of difference equations from expressions defining u_r as a function of r.

Example 1. If

$$u_r = A4^r, \quad (14)$$

where A is an arbitrary constant, then

$$u_{r+1} = A4^{r+1} = 4u_r. \quad (15)$$

Hence we see that the elimination of the arbitrary constant A leads to a first order difference equation. In general if the expression for

494

u_r contains m arbitrary constants it must satisfy an mth order difference equation. This is illustrated further by the next example.

Example 2. If

$$u_r = A + B3^r, \tag{16}$$

where A and B are arbitrary constants, then

$$u_{r+1} = A + B3^{r+1}, \tag{17}$$

and

$$u_{r+2} = A + B3^{r+2}. \tag{18}$$

Hence eliminating B from (17) and (18) using (16) we have

$$u_{r+1} = A + 3^{r+1}\left\{\frac{u_r - A}{3^r}\right\} = 3u_r - 2A, \tag{19}$$

and

$$u_{r+2} = A + 3^{r+2}\left\{\frac{u_r - A}{3^r}\right\} = 9u_r - 8A. \tag{20}$$

Finally eliminating A from (19) and (20) we get

$$u_{r+2} - 4u_{r+1} + 3u_r = 0. \tag{21}$$

Consequently (16) which contains two arbitrary constants must satisfy a second order difference equation.

It should now be clear that, as with ordinary differential equations, the general solution of an mth order linear difference equation must contain m arbitrary constants.

26.3 Solution of Difference Equations

The first type of equation we discuss here is the homogeneous linear first-order equation

$$u_{r+1} - a_r u_r = 0, \tag{22}$$

where a_r is a given function of r.

By putting $r = 1, 2, 3, \ldots$ in (22) we have

$$u_2 = a_1 u_1, \tag{23}$$

$$u_3 = a_2 u_2 = a_2 a_1 u_1, \tag{24}$$

$$u_4 = a_3 u_3 = a_3 a_2 a_1 u_1, \tag{25}$$

and so on, giving finally

$$u_r = a_{r-1} a_{r-2} \ldots a_3 a_2 a_1 u_1. \tag{26}$$

Hence, if u_1 has a known value, say c, the solution of (22) may be written as

$$u_r = c \prod_{p=1}^{r-1} a_p. \tag{27}$$

To solve the inhomogeneous linear first-order equation

$$u_{r+1} - a_r u_r = b_r, \tag{28}$$

where a_r and b_r are given functions of r, we define the general solution of (28) as

$$u_r = v_r + w_r, \tag{29}$$

where v_r is the general solution of the homogeneous equation

$$v_{r+1} - a_r v_r = 0, \tag{30}$$

and where w_r is any particular solution of the inhomogeneous equation

$$w_{r+1} - a_r w_r = b_r. \tag{31}$$

(This procedure is the same as that adopted in the solution of ordinary linear differential equations; see Chapter 21, 21.4.) The following examples illustrate these methods.

Example 3. To solve the homogeneous equation

$$u_{r+1} - 4^r u_r = 0, \quad \text{given } u_1 = 2, \tag{32}$$

we write

$$u_2 = 4 u_1, \tag{33}$$

$$u_3 = 4^2 . u_2 = 4^2 . 4 . u_1, \tag{34}$$

$$u_4 = 4^3 . u_3 = 4^3 . 4^2 . 4 u_1, \tag{35}$$

and so on.

Consequently

$$u_r = 4^{r-1}4^{r-2} \ldots 4^3 \cdot 4^2 \cdot 4 \cdot u_1$$

$$= 2 \prod_{p=1}^{p=r-1} 4^p = 2(4^{[r(r-1)]/2}). \tag{36}$$

Example 4. To solve the inhomogeneous equation

$$u_{r+1} - 2u_r = r \tag{37}$$

we use (29) to obtain the homogeneous equation

$$v_{r+1} - 2v_r = 0, \tag{38}$$

and the inhomogeneous equation

$$w_{r+1} - 2w_r = r. \tag{39}$$

The general solution of (38) is, by (27),

$$v_r = v_1 2^{r-1}, \tag{40}$$

where v_1 is arbitrary. Hence the solution of (37) now depends on our ability to find any one solution of (39). To do this we adopt a trial solution of the form

$$w_r = \alpha + \beta r, \tag{41}$$

where α and β are to be determined such that (41) satisfies (39) for all r. Substituting (41) in (39) we find

$$(\alpha + \beta(r+1)) - 2(\alpha + \beta r) = r, \tag{42}$$

which gives comparing coefficients

$$\beta - \alpha = 0, \quad \text{(coefficient of } r^0\text{)},$$

$$-\beta = 1, \quad \text{(coefficient of } r\text{)}.$$

Hence $\alpha = \beta = -1$ and therefore

$$w_r = -(1+r). \tag{43}$$

497

Consequently the solution of (37) is

$$u_r = v_1 2^{r-1} - (1+r). \tag{44}$$

If in addition we are given that $u_1 = c$, then from (44)

$$c = v_1 - 2,$$

and therefore

$$u_1 = (2+c)2^{r-1} - (1+r). \tag{45}$$

As with inhomogeneous differential equations we see that the solution of an inhomogeneous difference equation depends essentially on choosing a sensible trial function for the particular integral. With experience this choice is easily made for simple equations, and although an alternative method exists which does not require this trial and error approach it is not of any great importance here.

We now come to the second and last type of difference equation to be discussed here. This is the second-order constant coefficient equation

$$a_0 u_{r+2} + a_1 u_{r+1} + a_1 u_r = f(r), \tag{46}$$

where a_0, a_1, a_2 are numbers independent of r and $f(r)$ is a given function of r. As with the first-order equation, the general solution of (46) is defined as

$$u_r = v_r + w_r, \tag{47}$$

where v_r is the general solution of the homogeneous equation

$$a_0 v_{r+2} + a_1 v_{r+1} + a_2 v_r = 0, \tag{48}$$

and w_r is any solution of the inhomogeneous equation

$$a_0 w_{r+2} + a_1 w_{r+1} + a_2 w_r = f(r). \tag{49}$$

Homogeneous equations of the type (48) may always be solved by following the methods of the next three examples.

Example 5. To solve

$$u_{r+2} - 5u_{r+1} + 6u_r = 0 \tag{50}$$

we put $u_r = x^r$ to get

$$x^{r+2} - 5x^{r+1} + 6x^r = 0 \tag{51}$$

or

$$x^r(x^2 - 5x + 6) = 0. \tag{52}$$

Hence, assuming $x \neq 0$, (50) reduces to a quadratic equation in x whose roots are $x = 2$ and $x = 3$.

Consequently

$$u_r = 2^r \quad \text{and} \quad u_r = 3^r \tag{53}$$

are both possible solutions of (50). The general solution is now taken as an arbitrary linear combination of these solutions with the form

$$u_r = A2^r + B3^r, \tag{54}$$

where A and B are arbitrary constants. This is consistent with the condition that the solution of a second-order equation must contain two arbitrary constants. A and B are, of course, determined once particular values are given to any two terms of the sequence u_r, say u_1 and u_2.

Example 6. To solve

$$u_{r+2} - 4u_{r+1} + 5u_r = 0 \tag{55}$$

we again put $u_r = x^r$ and obtain the equation

$$x^2 - 4x + 5 = 0. \tag{56}$$

Since (56) gives $x = 2 \pm i$, we write

$$\left. \begin{array}{l} 2 + i = Re^{i\theta}, \\ 2 - i = Re^{-i\theta}, \end{array} \right\} \tag{57}$$

whence

$$R = \sqrt{5}, \quad \sin \theta = \frac{1}{\sqrt{5}}, \quad \cos \theta = \frac{2}{\sqrt{5}}. \tag{58}$$

Consequently the two possible solutions of (55) are

$$u_r = (\sqrt{5}e^{i\theta})^r = 5^{r/2} (\cos r\theta + i \sin r\theta), \tag{59}$$

and

$$u_r = (\sqrt{5}e^{-i\theta})^r = 5^{r/2} (\cos r\theta - i \sin r\theta). \tag{60}$$

The general solution of (55) is therefore an arbitrary linear combination of (59) and (60) with the form

$$u_r = A5^{r/2}(\cos r\theta + i \sin r\theta) + B5^{r/2}(\cos r\theta - i \sin r\theta), \qquad (61)$$

where A and B are arbitrary constants.

This is more conveniently written as

$$u_r = 5^{r/2}\{C \cos r\theta + D \sin r\theta\}, \qquad (62)$$

where C and D are new arbitrary constants.

Example 7. In some cases, as with the equation

$$u_{r+2} - 6u_{r+1} + 9u_r = 0, \qquad (63)$$

the substitution of $u_r = x^r$ leads to a quadratic equation with equal roots. For instance (63) becomes

$$x^2 - 6x + 9 = 0, \qquad (64)$$

which gives $x = 3$ (twice).

Hence we only obtain one independent solution of (63), namely $u_r = 3^r$. Consequently as with differential equations we attempt to find another solution by writing

$$u_r = v_r 3^r, \qquad (65)$$

where v_r is a function of r.

Substituting in (63) we find

$$v_{r+2} - 2v_{r+1} + v_r = 0, \qquad (66)$$

whose solution is easily seen to be

$$v_r = Ar + B, \qquad (67)$$

where A and B are arbitrary constants.

Hence the general solution of (63) is (using (65))

$$u_r = (Ar + B)3^r, \qquad (68)$$

which contains the correct number of arbitrary constants.

In fact, the reader will easily verify that if a homogeneous second-order difference equation leads to a quadratic equation with equal roots, say $x = c$, then the general solution is always

$$u_r = (Ar+B)c^r. \tag{69}$$

(The same factor arises in similar circumstances in solving ordinary linear second-order differential equations with constant coefficients.)

Finally the following example illustrates the method of solving an inhomogeneous second-order equation with constant coefficients.

Example 8. To solve

$$u_{r+2}-4u_{r+1}+3u_r = 1, \tag{70}$$

we write $u_r = v_r+w_r$ as in (47) and obtain

$$v_{r+2}-4v_{r+1}+3v_r = 0, \tag{71}$$

and

$$w_{r+2}-4w_{r+1}+3w_r = 1. \tag{72}$$

Following Example 5, the solution of (71) is easily seen to be

$$v_r = A3^r+B1^r = A3^r+B. \tag{73}$$

The solution of (72) again requires the use of a trial function. Putting

$$w_r = \alpha r, \tag{74}$$

and substituting in (72) and comparing coefficients, we find

$$\alpha = -\tfrac{1}{2}. \tag{75}$$

Hence one solution of (72) (which is all that is needed) is

$$w_r = -\frac{r}{2}. \tag{76}$$

Consequently the general solution of (70) is

$$u_r = A3^r+B-\frac{r}{2}. \tag{77}$$

As with inhomogeneous first-order equations, the correct choice of a trial function for the particular integral is largely a matter of commonsense and experience.

PROBLEMS 26

1. Solve the following equations subject to the given initial conditions

   (a) $u_{r+2} - 5u_{r+1} + 6u_r = 0$, $u_1 = 1$, $u_2 = 3$.

   (b) $u_{r+2} - 2u_{r+1} + 4u_r = 0$, $u_1 = 1$, $u_2 = 2$.

   (c) $4u_{r+2} - 4u_{r+1} + u_r = 2^r$.

   (d) $u_{r+1} - 3u_r = 4^r + r$.

2. Show that $u_r = 4^r/6$ is a solution of the difference equation

$$u_{r+3} - 6u_{r+2} + 11u_{r+1} - 6u_r = 4^r,$$

and find the general solution.

3. Show that the general solution of the recurrence relation

$$u_n - 2(\cosh \alpha)u_{n-1} + u_{n-2} = 0$$

may be written as

$$u_n = Ae^{(n+1)\alpha} + Be^{-(n+1)\alpha},$$

where A and B are arbitrary constants.

Prove that the nth order determinant

$$\Delta_n = \begin{vmatrix} k & 1 & 0 & 0 & . & . & . \\ 1 & k & 1 & 0 & . & . & . \\ 0 & 1 & k & 1 & . & . & . \\ . & . & . & . & . & . & . \\ . & . & . & . & 1 & k & 1 \\ . & . & . & . & 0 & 1 & k \end{vmatrix}$$

(where $k = 2 \cosh \alpha$) satisfies this recurrence relation.

Prove that

$$\Delta_n = \frac{\sinh (n+1)\alpha}{\sinh \alpha}. \tag{C.U.}$$

4. If $u_0, u_1, u_2, \ldots$ satisfy the recurrence relation

$$u_{r+2} - 2\alpha u_{r+1} + (\alpha^2 + \lambda^2)u_r = 0,$$

$(r = 0, 1, \ldots)$ and $u_0 = 0$, $u_1 = 1$, prove that, for $\lambda > 0$,

$$u_n = \frac{1}{\lambda}(\alpha^2 + \lambda^2)^{n/2} \sin \left(n \tan^{-1} \frac{\lambda}{\alpha} \right).$$

Find the corresponding formula for u_n, subject to the same initial conditions, when $\lambda = 0$. (L.U.)

502

CHAPTER 27

Simple Integral Equations

27.1 Types of Integral Equations

In certain problems it is not unusual to meet equations in which the unknown function occurs under an integral sign. Such equations are called integral equations and are usually difficult to solve analytically. Two important types of integral equation which we shall discuss here are (a) the linear integral equation of the first kind defined by

$$f(x) = \int_a^b K(x, y)u(y)\, dy, \tag{1}$$

and (b) the linear integral equation of the second kind defined by

$$u(x) = f(x) + \lambda \int_a^b K(x, y)u(y)\, dy, \quad (\lambda = \text{const.}). \tag{2}$$

In both these equations u is the unknown function, whilst the functions $f(x)$, $K(x, y)$ and the limits of integration a and b are assumed known. The function $K(x, y)$ is called the kernel of the equation.

Various special forms of (1) and (2) are known by particular names. For example, if a and b are constants (2) is called Fredholm's equation and is said to be homogeneous if $f(x) = 0$, whilst if $a = 0$, $b = x$ (i.e. variable limit of integration) (2) is known as Volterra's equation (again homogeneous if $f(x) = 0$). If however, in (1) or (2) either a or b or both are infinite, or if the kernel $K(x,y)$ becomes infinite in the range of integration, the equation is said to be singular, and special techniques are required for its solution.

It is not possible in this book to enter into the general theory of these various types of equations but the following examples illustrate a few methods of obtaining exact solutions of some simple equations The reader should realise, however, that these simple methods are necessarily restrictive and that when faced with an integral equation in practice a numerical approach is usually the best, if not the only, way of obtaining a solution.

27.2 Some Methods of Solution

Example 1. When the kernel $K(x, y)$ can be written as

$$K(x, y) = g(x)h(y), \tag{3}$$

where $g(x)$ and $h(y)$ are respectively functions of x only and y only, the Fredholm equation may be solved in a simple manner. This follows by putting (3) into (2) and remembering that $g(x)$ is a constant as far as the y-integration is concerned; consequently it may be taken outside the integral sign to give

$$u(x) = f(x) + \lambda g(x) \int_a^b h(y)u(y)\, dy. \tag{4}$$

Hence since

$$\int_a^b h(y)u(y)\, dy = C(= \text{const.}), \tag{5}$$

we have the solution

$$u(x) = f(x) + \lambda C g(x). \tag{6}$$

The value of C may now be obtained by putting (6) into (5).

As an example of this method, we consider the equation

$$u(x) = \cosh x - \frac{x}{2} + \frac{1}{3}\int_0^1 xyu(y)\, dy \tag{7}$$

which clearly has the solution

$$u(x) = \cosh x - \frac{x}{2} + \frac{Cx}{3}, \tag{8}$$

where

$$C = \int_0^1 yu(y)\, dy. \tag{9}$$

Hence from (8) and (9) we have

$$C = \int_0^1 y\left(\cosh y - \frac{y}{2} + \frac{Cy}{3}\right) dy, \tag{10}$$

which after some simplification gives

$$C = \frac{15}{16} - \frac{9e^{-1}}{8}.$$ (11)

Consequently

$$u(x) = \cosh x - \frac{x}{2} + \left(\frac{5}{16} - \frac{3e^{-1}}{8}\right)x$$ (12)

$$= \cosh x - \frac{3}{16}(1 + 2e^{-1})x.$$ (13)

Example 2. The homogeneous Fredholm equation

$$u(x) = \lambda \int_0^{\pi/2} \sin x \sin y u(y)\, dy$$ (14)

only has a solution for a particular value of λ. To find λ (the eigen-value) and the corresponding solution for u (the eigenfunction) we write

$$u(x) = \lambda \sin x \int_0^{\pi/2} \sin y u(y)\, dy = C\lambda \sin x,$$ (15)

where

$$C = \int_0^{\pi/2} \sin y u(y)\, dy.$$ (16)

Hence putting (15) into (16) we have

$$C = C\lambda \int_0^{\pi/2} \sin^2 y\, dy = \frac{C\pi\lambda}{4}$$ (17)

which, assuming $C \neq 0$, gives $\lambda = 4/\pi$. The solution corresponding to this value of λ is therefore (from (15))

$$u(x) = A \sin x,$$ (18)

where A is an arbitrary constant.

Example 3. The Volterra equation can sometimes be transformed into an ordinary differential equation which may be easier to solve than the integral equation. For example, consider the equation

$$u(x) = 2x + 4\int_0^x (y-x)u(y)\,dy. \tag{19}$$

Then differentiating with respect to x using the methods of Chapter 10, 10.2 (equation (16)), we have

$$\frac{d}{dx}u(x) = 2 + 4\left[\left\{(y-x)u(y)\right\}_{y=x} - \int_0^x u(y)\,dy\right], \tag{20}$$

$$= 2 - 4\int_0^x u(y)\,dy. \tag{21}$$

Differentiating again we get

$$\frac{d^2}{dx^2}u(x) = -4u(x), \tag{22}$$

which has the solution

$$u(x) = A\cos 2x + B\sin 2x, \tag{23}$$

where A and B are constants. The values of these constants may now be obtained by putting (23) into (19) and integrating. In this way we find $A = 0$ and $B = 1$. Consequently from (23) the solution of (19) is

$$u(x) = \sin 2x. \tag{24}$$

Example 4. The Laplace transformation (see Chapter 23) and the convolution theorem discussed in Chapter 23, Problem 5, are useful in solving the Volterra equation

$$u(x) = f(x) + \lambda \int_0^x K(x, y)u(y)\,dy \tag{25}$$

when the kernel has the form

$$K(x, y) = g(x-y). \tag{26}$$

For taking the Laplace transformation of (25) with a kernel given by (26) and using the convolution theorem

$$L\left\{\int_0^x g(x-y)u(y)\,dy\right\} = \bar{g}(p)\bar{u}(p), \tag{27}$$

we have

$$\bar{u}(p) = \bar{f}(p) + \lambda \bar{g}(p)\bar{u}(p), \tag{28}$$

or

$$\bar{u}(p) = \frac{\bar{f}(p)}{1 - \lambda \bar{g}(p)}. \tag{29}$$

Consequently inverting

$$u(x) = L^{-1} \left\{ \frac{\bar{f}(p)}{1 - \lambda \bar{g}(p)} \right\}. \tag{30}$$

In some simple equations this inversion is easily performed. For example, if

$$u(x) = x + \int_0^x u(y) \sin(x - y) \, dy \tag{31}$$

then, by (28),

$$\bar{u}(p) = \frac{1}{p^2} + \frac{\bar{u}(p)}{p^2 + 1}, \tag{32}$$

(since $L\{x\} = 1/p^2$ and $L\{\sin x\} = 1/[p^2 + 1]$). Hence

$$\bar{u}(p) = \frac{1}{p^2} + \frac{1}{p^4}. \tag{33}$$

This is easily inverted by inspection to give

$$u(x) = x + \frac{x^3}{6} \tag{34}$$

as the solution of (31).

PROBLEMS 27

1. Solve the equation

$$u(x) = 1 + \int_0^x u(y) \, dy.$$

2. By separating the kernel solve the equation

$$u(x) = \sin x + \int_0^\infty e^{-\alpha(x+y)} u(y) \, dy,$$

where α is a real constant, and show that the solution is not valid when $\alpha = 0$ and $\alpha = \frac{1}{2}$.

3. Obtain the eigenvalues and eigenfunctions of the following equations

(a) $u(x) = \lambda \int_{-1}^1 (x-y)^2 u(y) \, dy,$

(b) $u(x) = \lambda \int_{-\pi}^\pi \cos^2 (x+y) u(y) \, dy.$

4. By using the Laplace transformation and the convolution theorem solve the equation

$$u(x) = \sin 2x + \int_0^x u(x-y) \sin y \, dy.$$

Answers to Problems

Problems 1

1. Yes.

2. $(3x+1)/(4-x)$, $(7x-3)/(x+7)$.

4. $\dfrac{x^3-x^2+x-12}{-x^3+5x^2-2x-8}$, $x=-1,4,2$.

5. (a) odd, (b) neither, (c) even, (d) odd, (e) odd.

6. $\cosh ax \cos bx$, $-\sinh ax \cos bx$.
   $1/\sqrt{(1-x^2)}$, $-x/\sqrt{(1-x^2)}$.

7. (a) π, (b) not periodic, (c) π, (d) $\pi/2$, (e) $2\pi/\omega$, (f) π.

9. (a) increasing, (b) not, (c) not, (d) increasing, (e) increasing for $x<0$, decreasing for $x>0$.

Problems 2

1. (a) $x<1, x>2$, (b) all x, (c) no x, (d) $x<-2, x>5$,
   (e) $-5<x<\tfrac{1}{2}$.

10. Corners at $(1,\tfrac{3}{2})$, $(1,-1)$.

Problems 3

1. (a) 1, (b) $\tfrac{1}{2}$, (c) 1, (d) $1/(2\sqrt{2})$, (e) 0, (f) $\tfrac{1}{2}$, (g) 1, (h) 0, (i) $\tfrac{1}{2}$.

2. (a) $x=2,4$. (b) $x=n\pi/2, n=\pm1,\pm3,\pm5\dots$, (c) $x=0$.

3. $-1/[(1+x)\sqrt{(1-x^2)}]$, $-x/\sqrt{(a^2-x^2)}$.

4. (a) $(1/x^2)\tan(1/x)$,   (b) $6xe^{3x^2}$,   (c) $1/[(x+1)(2x+1)^{\frac{1}{2}}]$,
   (d) $e^{\sin^2 x}\sin 2x$, (e) $-\sin x \cos(\cos x)$, (f) $2x\tan[(\pi/4)-x^2]$,
   (g) $x^{\cos x}\{[(\cos x)/x]-\sin x \log_e x\}$, (h) $x^x(1+\log_e x)$,

   (i) $a^x\log_e a$,   (j) $\dfrac{2x}{\sin^{-1}x^2.(\sqrt{(1-x^4)})}$.

5. (a) $\dfrac{(-1)^n a^n (2n)!(ax+b)^{-(n+\frac{1}{2})}}{2^{2n}n!}$, (b) $24(-1)^{n+1}(n-5)!x^{4-n}, (n \geqslant 5)$

7. Twice.

13. (a) Function does not exist for real x.
    (b) Minimum value $= 1$ at $x = 0$.

14. Maximum at $x = 2$, minimum at $x = 0$.

Problems 4

1. (a) $\sin x - x \cos x$,   (b) $\frac{1}{3}x^3 + \frac{1}{3}\log_e (x^3 - 1)$,
   (c) $\frac{1}{4}\log_e [(1 + 2\tan \frac{1}{2}x)/(2 - \tan \frac{1}{2}x)]$,   (d) $\frac{1}{2}(x^2 - 1)e^{x^2}$,
   (e) $-1/\sqrt{(1 + 2ax + x^2)}$,   (f) $x\cos^{-1} x - \sqrt{(1 - x^2)}$,
   (g) $\log_e [(x - 2)/(x - 1)]$,   (h) $2\sqrt{x} - 2\tan^{-1} \sqrt{x}$.

2. (a) $\pi/4$, (b) $\alpha/(\sin \alpha)$, (c) $\frac{1}{2}$, (d) $\frac{1}{2}\log_e \sqrt{\frac{5}{3}}$, (e) $\frac{2}{3}$, (f) $(\pi^2/16) + \frac{1}{4}$,
   (g) $\frac{4}{3}$, (h) $\sqrt{3} - 1$, (i) $4 - 2\sqrt{e}$.

3. 1.

6. $\pi/3$, $\frac{3}{4} + \log_e 2$.

7. $(-1)^n [b/(a^2 + b^2)](e^{-n\pi a/b} + e^{-(n+1)\pi a/b})$.

10. (a) $\frac{5}{16} > I > \frac{5}{25}$,   (b) $\pi/4 < I < \frac{1}{8}\pi\sqrt{5}$,
    (c) $16\pi^2/45 > I > 10\pi^2/45$, (d) $\pi^2/8 > I > 0$, (e) $0.29 < I < 0.80$.

11. divergent, $-\frac{3}{4}$, $\frac{1}{2}$.

12. $\alpha < -1$; $\alpha > 0$, $\beta > 0$; $|\alpha| < 2$.

Problems 5

1. (a) convergent, (b) divergent, (c) divergent, (d) convergent,
   (e) convergent for $p > 1$, divergent for $0 \leq p \leq 1$, (f) divergent,
   (g) convergent for $0 < x < 1$, divergent for $x \geq 1$, (h) divergent,
   (i) divergent, (j) divergent, (k) convergent.

2. divergent, conditionally convergent, divergent.

4. $0 < x < 2$.

9. $2\left(x + \frac{1}{2}\frac{x^3}{3} + \frac{1 \cdot 3}{2 \cdot 4}\frac{x^5}{5} + \ldots\right)$.

14. (i) absolutely convergent,   (ii) conditionally convergent,
    (iii) divergent.

15. Convergent for $p > 1$, divergent for $p \leq 1$.

Problems 6

1. (a) $1 + \frac{1}{2}x - \frac{1}{8}x^2 + \ldots$, $|x| \leq 1$.

(b) $x - \dfrac{x^3}{3} + \dfrac{x^5}{5} - \ldots, \quad x^2 < 1.$

(c) $x + \dfrac{x^2}{2} + \dfrac{x^3}{3} + \ldots, \quad x^2 < 1, x = -1.$

(d) $1 + \dfrac{x^2}{2} + \dfrac{5x^4}{24} + \ldots, \quad x^2 < \dfrac{\pi^2}{4}.$

(e) $1 - \dfrac{x^2}{2} + \dfrac{5x^4}{24} - \ldots, \quad \text{all } x.$

(f) $x + x^2 + \dfrac{2x^3}{3} + \ldots, \quad x^2 < 1.$

3. $1 + \dfrac{1}{1!}\left(\dfrac{x-1}{2}\right) - \dfrac{1}{2!}\left(\dfrac{x-1}{2}\right)^2 + \dfrac{1 \cdot 3}{3!}\left(\dfrac{x-1}{2}\right)^3 - \ldots,$

4. $x^2 - x^3 + 1\frac{1}{12}x^4$, (i) maximum at $x = 0$, (ii) no limit at $x = 0$.

5. (a) $\frac{1}{2}$, (b) 2, (c) 0, (d) $-\pi$, (e) finitely oscillating, (f) 1, (g) 1.

8. $\sin 31° \simeq 0 \cdot 5150.$

10. $y = 1 + \dfrac{x^3}{2 \cdot 3} + \dfrac{x^6}{2 \cdot 3 \cdot 5 \cdot 6} + \ldots.$

Problems 7

1. $z_1 + z_2 = 2 + 3i$, $z_1 - z_2 = 4 + i$, $z_1 z_2 = -5 + i$, $z_1/z_2 = -\frac{1}{2} - \frac{3}{2}i$, $z_1 \bar{z}_1 = 13$, $\bar{z}_2 z_2 = 2.$

2. (a) $(12 + 5i)/13$,   (b) $0 + i(\pi/6)$,

    (c) $e^{-(\pi y/4)}\{\cos(\log_e 2^{y/2}) + i \sin(\log_e 2^{y/2})\}$,   (d) $0 - i$,

    (e) $-\frac{1}{2} - \frac{1}{2}i\sqrt{3}$,   (f) $\dfrac{\sin x \cos x + i \sinh y \cosh y}{\cosh^2 y \cos^2 x + \sinh^2 y \sin^2 x}.$

4. (a) $z = (\alpha + 1)/(\alpha - 1)$, where $\alpha = e^{(2n+1)\pi i/7}$, $n = 0, 1, 2, 3, 4, 5, 6$.

    (b) $z = 2^{\frac{1}{2}}e^{i\theta}$, where $\theta = (\pi/6) + (n\pi/2)$, $n = 0, 1, 2, 3$.

    (c) $z = e^2 e^{i\theta}$, where $\theta = (6n+1)\pi/3$, n integral or zero.

9. (a) $e^{\cos\theta} \sin(\sin\theta)$, (b) $\dfrac{4 - 2\cos\theta}{5 - 4\cos\theta}$, (c) $\dfrac{\sin\theta + \sin n\theta - \sin(n+1)\theta}{2(1 - \cos\theta)}.$

10. (a) $y + x = 0$, (b) $x^2 + y^2 + 2y \cot \alpha - 1 = 0$,
    (c) $x^2/(\sin^2 \alpha) - y^2/(\cos^2 \alpha) = 1$.

12. Region between line $u = \frac{3}{2}$ and the circle with $u = \frac{5}{3}$ to $u = 3$ as diameter ($w = u + iv$).

13. $\text{Log}\,(1+i) = \frac{1}{2} \log_e 2 + \frac{1}{4}\pi i$, $\quad \text{Log}\, i = \frac{1}{2}\pi i$,
    $(1+i)^i = e^{-[(\pi/4) + 2\pi n]}\{\cos (\log_e \sqrt 2) + i \sin (\log_e \sqrt 2)\}$, n integral.
    $i^i = e^{-(4n+1)\pi/2}$, n integral.

14. (a) $\left(\dfrac{\pi}{2} + 2\pi k\right) \pm i \log_e (2 + \sqrt 3)$,

    (b) $(2k+1)\dfrac{\pi}{2} + i \log_e \sqrt 3$, $k = 0$ or integral in each case.

15. $z_3 = (7 + 2\sqrt 3) + i(4 + 3\sqrt 3)$ or $(7 - 2\sqrt 3) + i(4 - 3\sqrt 3)$.
    Vertices $(7 + 2\sqrt 3) + i(4 + 3\sqrt 3)$, $(7 - 2\sqrt 3) + i(4 - 3\sqrt 3)$,
    $-2 + 10i$, $(1 + 2\sqrt 3) + i(8 + 3\sqrt 3)$, $(1 - 2\sqrt 3) + i(8 - 3\sqrt 3)$.

18. $\lambda = \pm 3$.

19. (a) analytic, (b) not analytic, (c) not analytic.

20. (a) $z = -1$, (b) $z = -3$, (c) $z = \pm i$.

Problems 8

4. (a) sech x, (b) sech x cosech x,
   (c) $\cos \sqrt x / \{2\sqrt{(x)}\sqrt{[1 + \sin^2 \sqrt{(x)}]}\}$, (d) $1/\{2(1-x)\sqrt{(2-x)}\}$,
   (e) $(\cosh x)^{\sin x}\{\cos x \log_e \cosh x + \sin x \tanh x\}$.

6. (a) -1, (b) 0, (c) e.

8. (a) $\frac{1}{2}(\cosh x \sin x - \sinh x \cos x) + c$,
   (b) $\tan^{-1} (\sinh x) + c\,(= 2\tan^{-1} e^x + c')$.
   (c) $\frac{7}{15}$, (d) $\frac{1}{2}e + \frac{5}{2}e^{-1} - 2$,
   (e) $\frac{1}{2}a\sqrt{(1+a^2)} \sinh^{-1} a + \frac{1}{4}(\sinh^{-1} a)^2 - \frac{1}{4}a^2$.

10. $x = 0$, $y = 2n\pi$.

Problems 9

3. (a) $e^{xyz}(1 + 3xyz + x^2y^2z^2)$, (b) $(2x-z)/(2x+z)^3$.

5. $nx^{n-1}y^{n-1}(bx \cos bt - ay \sin at)$.

6. $2xy + x - [1/(xy^2)]$.

7. $\partial V/\partial x = 2 + \log_e (x^2 + y^2)$, $\partial V/\partial y = -2 \tan^{-1} (y/x)$.

9. (a) $2xy + [1/(x+y)]$, (b) 0, (c) $2u(df/du)$, where $u = xy$.

17. $e^6\{1 + 3(x-2) + 2(y-3) + \frac{9}{2}(x-2)^2 + 7(x-2)(y-3)$
$$+ 2(y-3)^2 + ...\}.$$

18. $(0, 0)$, $(-1, -1)$.

19. (i) $(0, 0)$ saddle point, $(\frac{1}{6}, -\frac{1}{12})$ minimum.
(ii) $(\frac{1}{2}\sqrt{5}, 0, -\frac{1}{2})$, $(-\frac{1}{2}\sqrt{5}, 0, -\frac{1}{2})$.

20. $(0, 0)$ maximum, $(\frac{1}{3}, \frac{1}{3})$ saddle point.

22. $c^6/27$.

23. $(\frac{18}{13}, \frac{12}{13})$ minimum.

Problems 10

1. (a) $-e^{-x}/(2x)$, (b) $[4/(3x)][\sqrt{(1+8x^4)} - \sqrt{(1+x^4)}]$.

Problems 12

1. (a) $\frac{32}{3}$, (b) -2.

2. (i) $-\frac{278}{15}$, (ii) -19.

3. 1.

4. $3\pi a^2/4$.

5. $16(\pi - 2)$.

6. (i) 1, (ii) 1.

7. 1.

8. (i) $(\pi/4) - [\pi/(3\sqrt{3})]$, (ii) 4.

9. (a) $5\pi a^6/6$, (b) $\frac{1}{3}a^2b$.

10. $\pi/4$.

12. (i) $a^2/2$, (ii) $a^4/3$, (iii) $\frac{1}{35}$, (iv) πa^3.

14. $\pi b^2 a^3/16$.

15. (i) $\pi a^3/12$, (ii) $2 \log_e 2 - 1$.

16. $e^2 - 1$.

513

Problems 13

1. $\Gamma(4 \cdot 1) = 6 \cdot 810$, $\Gamma(-3 \cdot 9) = 0 \cdot 492$.

Problems 14

1. (a) $1 \cdot 178$, (b) $1 \cdot 060$, (c) $0 \cdot 658$, (d) $0 \cdot 028$, (e) $1 \cdot 202$, (f) $5 \cdot 403$.

2. $0 \cdot 597$.

3. $3 \cdot 54$.

4. $0 \cdot 1249$, (exact value $0 \cdot 1250$).

Problems 15

3. $\dfrac{1}{\pi}\left\{\dfrac{\sinh \pi}{2} + \sum_{r=1}^{\infty} \dfrac{(-1)^r \sinh \pi \cos rx}{1+r^2}\right.$
$$\left. + \sum_{r=1}^{\infty} \dfrac{[1-(-1)^r \cosh \pi] r \sin rx}{1+r^2}\right\}.$$

7. $\dfrac{2}{\pi} \sum_{r=1}^{\infty} \{1-(\pi+1)(-1)^r\} \dfrac{\sin rx}{r}$.

9. (i) $\sum_{r=1}^{\infty}\left(\dfrac{2 \sin (r\pi/2)}{\pi r^2} - \dfrac{1}{r}(-1)^r\right) \sin rx$.

(ii) $\dfrac{3\pi}{8} + \sum_{r=1}^{\infty} \dfrac{2}{\pi r^2}[\cos (r\pi/2)-1] \cos rx$.

12. $\dfrac{4}{\pi} \sum_{r=1}^{\infty}\left(\dfrac{\sin (r\pi/2)}{r^2}\right) \sin rx$.

Problems 16

1. $(\mu-\nu)(\nu-\lambda)(\lambda-\mu)(\lambda+\mu+\nu)$, -240.

3. $\lambda = 17$ and 5.

4. $\lambda = 0$, $\pm\sqrt{2}$.

8. (ii) (a) $x = 0, 0, -3$, (b) $x = 0$, three times.

9. $\partial(\xi, \eta)/\partial(x, y) = 1/[c^2(\cosh^2 \xi - \cos^2 \eta)]$.

10. $x = 5$, $y = 1$, $z = -5$.

12. $k = -\frac{1}{25}$ and 3. $x = 2$, $y = 1$, $z = -4$.

Problems 17

1. $A + B = \begin{pmatrix} -1 & 2 & 2 \\ 9 & 7 & 8 \\ 2 & -1 & 3 \end{pmatrix}$, $\quad A - B = \begin{pmatrix} 5 & 0 & 2 \\ -3 & 3 & 6 \\ 0 & 1 & -1 \end{pmatrix}$,

$B - A = \begin{pmatrix} -5 & 0 & -2 \\ 3 & -3 & -6 \\ 0 & -1 & 1 \end{pmatrix}$, $\quad AB = \begin{pmatrix} 2 & 2 & 5 \\ 28 & 6 & 19 \\ -2 & 0 & 2 \end{pmatrix}$,

$BA = \begin{pmatrix} -3 & 2 & 1 \\ 19 & 16 & 27 \\ 1 & -4 & -3 \end{pmatrix}$.

3. $\underline{A} = \begin{pmatrix} 1 & 1 & -3 \\ 1 & 0 & \frac{1}{2} \\ -3 & \frac{1}{2} & 2 \end{pmatrix}$, $\quad \overset{v}{A} = \begin{pmatrix} 0 & \frac{1}{2} & -2 \\ -\frac{1}{2} & 0 & \frac{1}{4} \\ 2 & -\frac{1}{4} & 0 \end{pmatrix}$.

7. $A^{-1} = \begin{pmatrix} \frac{1}{3} & -\frac{2}{3} & \frac{2}{3} \\ \frac{1}{6} & \frac{1}{6} & -\frac{1}{6} \\ 0 & 0 & \frac{1}{2} \end{pmatrix}$.

8. $A^{-1} = \begin{pmatrix} 1 & -2 & 1 & 0 \\ 0 & 1 & -2 & 1 \\ 0 & 0 & 1 & -2 \\ 0 & 0 & 0 & 1 \end{pmatrix}$.

9. (a) $x = 3$, $y = 1$, $z = 2$; (b) inconsistent.

16. 4, -1.

17. $\lambda^3 - 6\lambda^2 - 3\lambda + 18 = 0$.

18. $A = \pm \begin{pmatrix} 1 & \frac{1}{2} & 0 \\ 0 & 1 & 0 \\ 0 & -\frac{1}{4} & 1 \end{pmatrix}$.

19.

(a) $\lambda_1 = 1$, $X_1 = \begin{pmatrix} 1 \\ 0 \\ 0 \end{pmatrix}$; $\lambda_2 = 2$, $X_2 = \dfrac{1}{\sqrt{17}} \begin{pmatrix} 4 \\ 1 \\ 0 \end{pmatrix}$;

$\lambda_3 = 3$, $X_3 \dfrac{1}{\sqrt{989}} \begin{pmatrix} 29 \\ 12 \\ 2 \end{pmatrix}$.

(b) $\lambda_1 = 2$, $X_1 = \begin{pmatrix} 1 \\ 0 \\ 0 \end{pmatrix}$; $\lambda_2 = \lambda_3 = 1$, $X_2 = X_3 = \dfrac{1}{\sqrt{10}} \begin{pmatrix} -3 \\ 1 \\ 0 \end{pmatrix}$.

(c) $\lambda_1 = 0$, $X_1 = \dfrac{1}{\sqrt{2}} \begin{pmatrix} 1 \\ -1 \\ 0 \end{pmatrix}$; $\lambda_2 = 4$, $X_2 = \dfrac{1}{\sqrt{2}} \begin{pmatrix} 1 \\ 1 \\ 0 \end{pmatrix}$;

$\lambda_3 = 1$, $X_3 = \begin{pmatrix} 0 \\ 0 \\ 1 \end{pmatrix}$.

(d) $\lambda_1 = 0$, $X_1 = \dfrac{1}{\sqrt{3}} \begin{pmatrix} -(1+i) \\ 1 \end{pmatrix}$; $\lambda_2 = 3$, $X_2 = \dfrac{1}{\sqrt{6}} \begin{pmatrix} (1+i) \\ 2 \end{pmatrix}$.

Problems 19

1. $\mathbf{a} + 3\mathbf{b}$, $3\mathbf{b} - \mathbf{a}$, $2\mathbf{a} + 2\mathbf{b}$, $-5\mathbf{b} - \mathbf{a}$.

2. (i) $\mathbf{a} \cdot \mathbf{b} = 7$, $\mathbf{a} \wedge \mathbf{b} = 5\mathbf{i} - 3\mathbf{j} - \mathbf{k}$,
   (ii) $\mathbf{a} \cdot \mathbf{b} = 0$, $\mathbf{a} \wedge \mathbf{b} = -4\mathbf{i} + \mathbf{j} + 7\mathbf{k}$.

3. $3, 2\sqrt{2}, \sqrt{13}$. $\cos^{-1}(3/\sqrt{26})$, $\cos^{-1}[7/(3\sqrt{13})]$, $\cos^{-1}[1/(3\sqrt{2})]$.

5. $\mathbf{A} \cdot \mathbf{B} = \cos(\theta - \phi)$, $\mathbf{A} \wedge \mathbf{B} = \mathbf{k} \sin(\phi - \theta)$.

17. $R = \begin{pmatrix} u_1^2 - u_2^2 & 2u_1 u_2 \\ 2u_1 u_2 & u_2^2 - u_1^2 \end{pmatrix}$.

20. (a) $\mathbf{i}$, (b) $nr^{n-2}\mathbf{r}$, (c) $\mathbf{i} + \mathbf{j} + \mathbf{k}$, (d) $\mathbf{A}$.

21. 0.

22. (a) 3, (b) $yz + xz + yx$, (c) 0.

23. $2\pi a^2 \mathbf{k}$.

24. (a) 0, (b) 0, (c) $\mathbf{k}/(x+y)$, (d) $\mathbf{k}$.

Answers to Problems

Problems 20

2. $\frac{2}{3}$, 2, 6.

3. $k = 6$, $x = 1 \pm \sqrt{3}$, $1 \pm \sqrt{2}$.

5. -2, 3, 4, 6.

6. (a) $x = 2\sqrt{2} \cos \theta$, $\theta = \pi/12$, $7\pi/12$, $9\pi/12$.
   (b) $x = \frac{1}{2} \sec \theta$, $\theta = \pi/9$, $5\pi/9$, $7\pi/9$.

7. $x = 1$ twice, and $(1 \pm i\sqrt{3})/2$.

8. $x = 0.41$.

9. $x = 0.447$.

10. $\theta = -2.48$.

11. $x = 0.783$.

12. $x = 0.103$.

13. $x = 2.39$.

Problems 21

1. (a) $y = 2x + 1$,
   (b) $x^3(3y - 2) = 4$,
   (c) $y = 2x/(Cx^2 + 1)$,
   (d) $\frac{3}{5}(x + 2y) - \frac{4}{25} \log_e (5x + 10y + 13) = x + C$,
   (e) $y = [(\sin x)/x] - \cos x - (\pi/x)$.

2. (i) $y = \frac{1}{2}(x + 1)^5 + (x + 1)^3$,
   (ii) $y = (Ax + B)e^{-2x} + \frac{1}{4} + \frac{1}{8} \sin 2x$.

3. (i) $y = [x^2/(x + 1)](x^2 - 3x + C)$,
   (ii) $y = Ae^{-2x} + Be^{-x} + e^{-x}(\frac{1}{2}x^2 - x)$.

4. (i) $y = e^{3x}(A \cos x + B \sin x) + 2 - \frac{1}{2}e^{2x}$,
   (ii) $y = \left(\dfrac{x+1}{x+2}\right)\dfrac{x}{2} - \dfrac{(x+1)}{(x+2)^2} \log_e (1+x) + C\dfrac{(x+1)}{(x+2)^2}$.

5. (i) $y = (Ce^x/x^x) - (1/x^x)$,
   (ii) $y = A \cosh nx + B \sinh nx + (x/2n)e^{nx}$.

517

Answers to Problems

6. (a) exact; $(1+e^y) \sin x = \text{const.}$,
   (b) $y = e^{[(A/x)-(x^2/3)]}$,
   (c) $x^2 + y^2 = Ay$,
   (d) exact; $x^4 - 4x^3y + 10x^2y^2 - 28xy^3 + \frac{4}{5}y^5 + \frac{8}{3}y^3 = \text{const.}$

7. $y = (\frac{1}{4}x + \frac{5}{16})e^{-2x} - \frac{1}{3}e^{-4x} + \frac{1}{48}e^{2x}$.

8. (i) $y = Ae^{2x} + Be^{-x} + \frac{1}{10}(\cos x - 3 \sin x)$,
   (ii) $y = (Ax+B)e^{-x} + \frac{1}{2}e^x - x^2e^{-x}$,
   (iii) $y = e^x(A \cos 2x + B \sin 2x) + \frac{1}{3}e^x \sin x$,
   (iv) $y = e^{2x}(A \cos x + B \sin x) - (\frac{1}{2}e^{2x} \cos x)(\frac{1}{3}x^3 - \frac{1}{2}x)$
   $$+ (\frac{1}{4}x^2e^{2x}) \sin x.$$

9. $y = \sin x + \sin 2x + \cos 2x + 2x + 1$.

10. $y = A \cos (2 \sin \theta) + B \sin (2 \sin \theta) + 3 - 2 \sin^2 \theta$.

11. $y = A \cos (\log_e x) + B \sin (\log_e x)$.

12. $n = -2$,
    $y = (1/x^2)e^{-x}(A \cos x + B \sin x) + (1/2x)e^{-x} \sin x$.

14. (i) $y = Ax + (B/x)$,
    (ii) $y = Ax^2 + (B/x^2) + \frac{1}{12}x^4 + \frac{1}{4}x^2 \log_e x$,
    (iii) $y = (A/x^2) + x[B \cos (\sqrt{3} \log_e x) + C \sin (\sqrt{3} \log_e x)]$
    $$- \sin (\log_e x) + 8 \cos (\log_e x).$$

15. (i) $x = Ae^{-t} + Be^{3t}$, $y = Ae^{-t} - Be^{+3t}$,
    (ii) $x = At + B + Ce^{\sqrt{2}t} + De^{-\sqrt{2}t}$, $y = At + B - Ce^{\sqrt{2}t} - De^{-\sqrt{2}t}$.

16. $x = \frac{3}{2}t(e^t - e^{-t})$, $y = \frac{3}{2}(3+t)e^t - \frac{1}{2}(1-3t)e^{-t}$.

17. $x = (At+B)e^{-at} + (Ct+D)e^{at}$,
    $y = (At+B)e^{-at} - (Ct+D)e^{at} + (1/2a^2) \sin at$.

18. $x = Ae^{kt} + De^{-2kt}$, $y = Be^{kt} + De^{-2kt}$, $z = -(A+B)e^{kt} + De^{-2kt}$.

19. (a) $\frac{1}{4}\log_e\{[1 + \sin 2(x-2y)]/[1 - \sin 2(x-2y)]\} = x + C$,
    (b) $\tan y = (x^2/3) + (c/x)$.

20. (a) $4y = 1 + \sinh [(A \pm x)/\sqrt{2}]$,
    (b) $y = \tan (A \pm x)$,
    (c) $y = 4Ax^2/(1 - Ax^2)$,
    (d) $y = \cosh (\pm x + A)$.

22. (a) $x = be^{-2y}$, (b) $y^2 = b - 2x$, (c) $x^2 + y^2 - 2by = 0$, where in all cases b is an arbitrary constant.

518

27. $A^2 = \begin{pmatrix} 0 & 0 & 0 \\ 3 & 3 & 9 \\ -1 & -1 & -3 \end{pmatrix}$, $A^3 = \begin{pmatrix} 0 & 0 & 0 \\ 0 & 0 & 0 \\ 0 & 0 & 0 \end{pmatrix}$.

$$Y = \begin{pmatrix} 1+t & t & 3t \\ 5t+\frac{3}{2}t^2 & 2t+\frac{3}{2}t^2+1 & 6t+\frac{9}{2}t^2 \\ -2t-\dfrac{t^2}{2} & -t-\dfrac{t^2}{2} & 1-3t-\dfrac{3t^2}{2} \end{pmatrix} \begin{pmatrix} 0 \\ -1 \\ 1 \end{pmatrix}.$$

Problems 22

1. $y = \dfrac{x^3}{3} + \dfrac{x^4}{12} + \dfrac{x^5}{60} + \ldots$; exact solution $y = 2e^x - x^2 - 2x - 2$.

2. $y = x - \dfrac{x^2}{4} + \dfrac{x^3}{12} - \dfrac{3x^4}{64} + \dfrac{61x^5}{1920} + \ldots$.

3. $y = x - \dfrac{x^2}{3} + \dfrac{x^3}{3 \cdot 8} - \dfrac{x^4}{3 \cdot 8 \cdot 15} + \ldots$.

4. (a) $y = A\left(1 - \dfrac{4x}{3} + \dfrac{4x^2}{15}\right) + \dfrac{B}{\sqrt{x}}\left(1 - 5x + \dfrac{5x^2}{2} - \dfrac{x^3}{6} + \ldots\right)$,

   (b) $y = A(1 + 3x + 6x^2 + \ldots) + B\sqrt{x}\left(1 + \dfrac{7x}{3} + \dfrac{63x^2}{15} + \ldots\right)$,

   (c) $y = A\left(1 - \tfrac{1}{2}x + \dfrac{x^2}{20} - \dfrac{x^3}{480} + \ldots\right)$

   $\qquad\qquad\qquad + Bx^{1/3}\left(1 - \tfrac{1}{4}x + \dfrac{x^2}{56} - \dfrac{x^3}{1680} + \ldots\right)$.

5. $y_1 = Ax\left(1 - \dfrac{x^2}{2 \cdot 4} + \dfrac{x^4}{2 \cdot 4 \cdot 4 \cdot 6} - \ldots\right)$.

6. $b = 2a, c = 2a/3$. $\quad |x| < 2$.

7. Second solution:
$$y = x + \sum_{n=1}^{\infty} \frac{(2^2-k^2)(4^2-k^2)\ldots\{(2n)^2-k^2\}}{(2n+1)!}x^{2n+1}.$$

8. $b_2 = \tfrac{1}{4}$.

12. $v = A\left(1 - \dfrac{2n}{2!}x^2 + \dfrac{2n(2n-1)}{4!}x^4 - \ldots\right)$.

Problems 23

2. (a) $-\sin x + \sqrt{2}\sin(x\sqrt{2})$,
   (b) $-1 + e^x - xe^x + \frac{1}{2}x^2 e^x$.

3. (a) $y = (\frac{1}{20}e^{-2x})(39\cos 2x + 47\sin 2x) + \frac{1}{10}\sin 2x + \frac{1}{20}\cos 2x$,
   (b) $y = e^{-3x} - \frac{1}{2}e^{-x}$, $z = \frac{2}{3}(1 - e^{-3x})$.

Problems 24

1. (a) $\dfrac{\partial u}{\partial x} - \dfrac{\partial u}{\partial y} = 0$,

   (b) $x\dfrac{\partial u}{\partial x} - y\dfrac{\partial u}{\partial y} = 0$,

   (c) $\dfrac{\partial^2 u}{\partial x^2} = \dfrac{\partial^2 u}{\partial y^2}$,

   (d) $x\dfrac{\partial u}{\partial x} + y\dfrac{\partial u}{\partial y} = nu$.

2. (a) Hyperbolic; $u = F(x + (2 + \sqrt{7})y) + G(x + (2 - \sqrt{7})y)$,
   (b) Parabolic; $u = F(x + y) + (rx + sy)G(x + y)$, r and s arbitrary constants,
   (c) Elliptic; $u = F(x + 2iy) + G(x - 2iy)$.

4. $f(\theta) = C\cos\theta + D\sin\theta$.

5. $w = \alpha^2$, $V = 2e^{-x}\sin(t - x)$.

6. $V = e^{-4y}\cos 2x$.

7. $V = C\sec l\sin(l - x)e^{-at}$.

8. $\tan(wa/c) = wa/c$.

10. $V = (A\cos pt + B\sin pt)e^{(1 - p^2)x/4}$.

11. $z = \displaystyle\sum_k e^{-(k^2 + 1)y}(A_k\cos kx + B_k\sin kx)$,

    $z = 2\displaystyle\sum_{n=1}^{\infty}\frac{(-1)^{n-1}}{n}e^{(n^2 + 1)(1 - y)}\sin nx$.

12. $V = \displaystyle\sum_{n=1}^{\infty}B_n e^{-(kn^2\pi^2 t)/a^2}\sin\frac{n\pi x}{a}$, where

    $$B_n = \frac{4V_0}{\pi^2}\left(\frac{1}{n^2}\sin\frac{n\pi}{2} - \frac{\pi}{2n}\cos\frac{n\pi}{2}\right).$$

15. $T = T_0 \left(1 + \cos \dfrac{2\pi x}{a} e^{-4\pi^2 kt/a^2} \right).$

16. $\theta = 100 + \sin \left(\dfrac{\pi x}{2l} \right) e^{-k\pi^2 t/4l^2}.$

Problems 25

1. $I = 5\pi/2.$

2. $y = 2x.$

3. $y = x^3.$

4. $y = \sin x.$

Problems 26

1. (a) $u_r = 3^{r-1},$
   (b) $u_r = (2^r/\sqrt{3}) \sin (\pi r/3),$
   (c) $u_r = (Ar + B)(\tfrac{1}{2})^r + (2^r/9),$
   (d) $u_r = A3^r + 4^r - \tfrac{1}{2}r - \tfrac{1}{4}.$

2. $u_r = A + B2^r + C3^r + (4^r/6).$

3. $u_n = n\alpha^{n-1}.$

Problems 27

1. $u = e^x.$

2. $u = \sin x + \dfrac{2\alpha}{(\alpha^2 + 1)(2\alpha - 1)} e^{-\alpha x}.$

3. (a) $\lambda = 0, \quad u = 0; \quad \lambda = -\tfrac{3}{4}, \quad u = Ax; \quad 1 - \tfrac{2}{3}\lambda = \pm 2\lambda/\sqrt{5},$
   $u = B(1 \pm \sqrt{5}x^2),$
   (b) $\lambda = 0, u = 0; \lambda = 1/\pi, u = A;$
   $\lambda = 2/\pi, u = B \cos 2x; \lambda = -2/\pi, u = C \sin 2x,$

   where A, B and C are arbitrary constants.

4. $u = \tfrac{1}{2}x + \tfrac{3}{4} \sin 2x.$

INDEX

Index

Index

Index

Linear differential equations, 381, 394, 461
  equations, 276, 311
  inequalities, 16
Logarithmic function, 5, 98, 116

M

Maclaurin series, 92
Magnitude of a vector, 333
Many-valued functions, 3, 116
Matrix, adjoint, 307
  diagonal, 301
  Hermitian, 306
  inverse, 309
  null, 301
  orthogonal, 309
  reciprocal, 307
  square, 301
  transposed, 303
  unit, 302
  unitary, 309
Matrices, addition of, 297
  inversion of, 307
  multiplication of, 299
  symmetric and skew-symmetric, 305
Maxima and minima, 38, 168
Maxwell's equations, 367
Means, 12
Mean-value theorems for functions, 30
  for integrals, 57
Minor, 277
Modulus of a real number, 2
  of a complex number, 108
Monotonic function, 7
Multiple integrals, 210
Multiplication of complex numbers, 110
  of determinants, 288
  of matrices, 299
  of series, 81
  of vectors, 336, 339, 342
Multiplicity of roots, 368
Multiply connected region, 205

N

Negative vector, 332
Neutral element, 324
Newton's method, 376
Non-differentiable functions, 30

Non-linear ordinary differential equations, 420
  partial differential equations, 461
Normalised set of functions, 275
Null matrix, 301
Number, complex, 2, 107
  irrational, 1
  rational, 1
  real, 2
Numbers, Bernoulli's 91
Numerical integration, 245
  value, 2

O

Odd function, 5, 49
Open interval, 3
Operator, D-, 27, 406
  difference, 493
  linear, 401, 449
  vector, 352
Order of a determinant, 276
  of a differential equation, 381
  of a matrix, 297
Ordinary differential equation, 380
  point, 434
Orthogonal curves, 179, 427
  matrix, 309
  set of functions, 275
Orthonormal set of functions, 275

P

Parabolic partial differential equation, 463
Partial differential coefficients, 147
  differential equations, 461
  fractions, 51
  sums, 68
Particular integral, 396
Pathological functions, 30
Period, 6
Periodic function, 6, 254
Permutation group, 329
Piecewise regular function, 255
Plane, equation of, 348
Point of discontinuity, 25, 255
  of inflection, 39
Polar coordinates, 108, 162, 220
Polynomial, degree of, 4, 368
  factors of, 368
Polynomials, Legendre, 443
Positive definite, 18

Index

Index